The LEGO® MINDSTORMS® EV3 Idea Book

THE LEGO® MINDSTORMS® EV3 IDEA BOOK

181 Simple Machines and Clever Contraptions

YOSHIHITO ISOGAWA

Printed in China
First Printing

18 17 16 15 14   1 2 3 4 5 6 7 8 9

ISBN-10: 1-59327-600-1
ISBN-13: 978-1-59327-600-3

Publisher: William Pollock
Production Editor: Riley Hoffman
Cover Design: Beth Middleworth
Photographer: Yoshihito Isogawa
Author Photo: Sumiko Hirano
Developmental Editor: Tyler Ortman
Technical Reviewer: Sumiko Hirano
Proofreader: Fleming Editorial Services

For information on distribution, translations, or bulk sales, please contact No Starch Press, Inc. directly:
No Starch Press, Inc.
245 8th Street, San Francisco, CA 94103
phone: 415.863.9900; info@nostarch.com
www.nostarch.com

Library of Congress Cataloging-in-Publication Data
Isogawa, Yoshihito, 1962-
The LEGO Mindstorms EV3 idea book : 181 Simple Machines and Clever Contraptions / Yoshihito Isogawa.
pages cm
ISBN 978-1-59327-600-3 -- ISBN 1-59327-600-1
1. Machinery--Models. 2. LEGO Mindstorms toys. I. Title.
TJ248.I86 2015
621.8022'8--dc23
2014027048

Production Date: 8/5/14
Plant & Location: Printed by Everbest Printing (Guangzhou, China), Co. Ltd
Job / Batch #: 42088-0 / 703189

Contents

Introduction · · · · · 1

PART 1 • Basic Mechanisms

Gear ratios · · · · · 4

Compound gear systems · · · · · 18

Changing the angle of rotation · · · · · 22

Using worm drives · · · · · 30

Swinging mechanisms · · · · · 36

Reciprocating mechanisms · · · · · 42

Cam mechanisms · · · · · 48

Intermittent motion · · · · · 52

Transmitting rotation with rubber bands · · · · · 56

Transmitting rotation with caterpillar treads · · · · · 60

Transmitting rotation over a long distance 62

Off-center axes of rotation 64

Changeover mechanisms using rotational direction 68

Universal joints 74

PART 2 • Vehicles

Driving wheels with a motor 78

Driving wheels with two motors 82

Caster wheels 90

Crawlers 94

Suspended wheels 100

Steering 104

PART 3 • Moving Without Tires

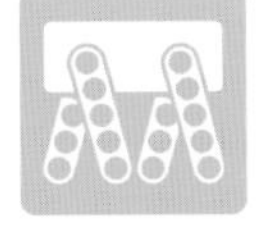

Walking machines · · · 110

Moving like an inchworm · · · 122

Moving through vibration · · · 126

PART 4 • Arms, Wings, and Other Movements

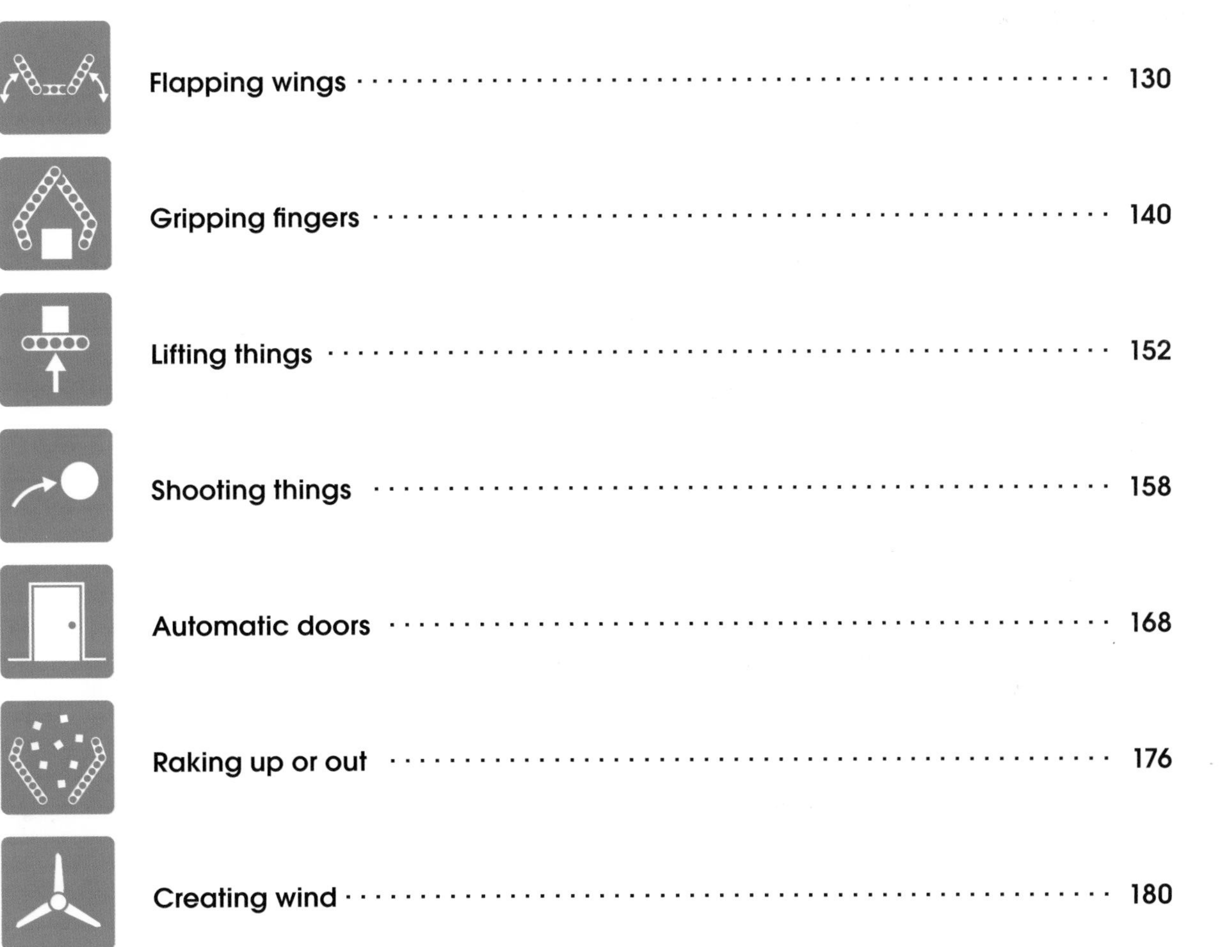

Flapping wings · · · 130

Gripping fingers · · · 140

Lifting things · · · 152

Shooting things · · · 158

Automatic doors · · · 168

Raking up or out · · · 176

Creating wind · · · 180

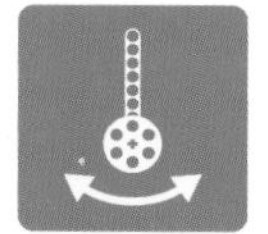

Swinging a pendulum · · · · · · · · · · 184

Using attachments to change motion · · · · · · · · · · 188

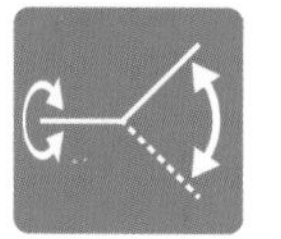

Meshing gears diagonally · · · · · · · · · · 194

Changing the angle of rotation freely · · · · · · · · · · 198

PART 5 • Sensors

Ideas for using the touch sensor · · · · · · · · · · 206

Ideas for using the buttons of the Intelligent EV3 Brick · · · · · · · · · · 214

Ideas for using the color sensor · · · · · · · · · · 216

PART 6 • Something Extra

Using the Pythagorean theorem · · · · · · · · · · 220

Try building something handy! · · · · · · · · · · 222

Introduction

The LEGO MINDSTORMS EV3 set is designed to allow builders of any age to create robots, vehicles, and other contraptions with moving parts.

Each model in this book is only a small mechanism, but you can make an infinite variety of larger models by combining these ideas. LEGO bricks aren't designed to fit in just one specific place or in one particular way. Your imagination is your guide when building with LEGO, and I hope that you will create your own wonderful masterpieces using this book as inspiration.

To build the models in this book, all you need is the LEGO MINDSTORMS EV3 set (home edition #31313).

Where Are the Words?

Other than this brief introduction and the table of contents, this book has almost no words. Instead, you'll find a series of photographs of increasingly complex models, each designed to demonstrate a mechanical principle or building technique.

While the book lists the pieces needed for each model, it does not include step-by-step building instructions. Look at the photographs taken from various angles and try to reproduce the model. Building in this way is something like putting together a puzzle. You'll get the hang of it after a little practice.

This is an idea book; it's about imagination. Rather than tell you what to see or think when you study the models, I encourage you to interpret them in your own way. Your interpretations will lead you to invent your very own models or use my mechanisms in entirely new ways!

What About Programs?

This book is about exploring the mechanical side of EV3 and robotics, rather than programming. You'll need only a few simple programs to test out your mechanical creations. Prepare these three simple programs in advance.

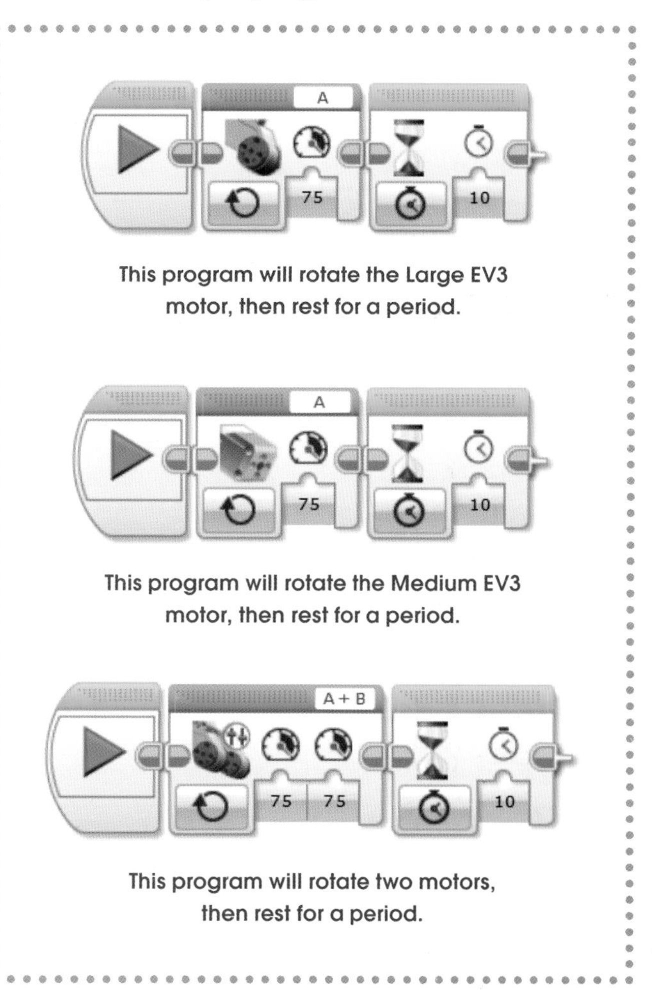
A
75
10
A
75
10
A + B
75
75
10

This program will rotate the Large EV3 motor, then rest for a period.

This program will rotate the Medium EV3 motor, then rest for a period.

This program will rotate two motors, then rest for a period.

A few of the models in this book require special programs. Take note of these programs, as they are written to avoid damaging parts by overextending a mechanism's range of motion.

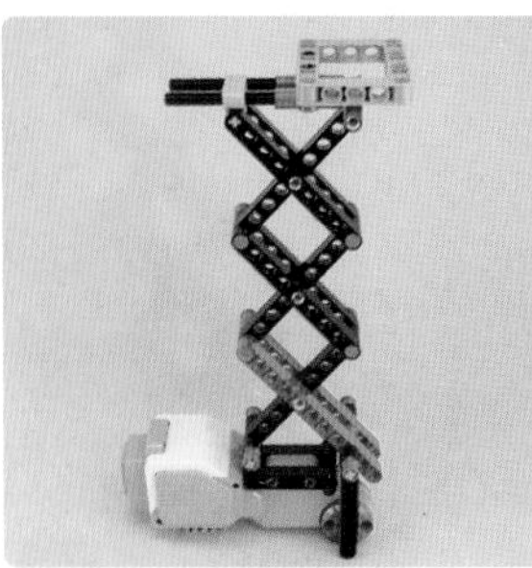
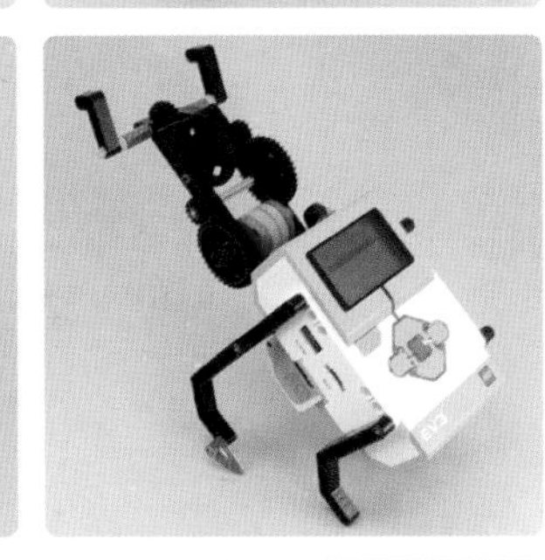

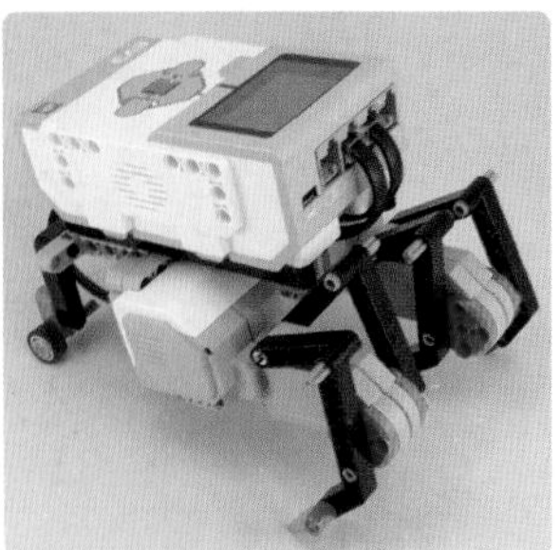

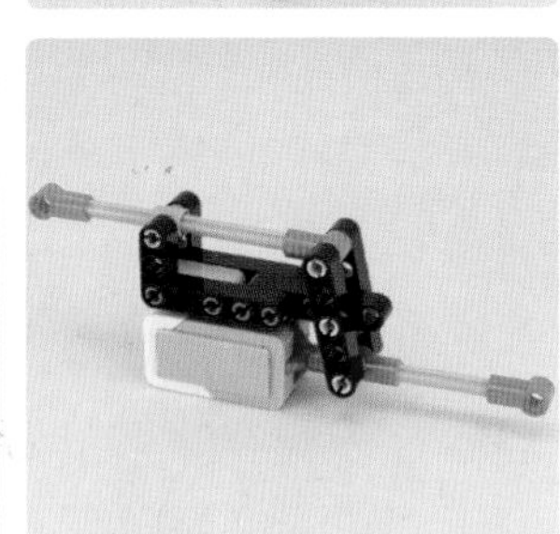

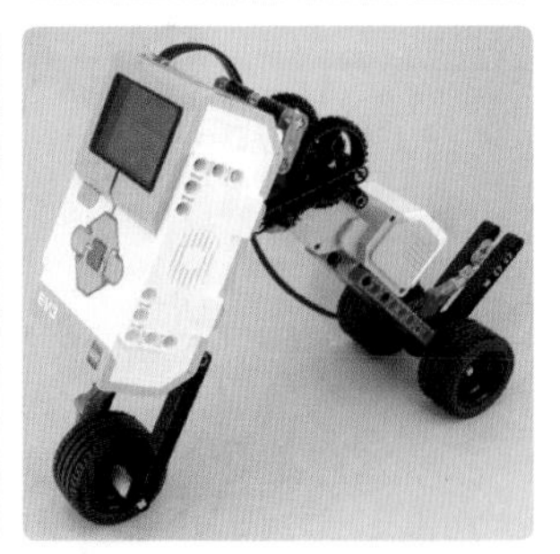

PART 1

Basic Mechanisms

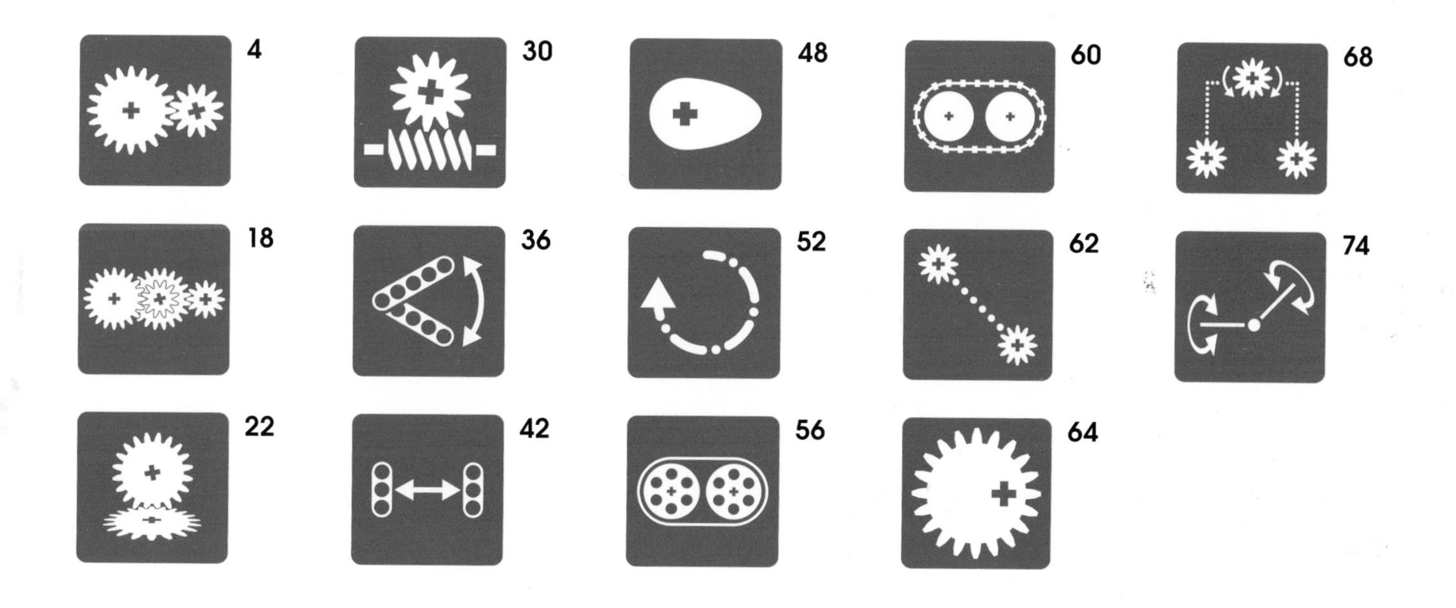
4
30
48
60
68
18
36
52
62
74
22
42
56
64

Gear ratios

#1

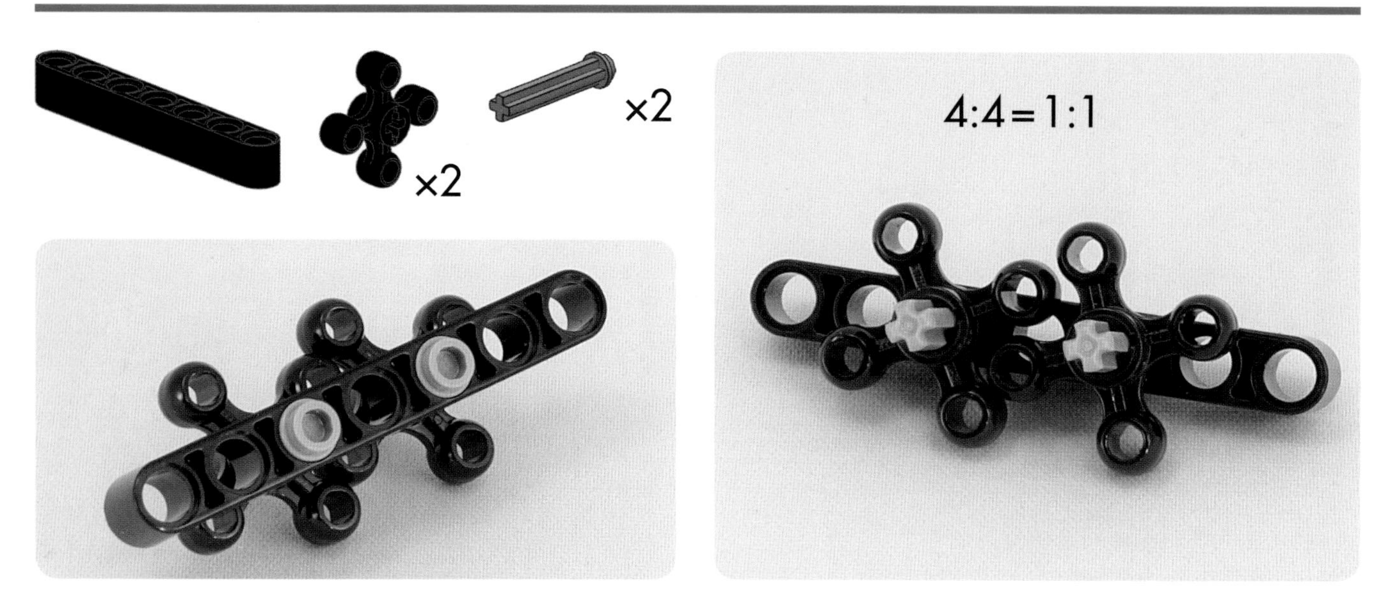
×2
×2
4:4=1:1

#2

24:24=1:1
×2
×2

#3

36:36 = 1:1
×2
×2
#4
×2
×2
×2
20:20 = 1:1

#5

#6

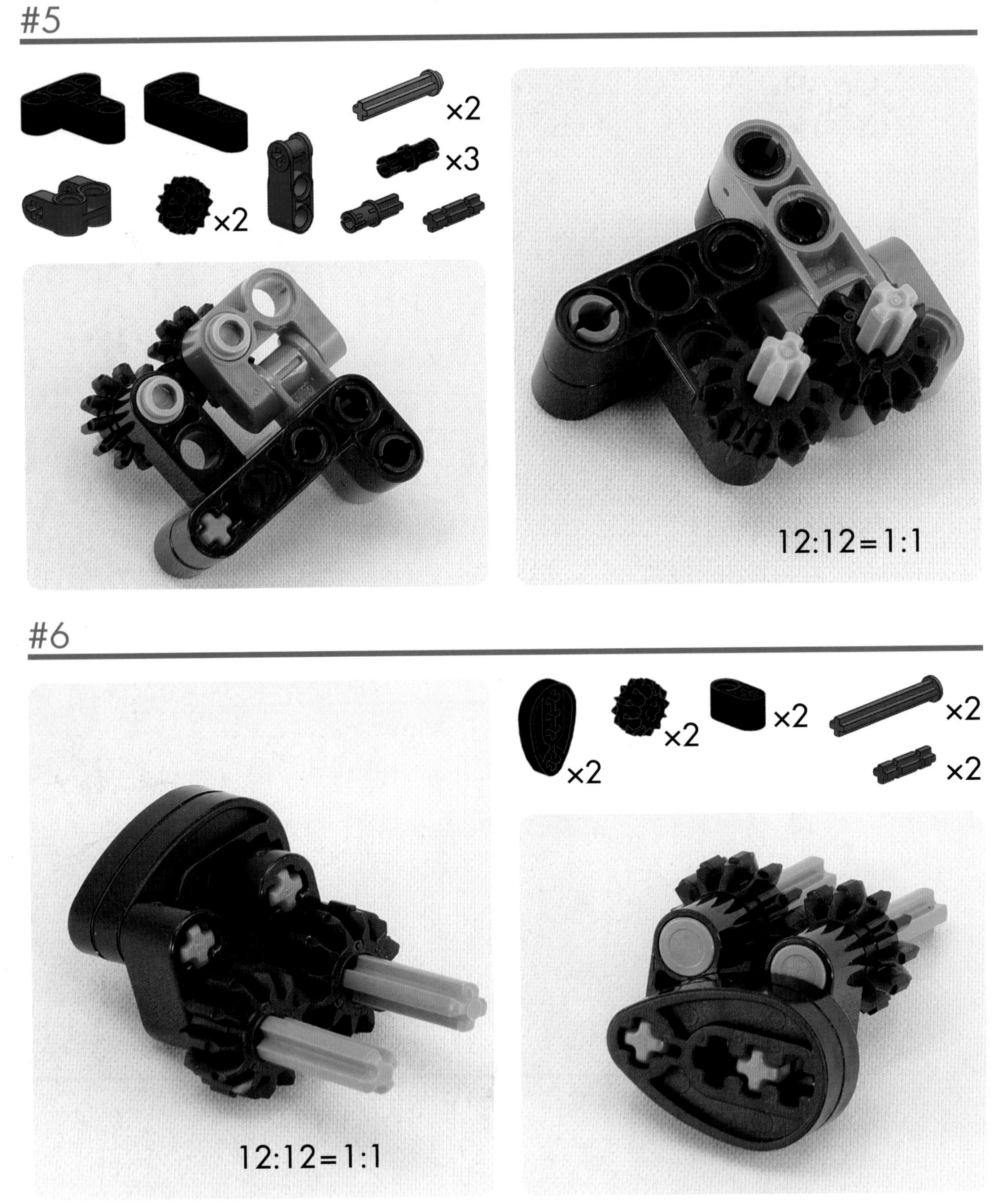
×2
×3
×2
12:12=1:1
×2
×2
×2
×2
×2
12:12=1:1

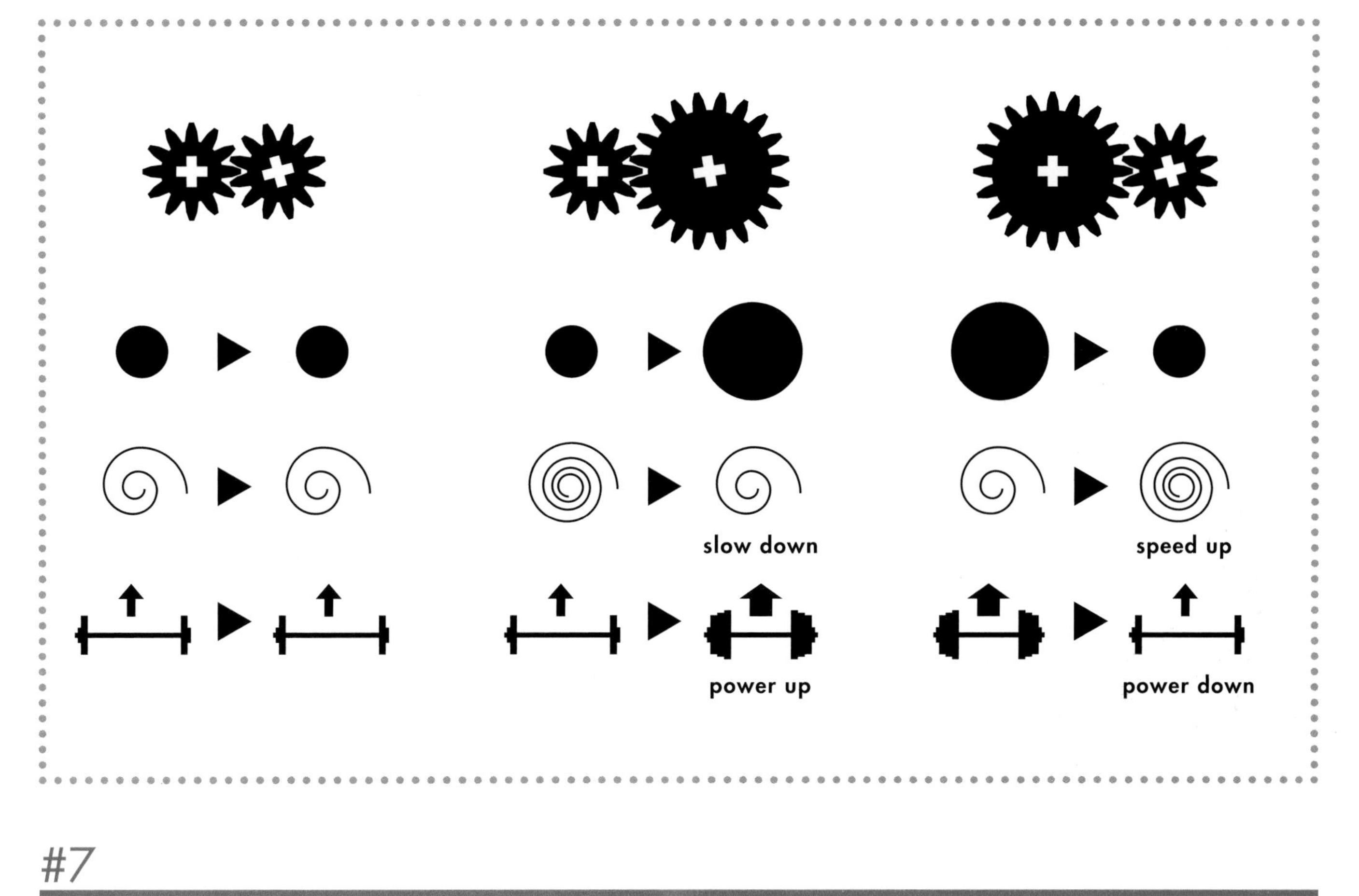
slow down
speed up
power up
power down

#7

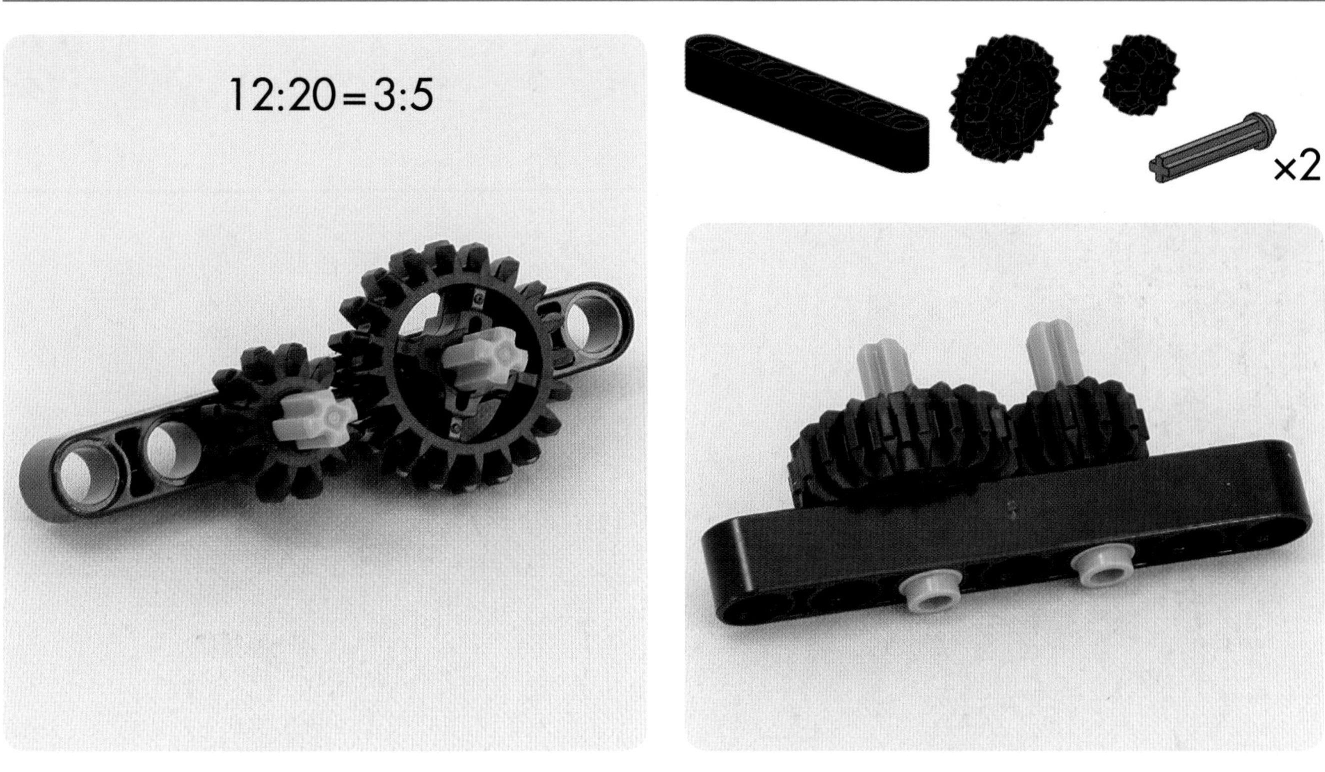
12:20 = 3:5
×2

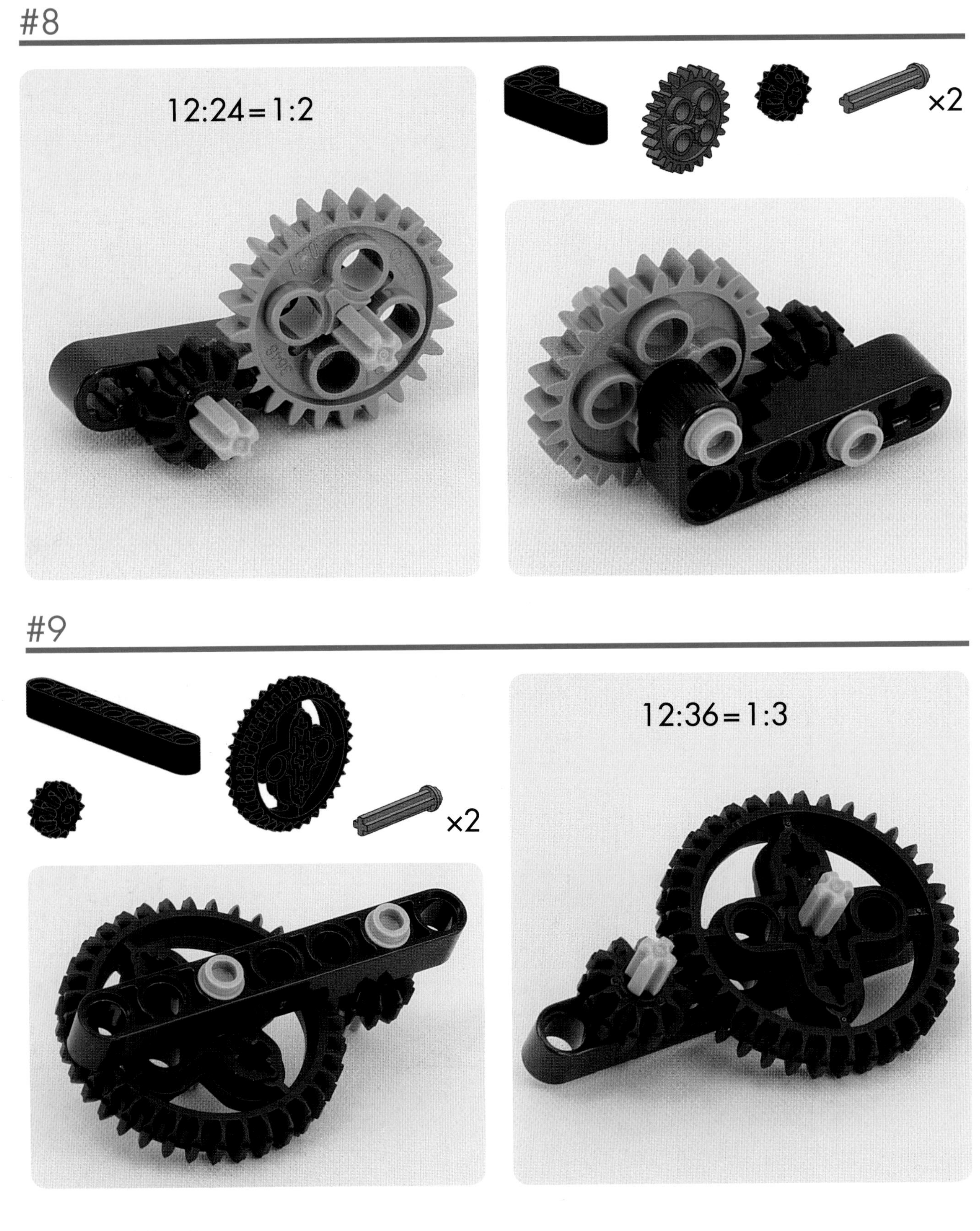

#8
12:24=1:2
×2
#9
×2
12:36=1:3

#10

#11

×2
×2
20:36 = 5:9
20:24 = 5:6
×2

#12

×2
24:36=2:3
12:20:12=3:5:3
×2
×3

#13

#14

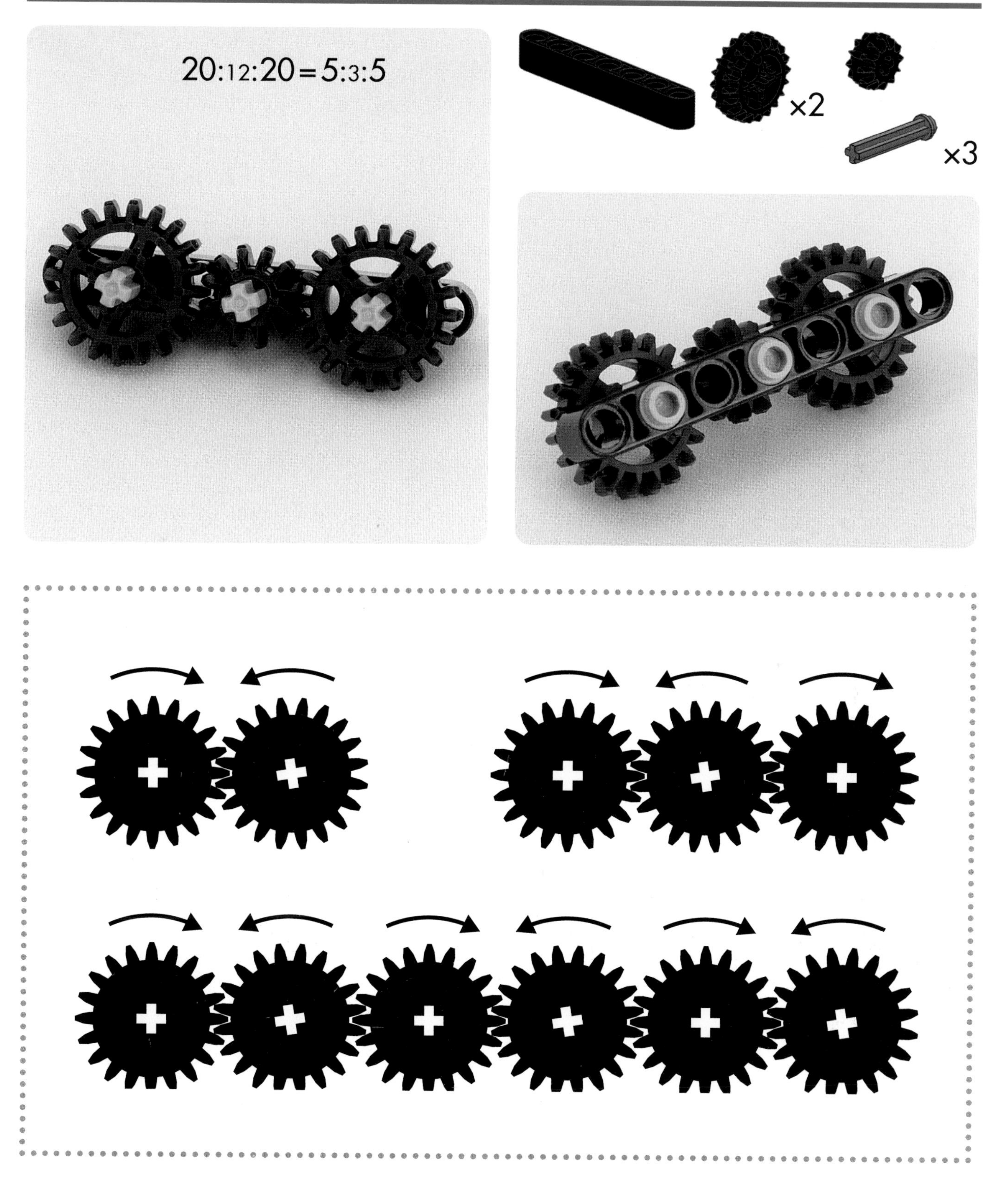
20:12:20 = 5:3:5
×2
×3

#15

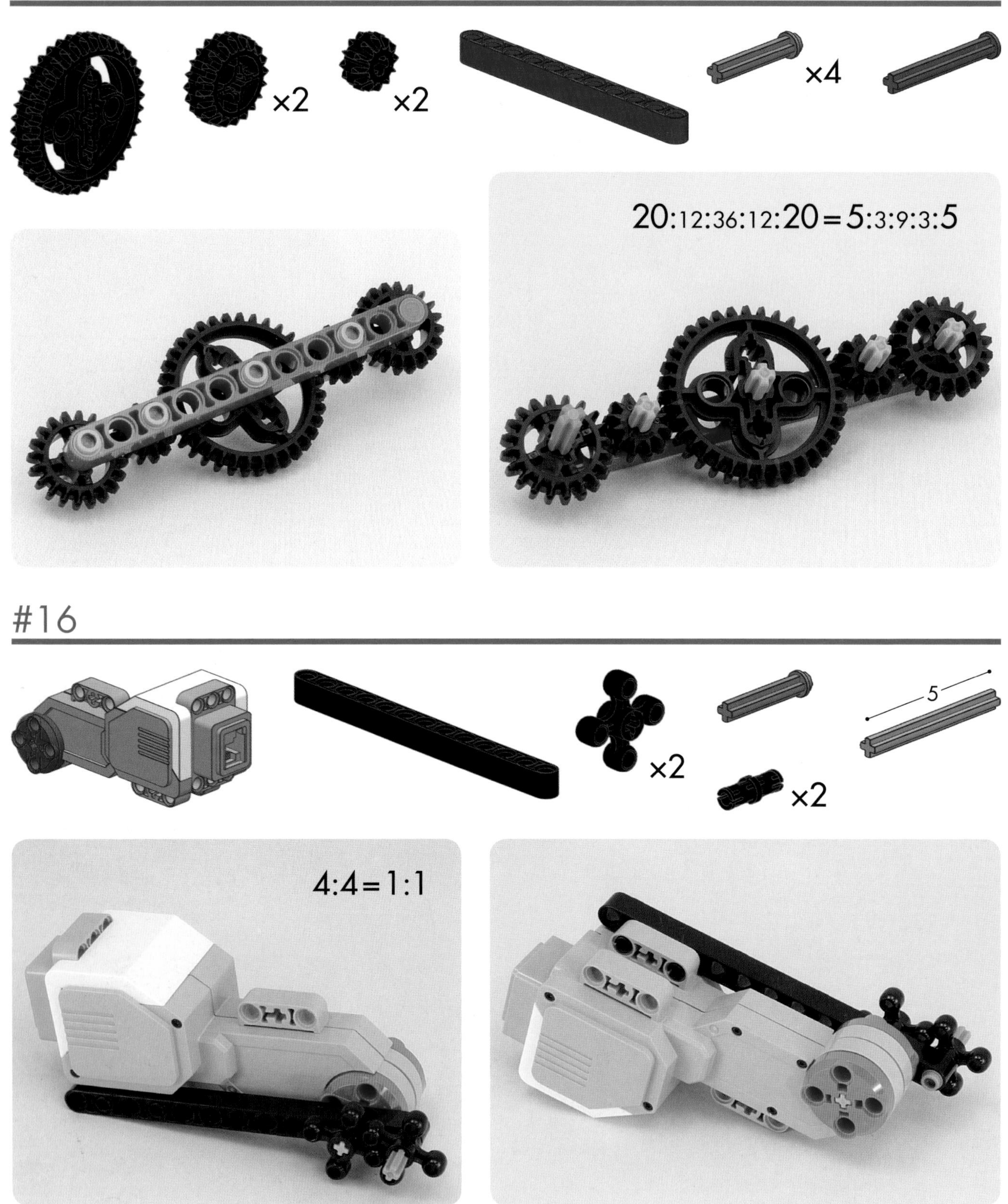
×2
×2
×4
20:12:36:12:20 = 5:3:9:3:5
#16
5
×2
×2
4:4 = 1:1

#17

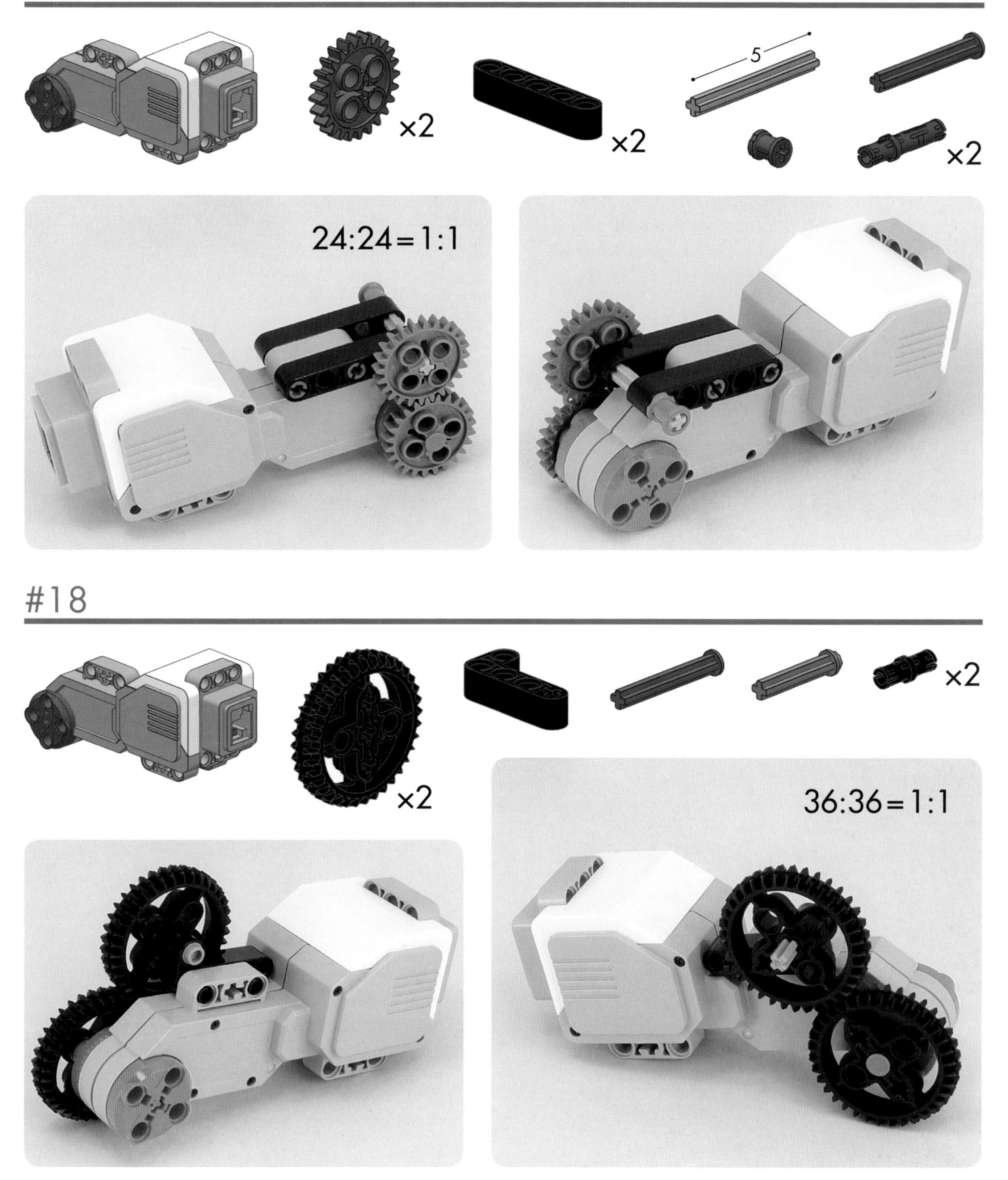
×2
×2
5
×2
24:24=1:1
#18
×2
×2
36:36=1:1

#19

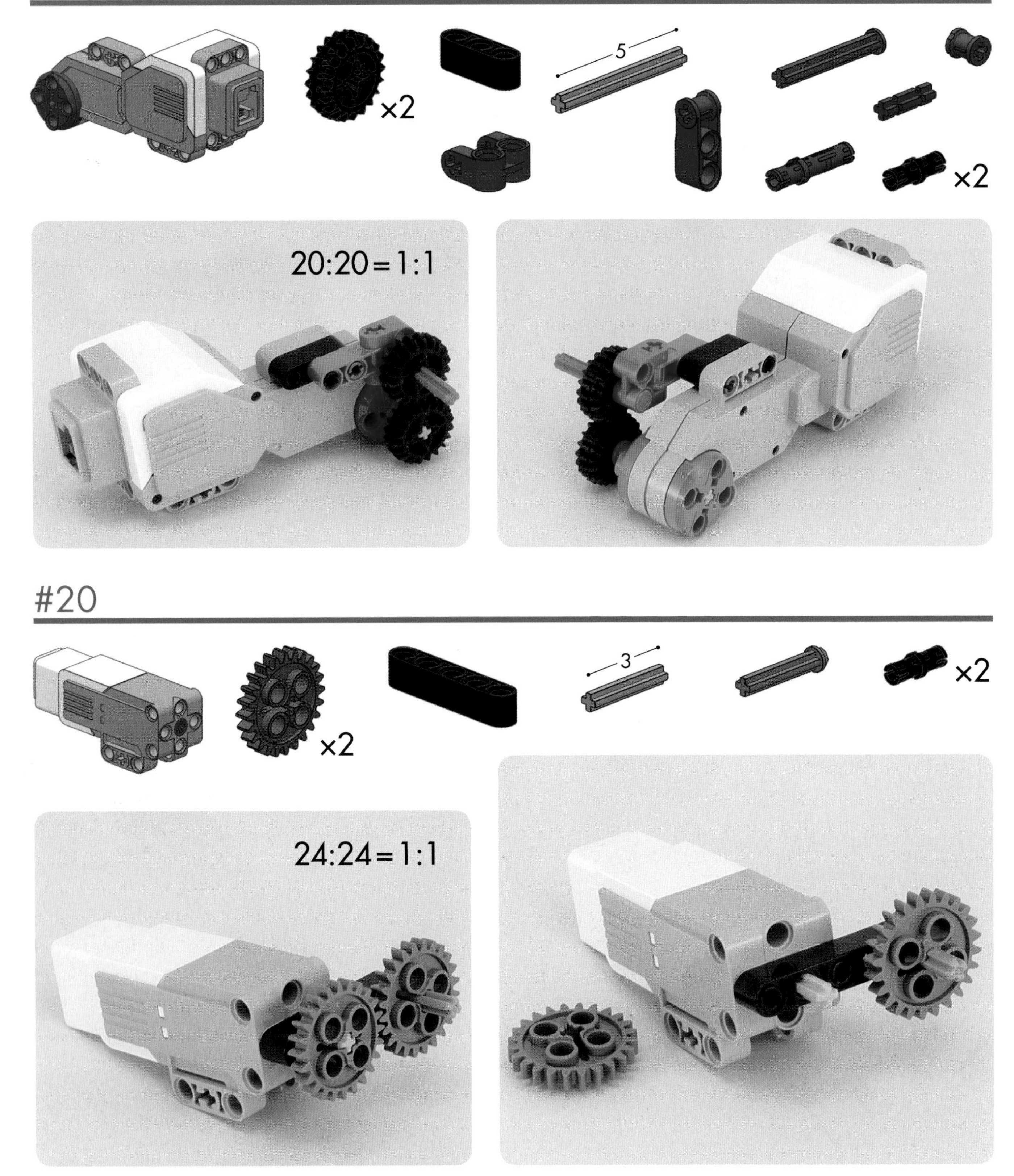
×2
5
×2
20:20 = 1:1
#20
3
×2
×2
24:24 = 1:1

#21

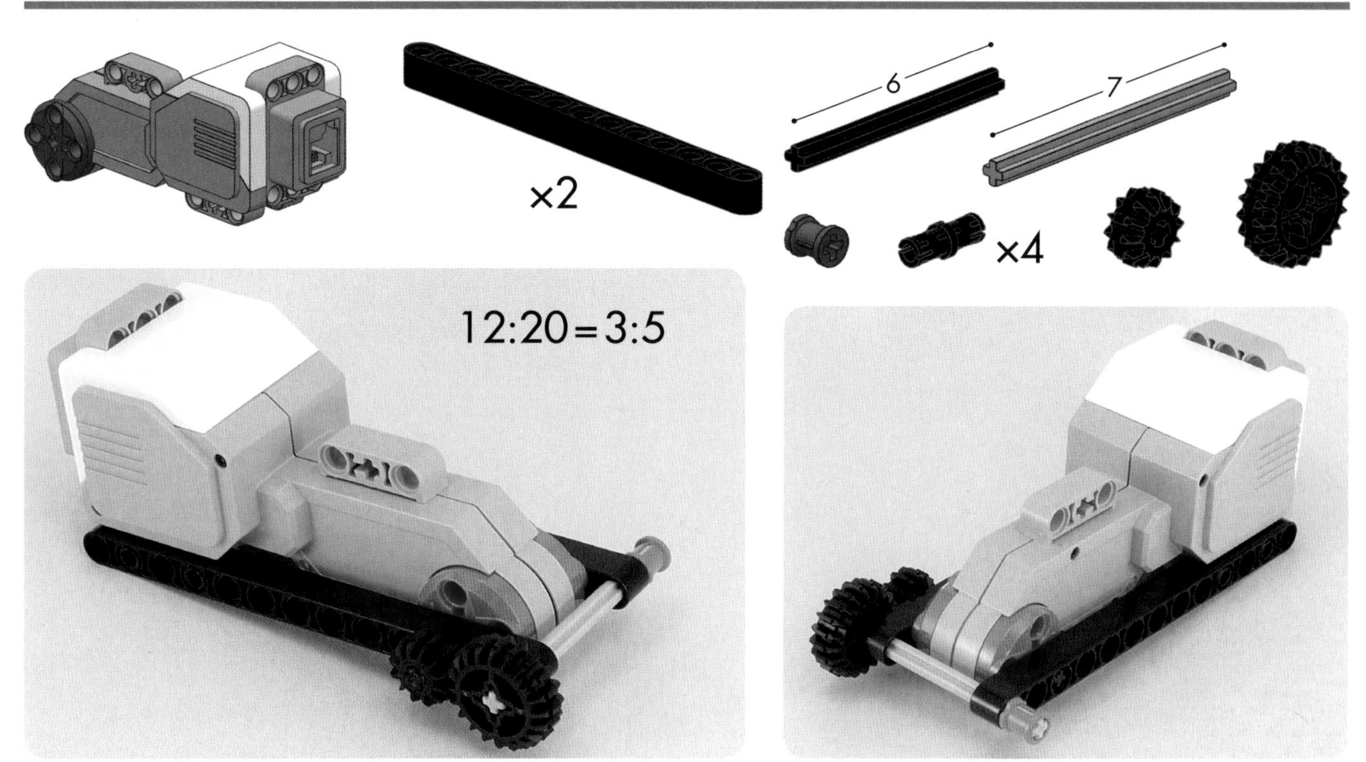
6
7
×2
×4
12:20=3:5

#22

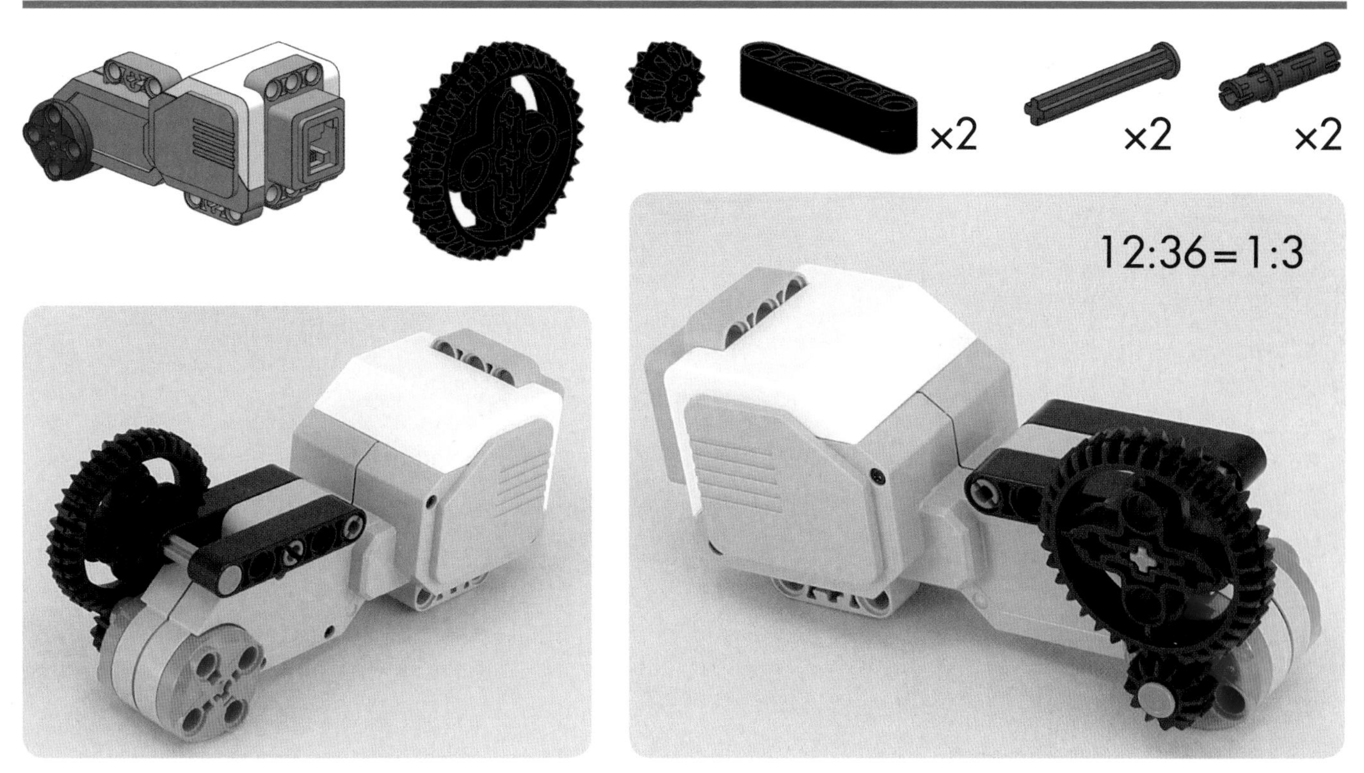
×2
×2
×2
12:36=1:3

#23

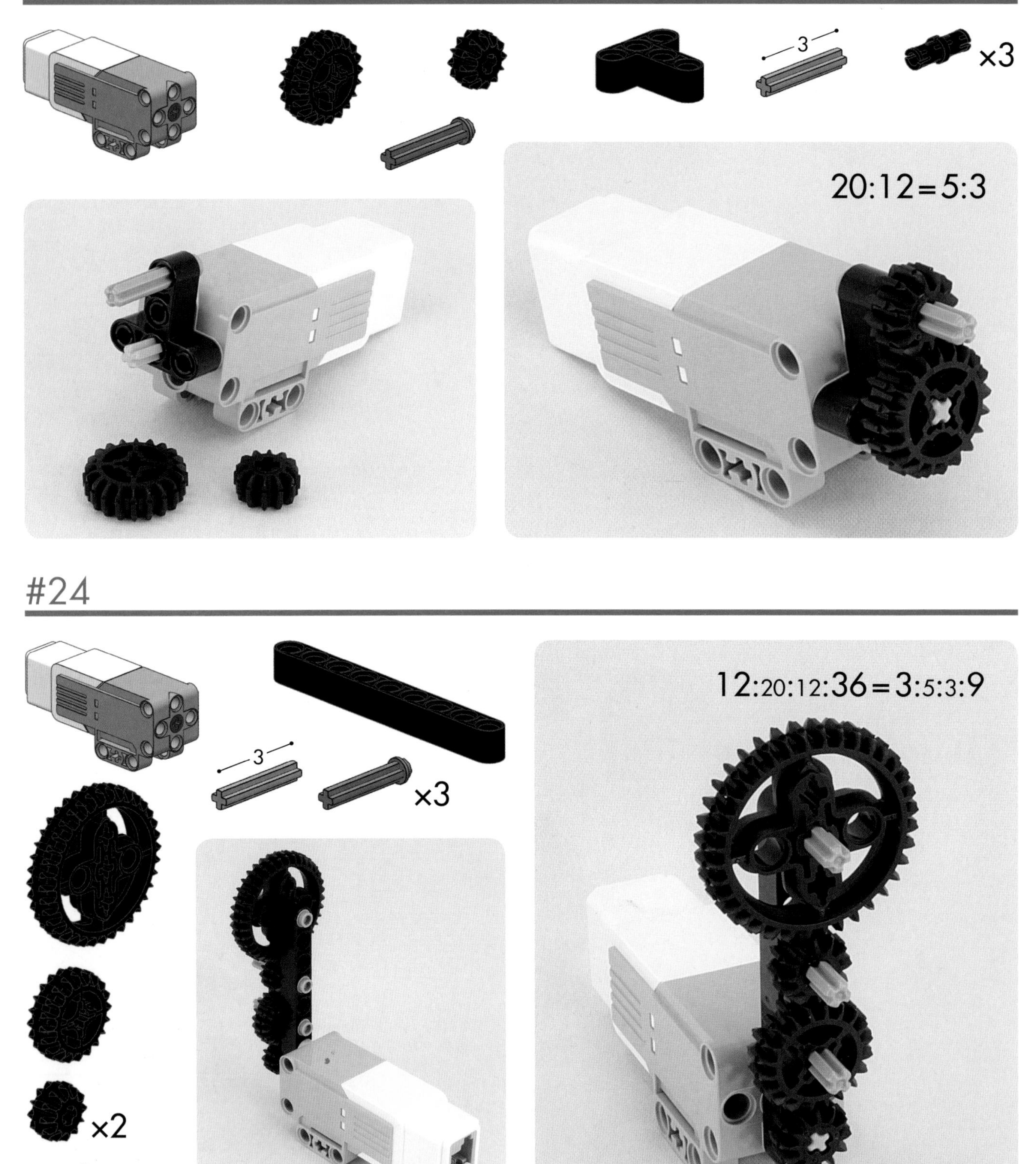
3
×3
20:12=5:3
#24
3
×3
12:20:12:36=3:5:3:9
×2
×2

#25

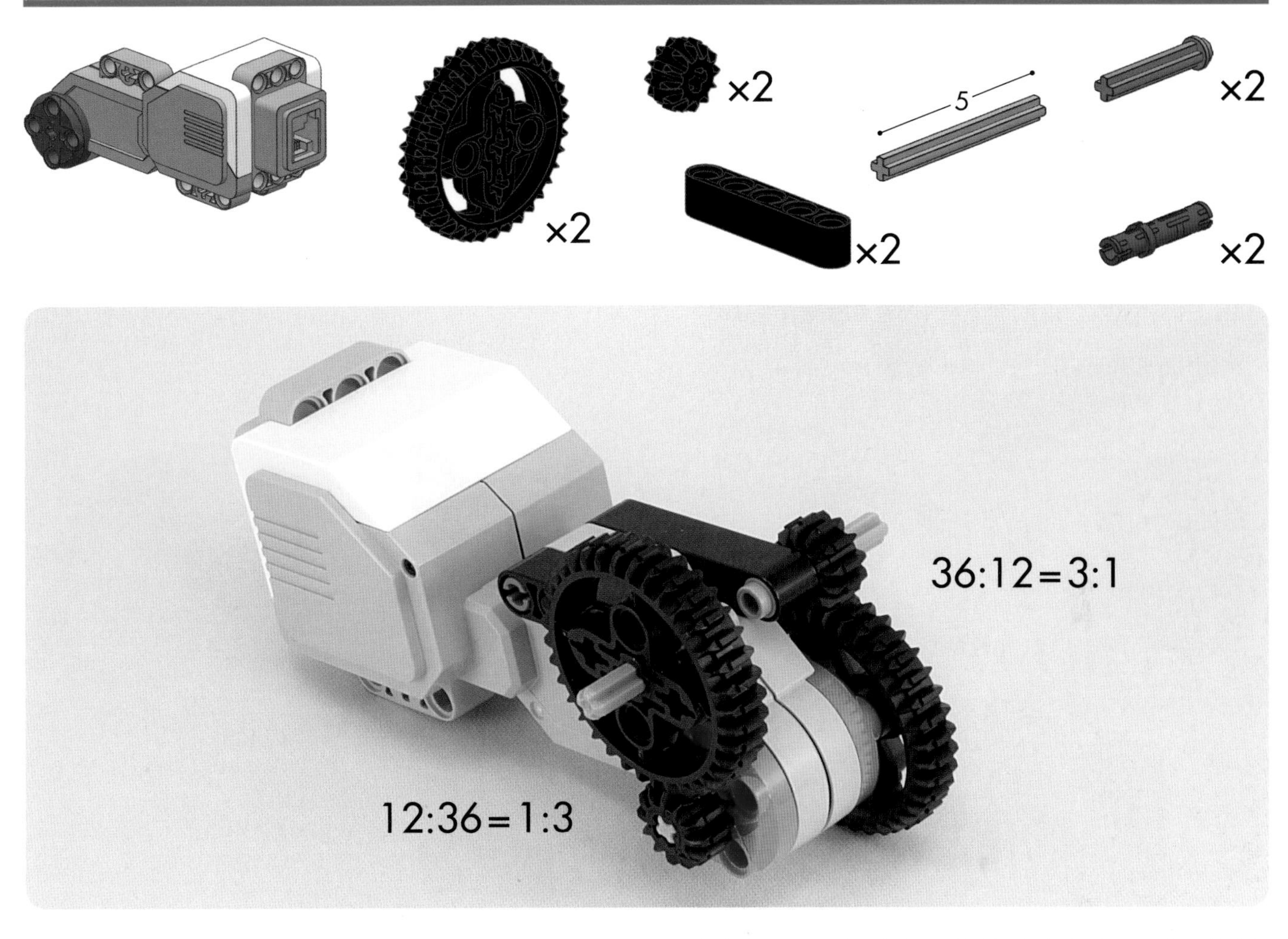
×2
×2
5
×2
×2
×2
×2
36:12=3:1
12:36=1:3

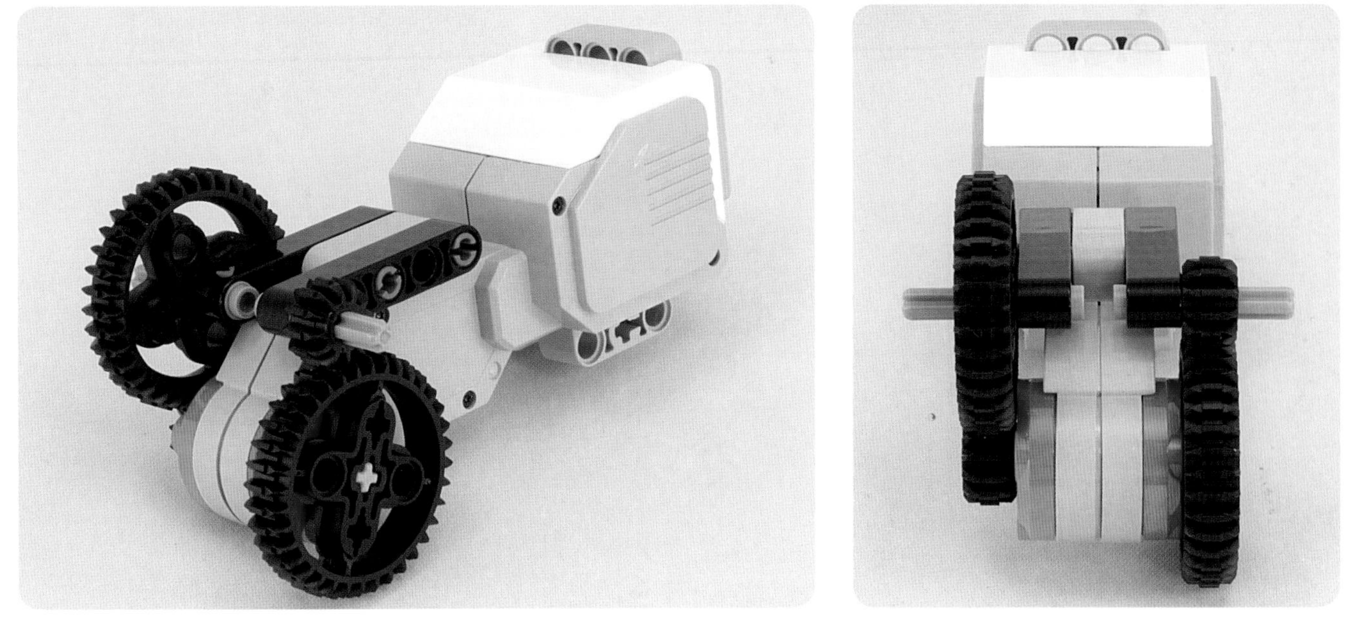

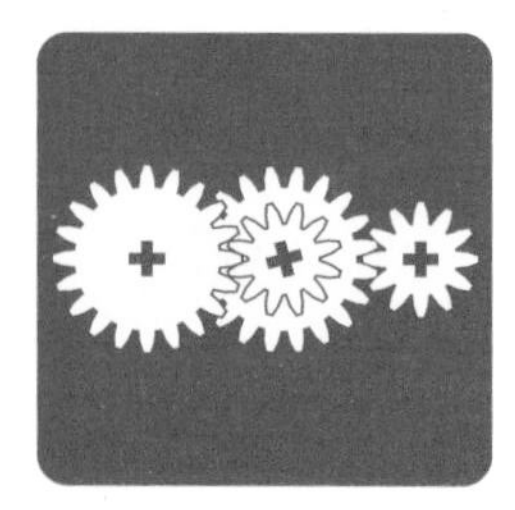

Compound gear systems

#26

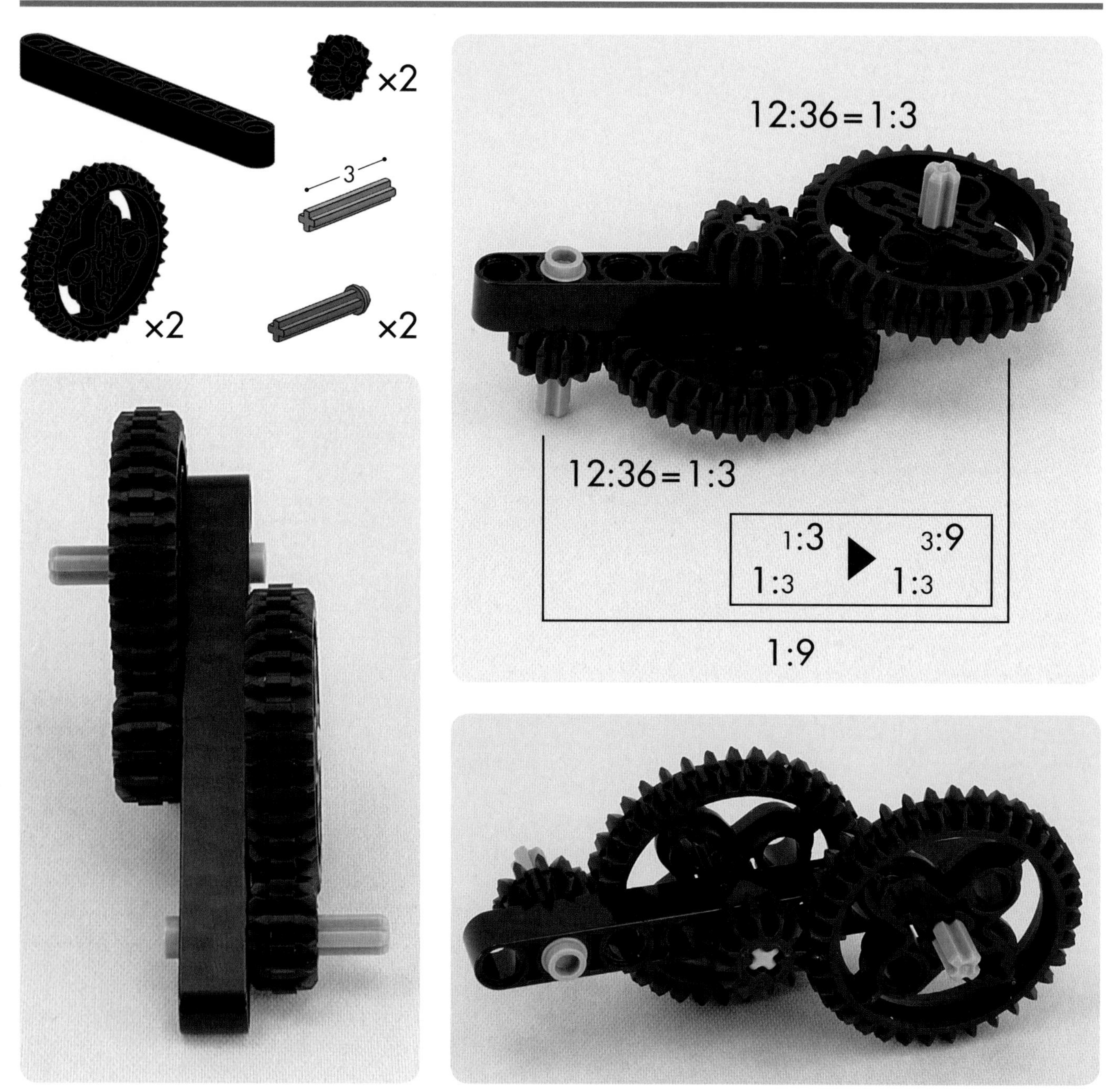

×2
×2
3
×2
12:36=1:3
12:36=1:3
1:3
3:9
1:3
1:3
1:9

#27

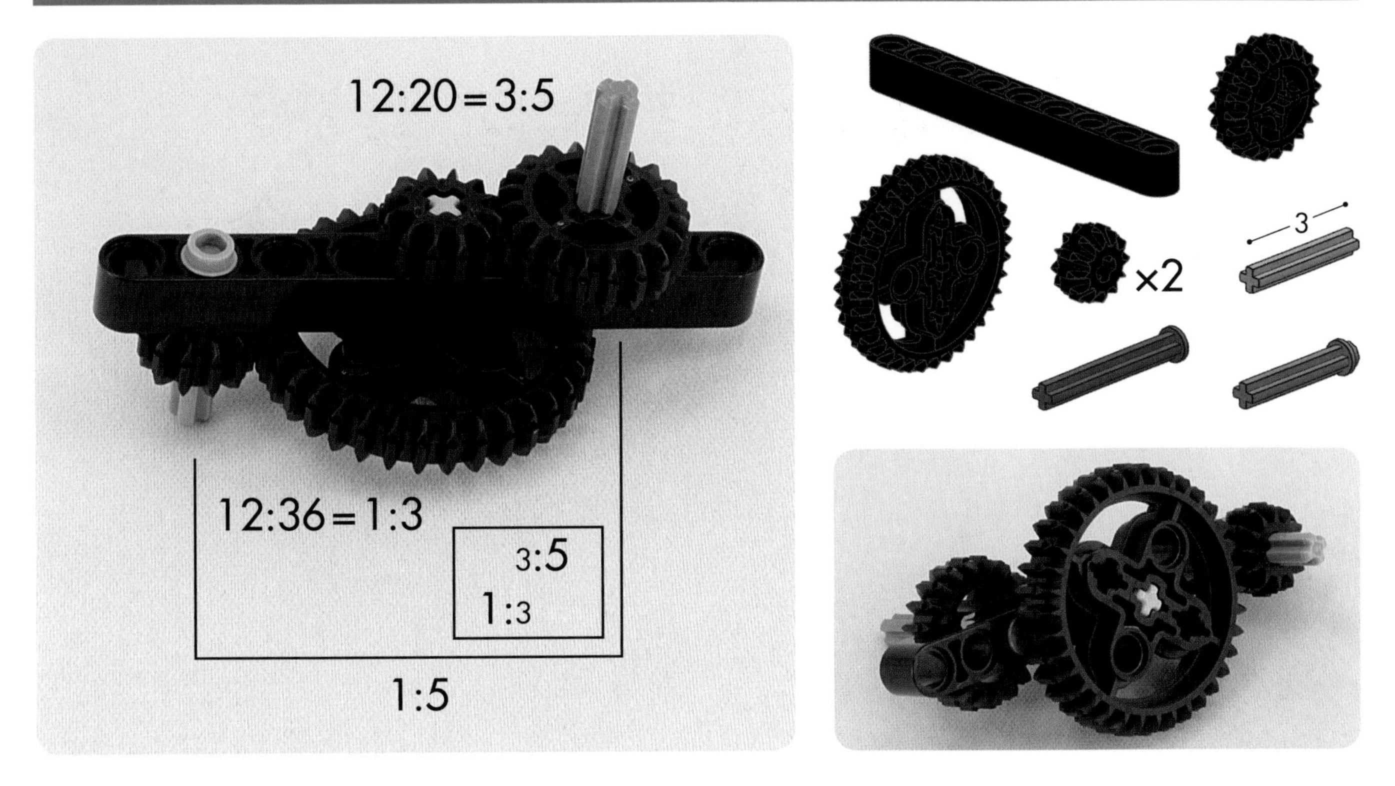
12:20=3:5
12:36=1:3
3:5
1:3
1:5
×2
3

#28

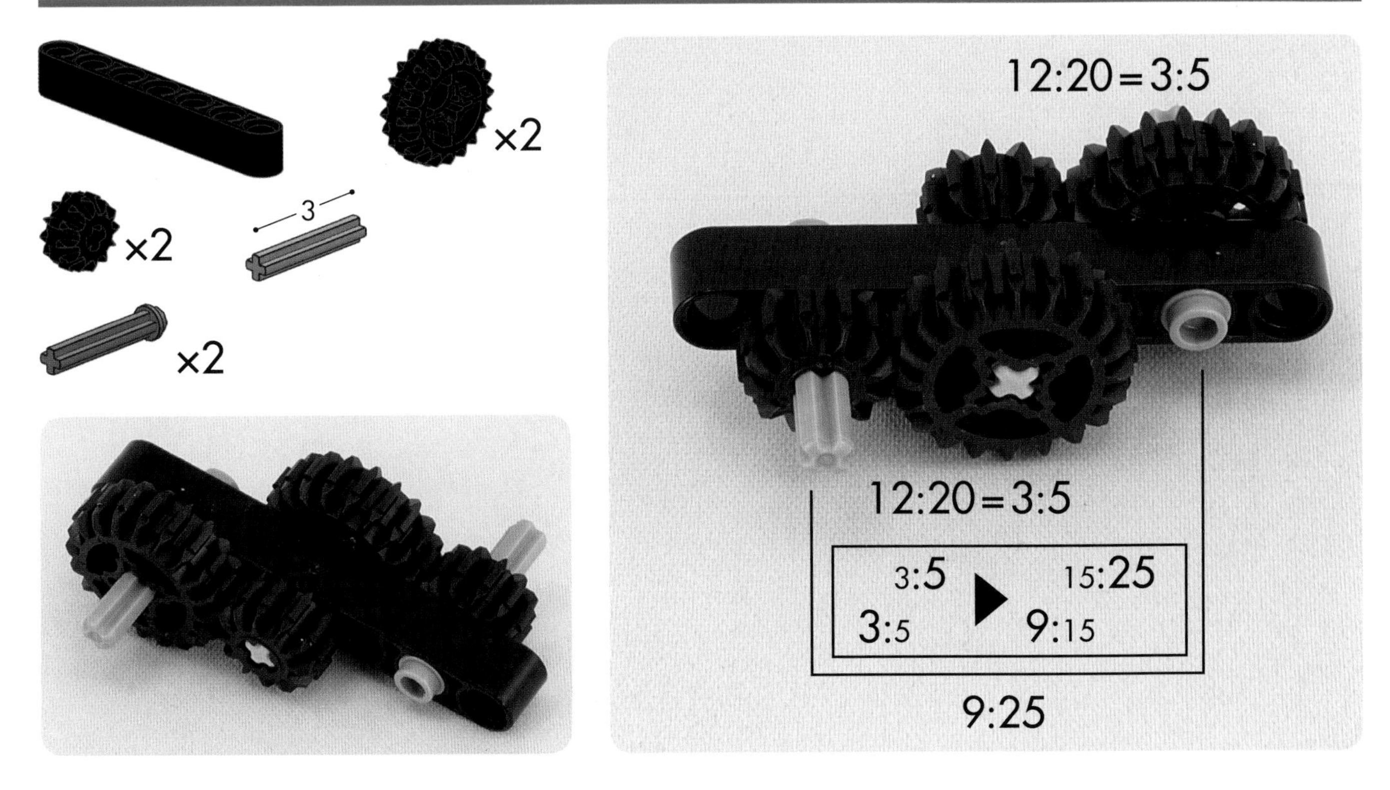
×2
×2
3
×2
12:20=3:5
12:20=3:5
3:5
3:5
15:25
9:15
9:25

#29

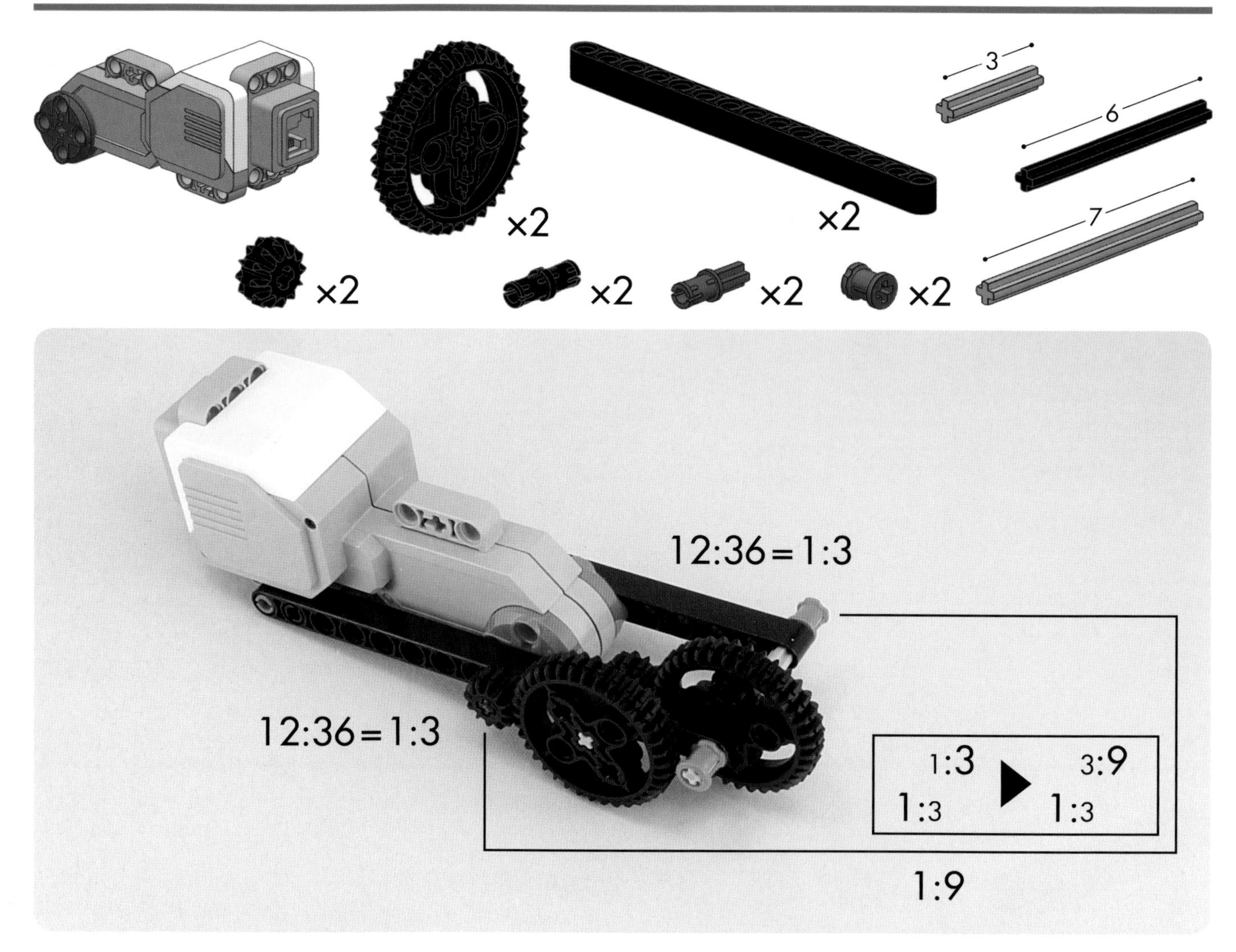
3
6
7
×2
×2
×2
×2
×2
×2
12:36=1:3
12:36=1:3
1:3
1:3
3:9
1:3
1:9

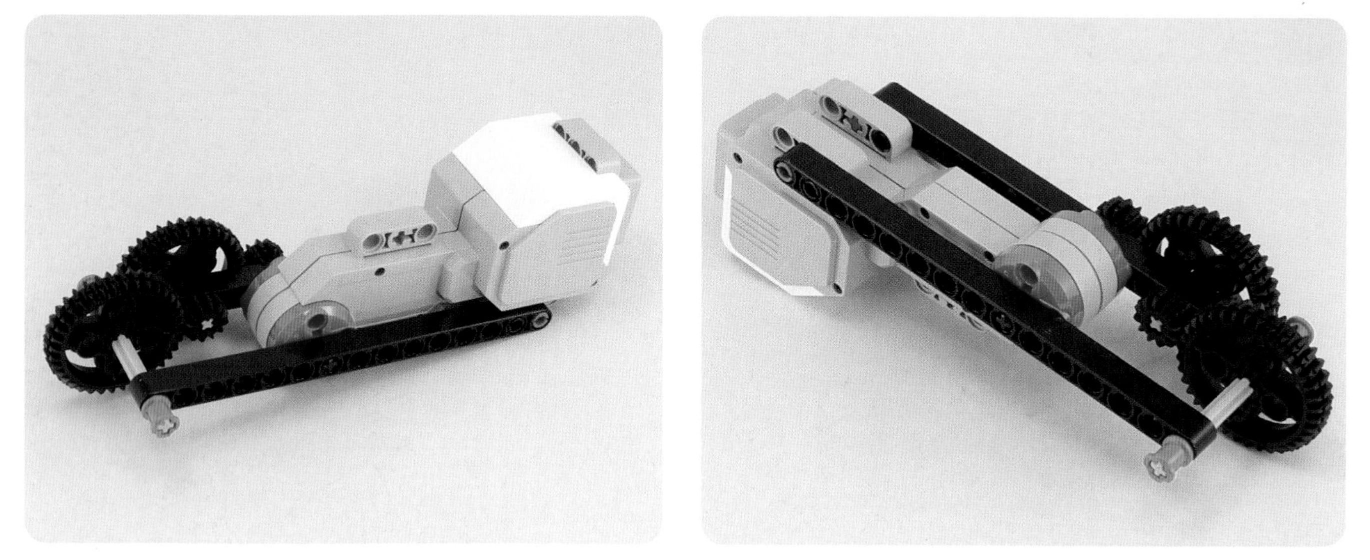

#30

×2 ×2 ×2 ×2 ×2 ×2

3 5

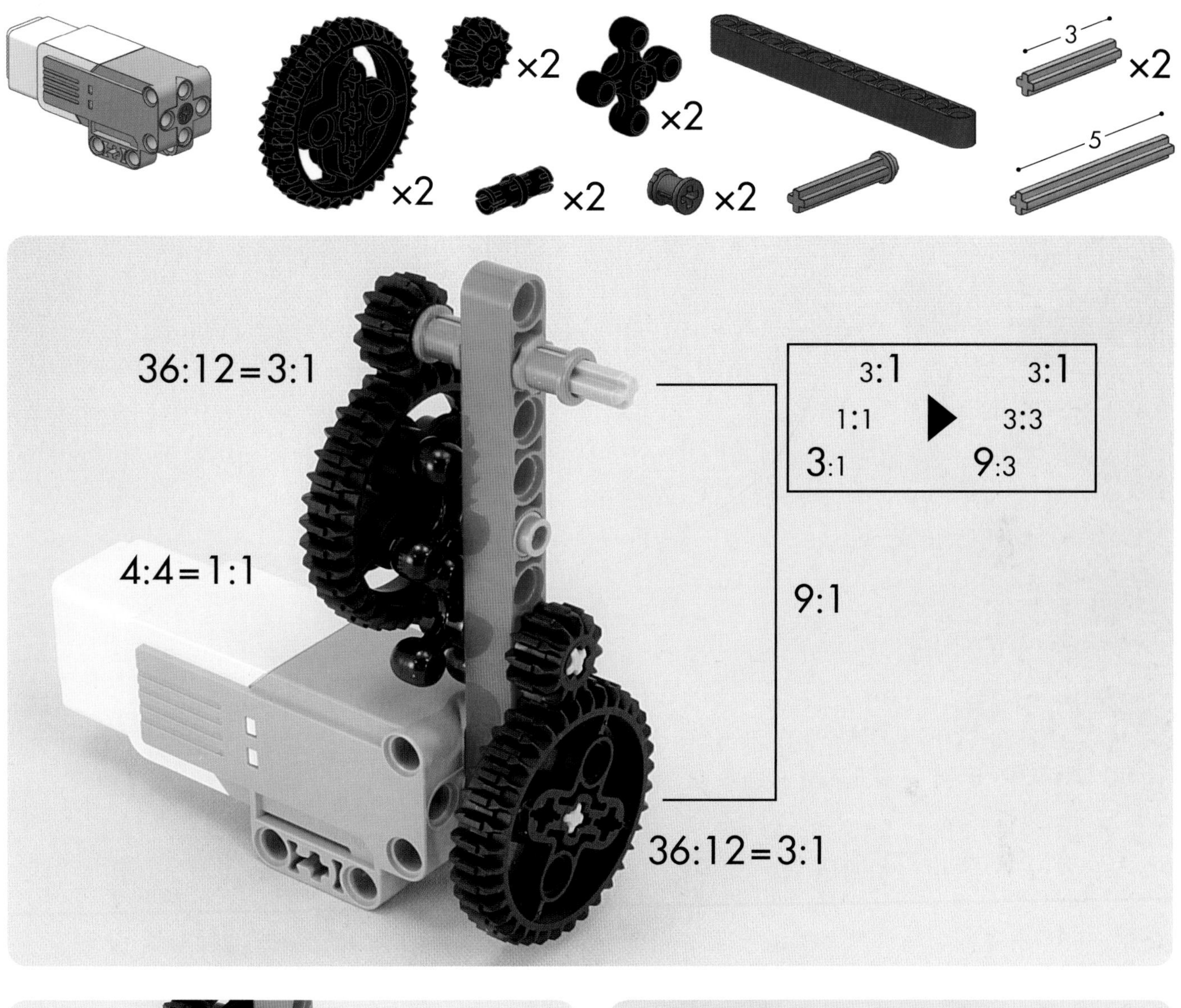

36:12=3:1
4:4=1:1
9:1
36:12=3:1
3:1 3:1
1:1 3:3
3:1 9:3

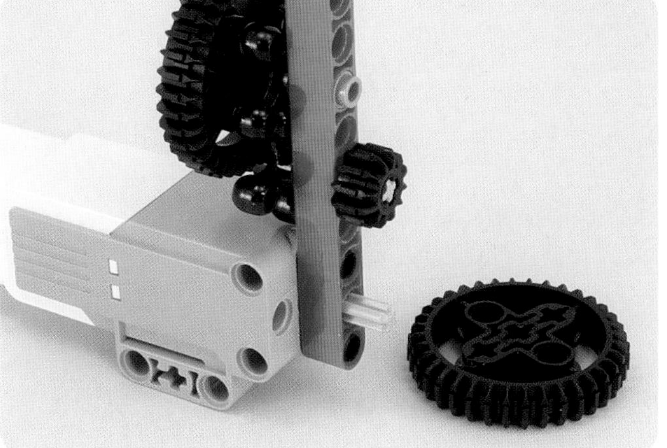

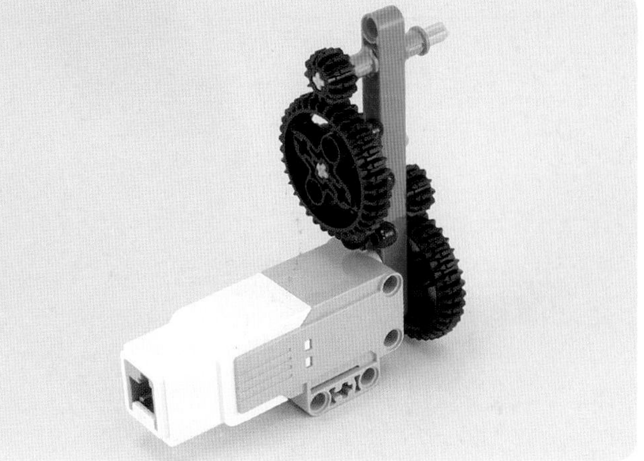

Changing the angle of rotation

#31

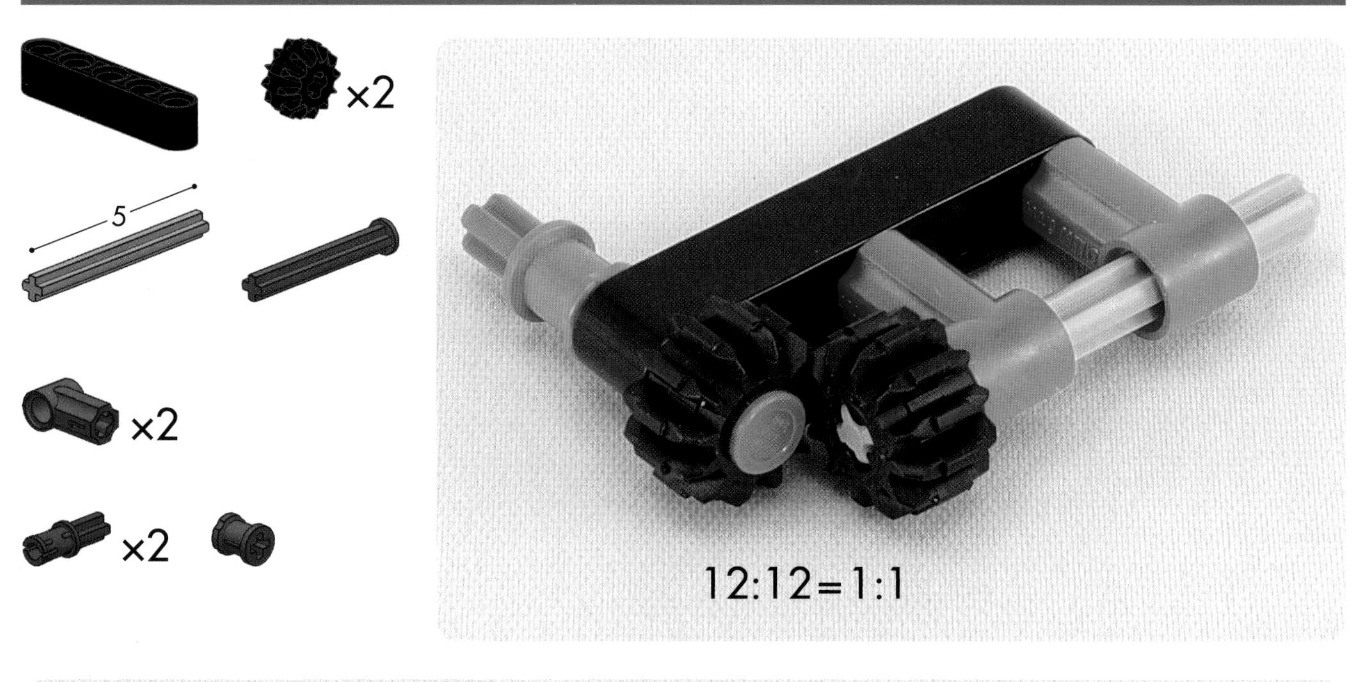
×2
5
×2
×2
12:12=1:1

#32

×2
×2
×2
5
7
4:4 = 1:1
3
×2
12:20 = 3:5

#33

#34

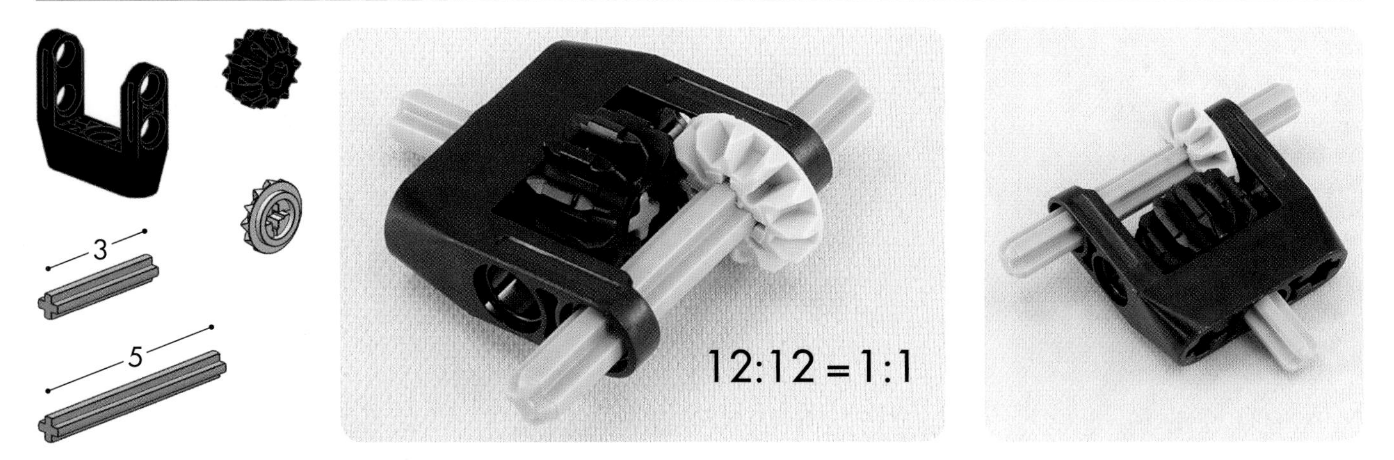

3
5
12:12 = 1:1

#35

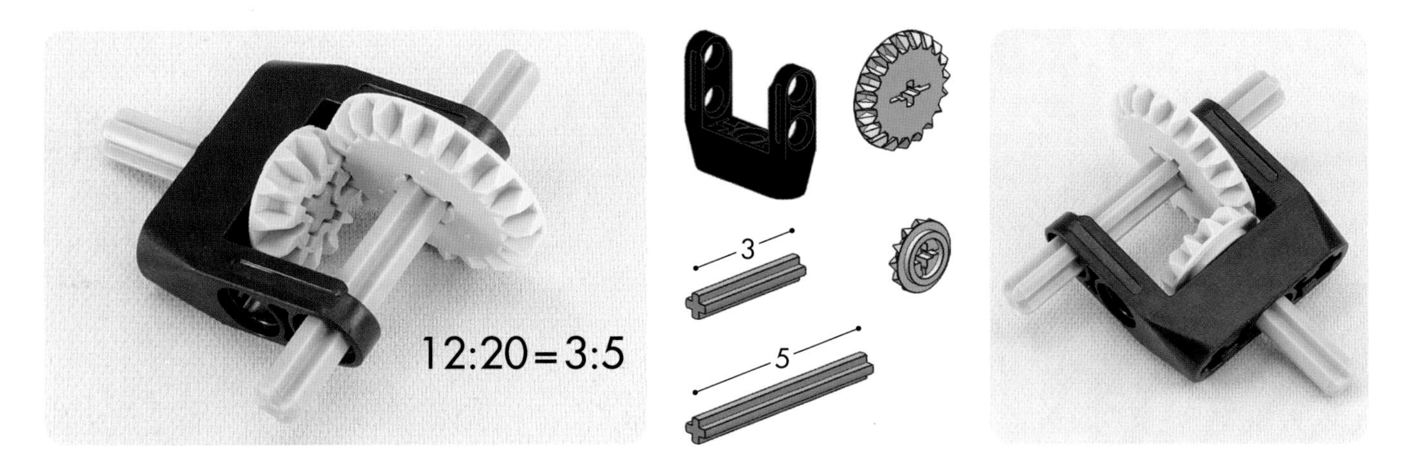

12:20 = 3:5
3
5

#36

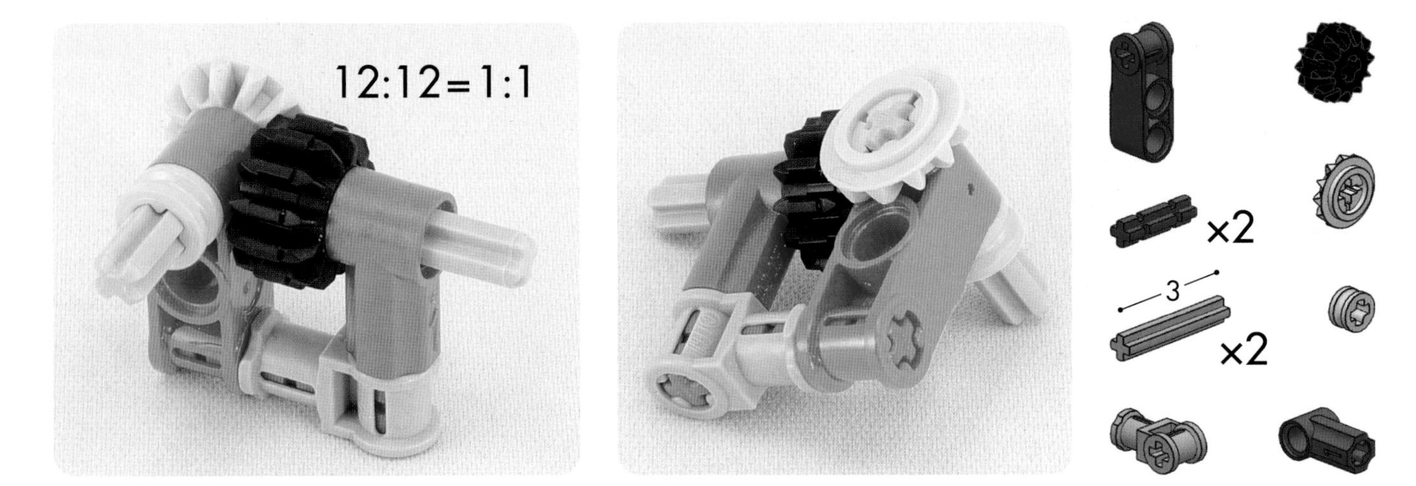

12:12 = 1:1
×2
3
×2

#37

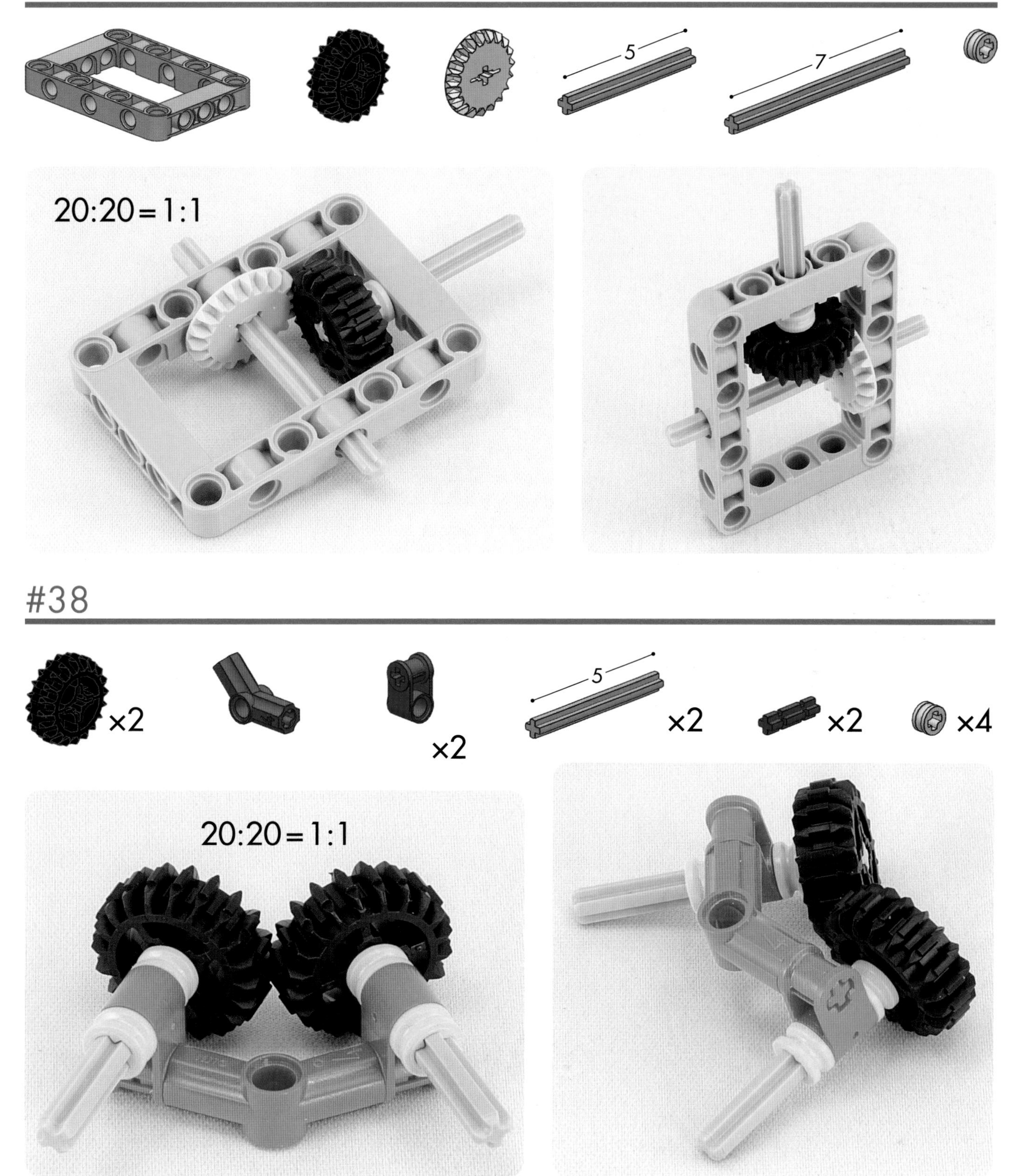
5
7
20:20 = 1:1
#38
×2
×2
5
×2
×2
×4
20:20 = 1:1

#39

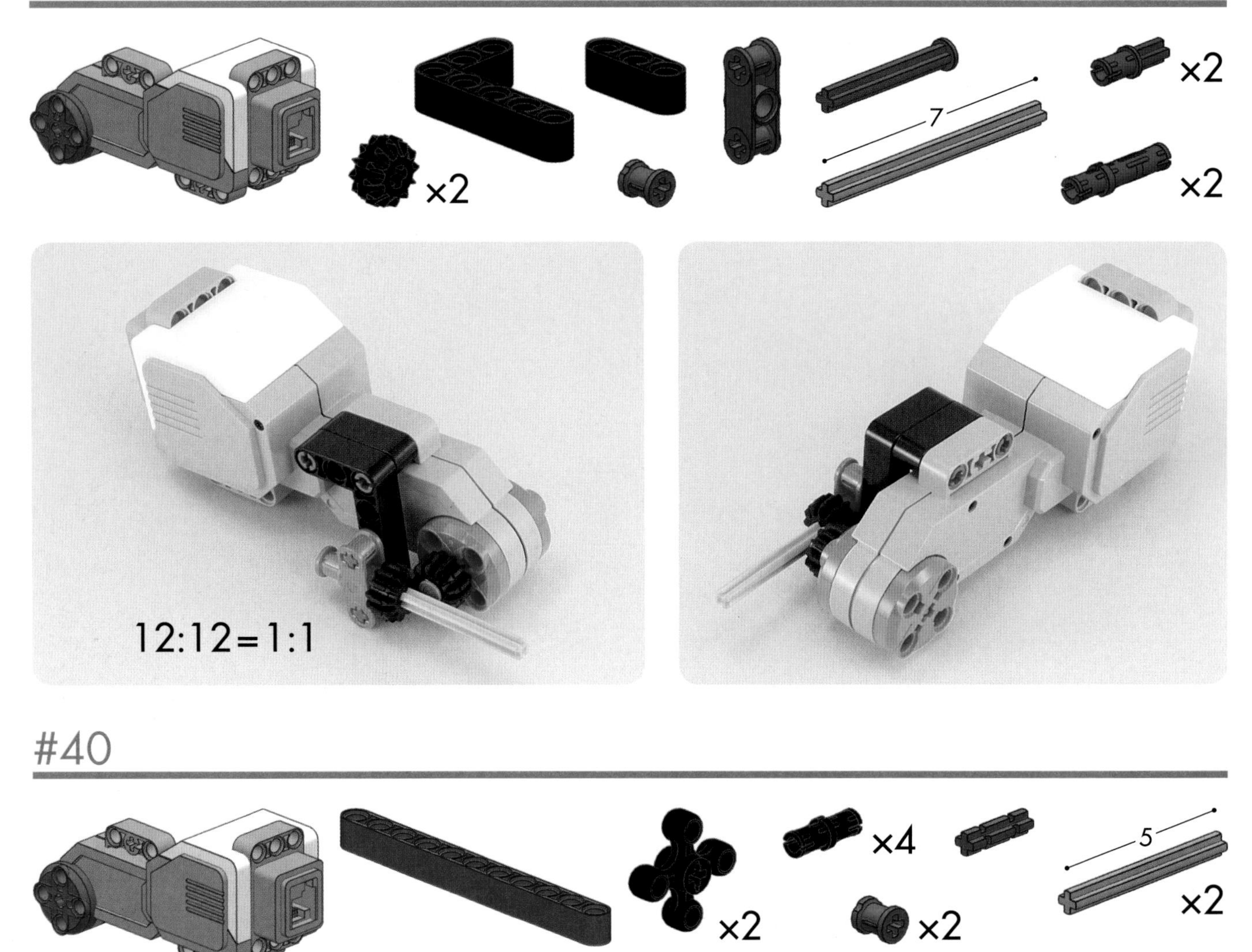
×2
7
×2
×2
12:12=1:1

#40

×4

×2

5

×2

×2

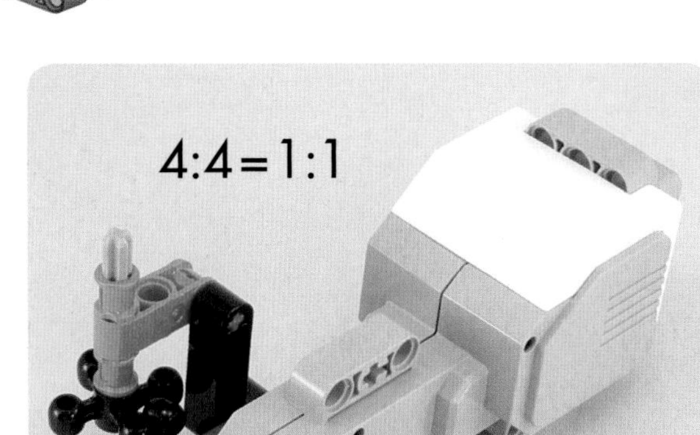
4:4=1:1

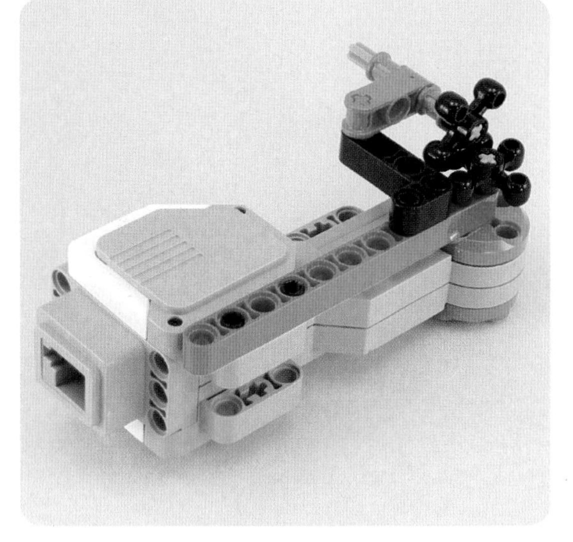

#41

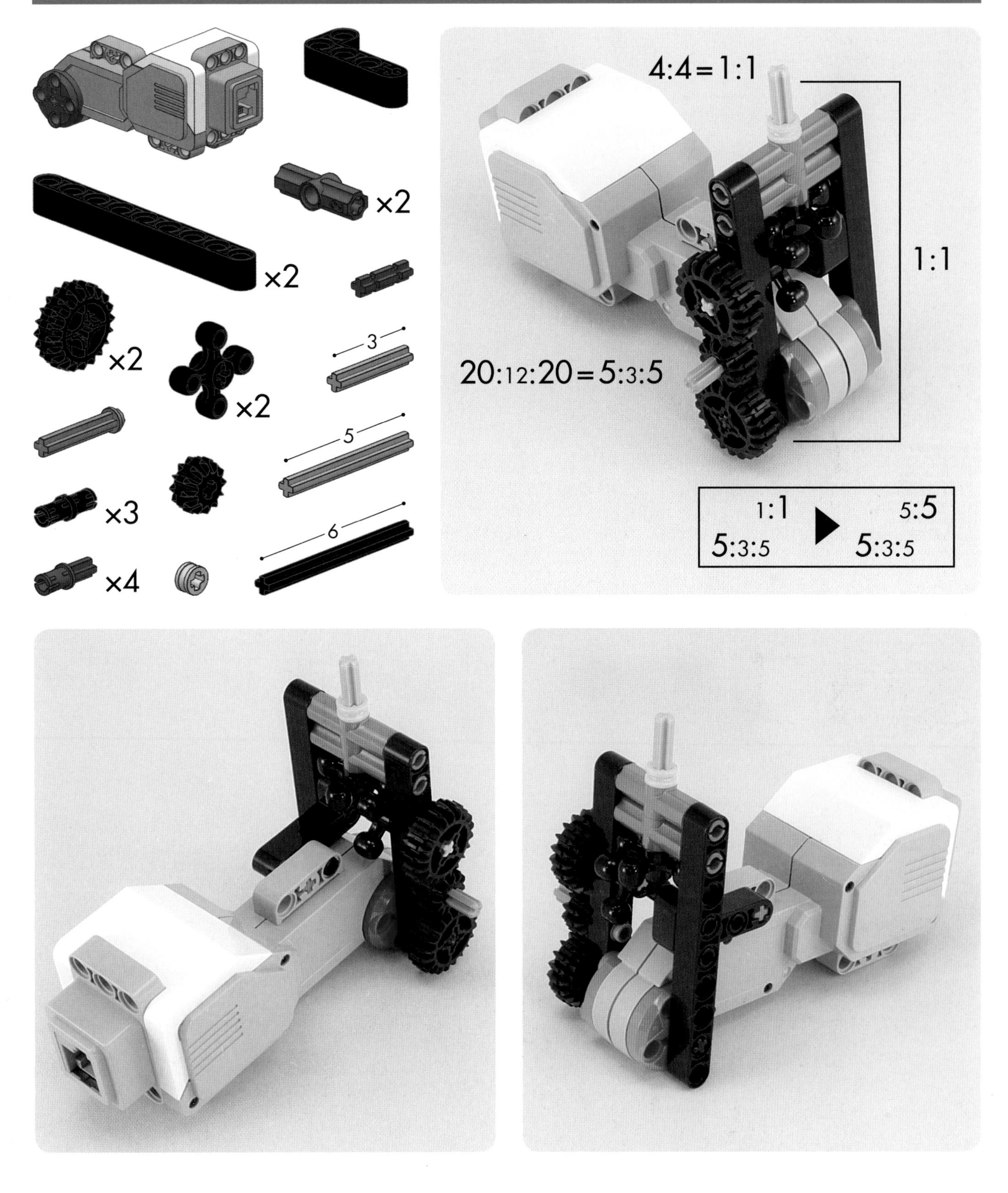
×2
×2
×2
×2
3
×2
5
×3
6
×4
4:4=1:1
1:1
20:12:20=5:3:5
1:1
5:3:5
5:5
5:3:5

#42

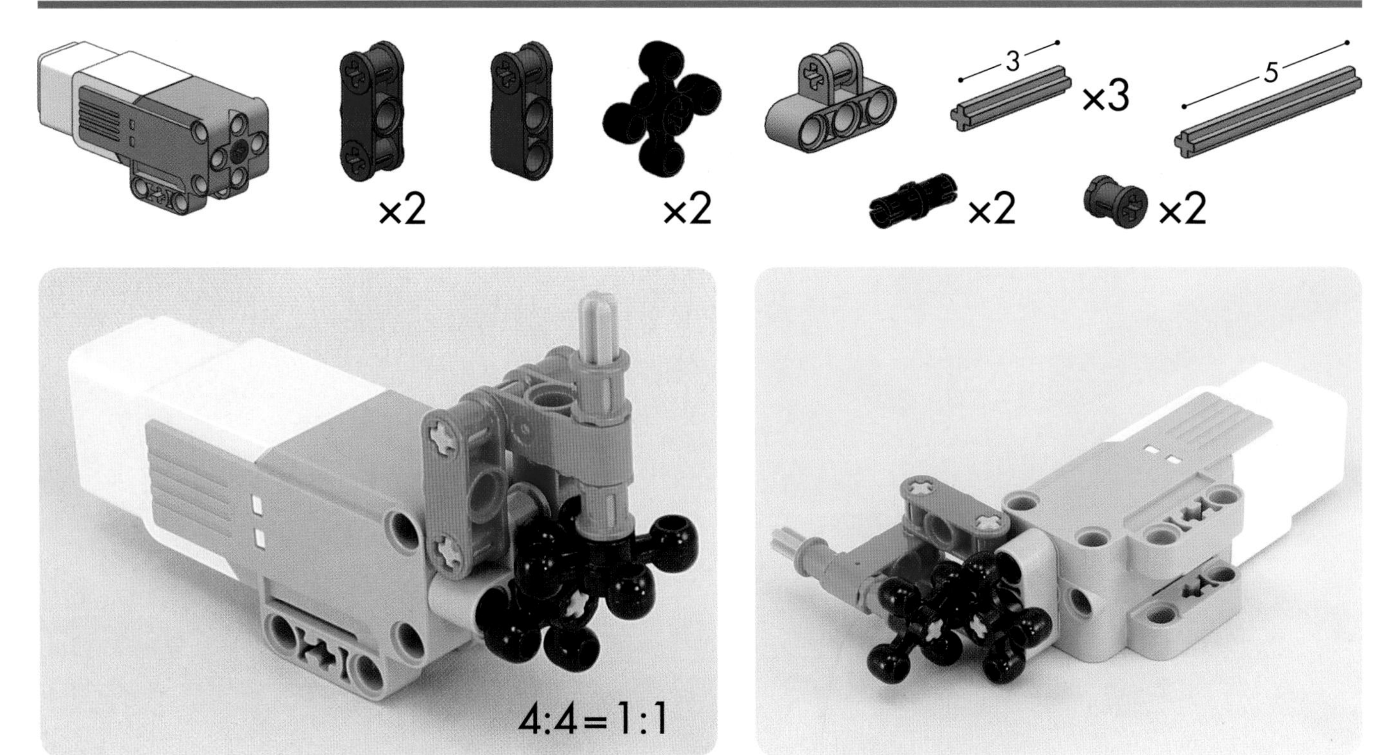

3
×3
5
×2
×2
×2
×2
4:4=1:1

#43

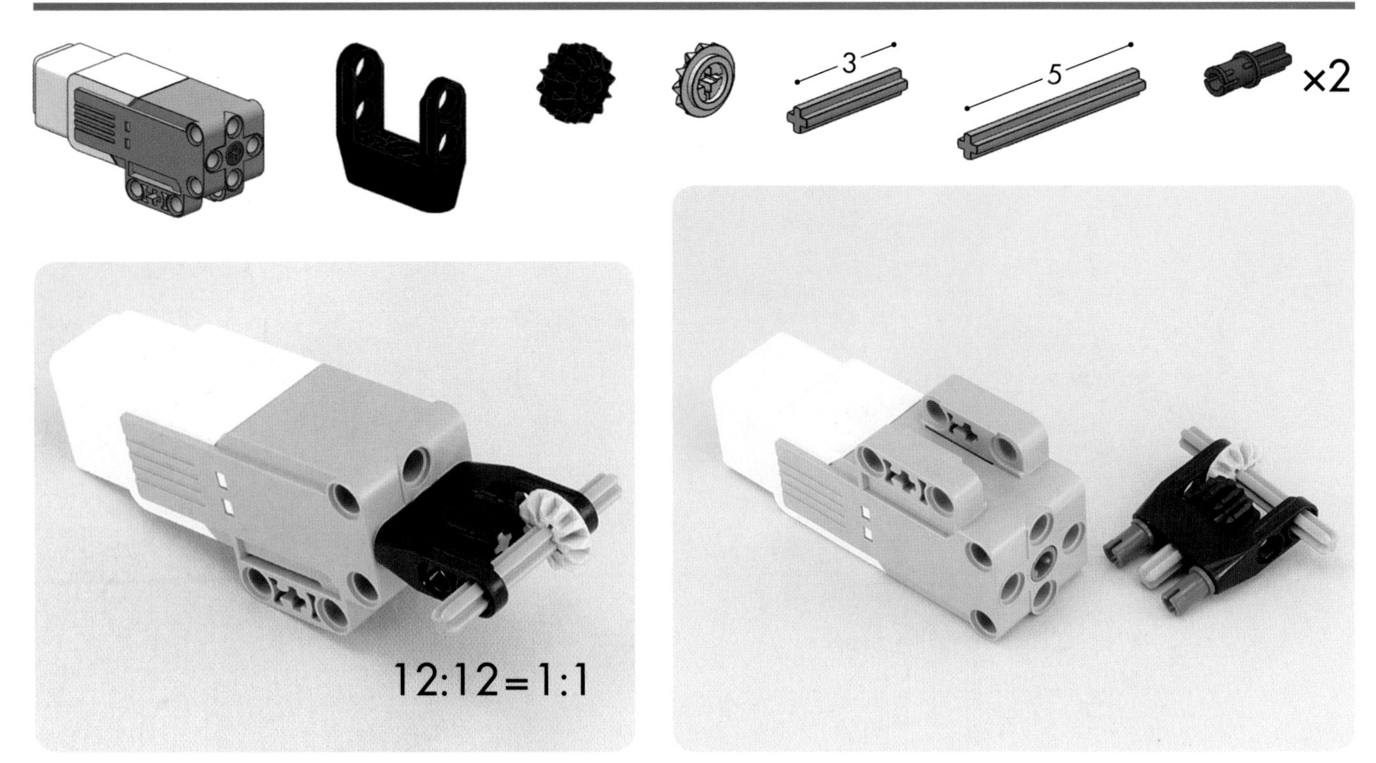

3
5
×2
12:12=1:1

#44

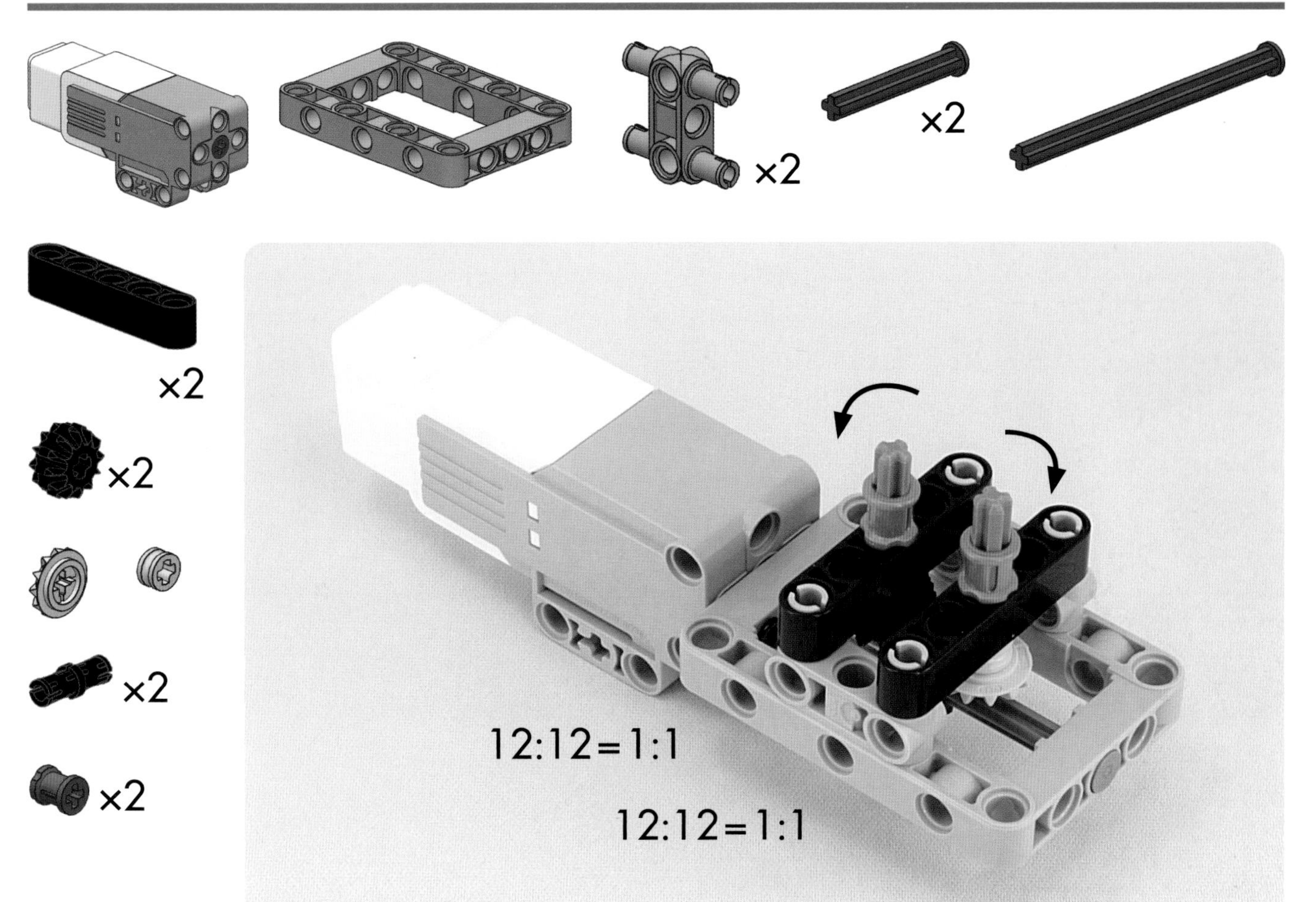
×2
×2
×2
×2
×2
×2
×2
12:12=1:1
12:12=1:1

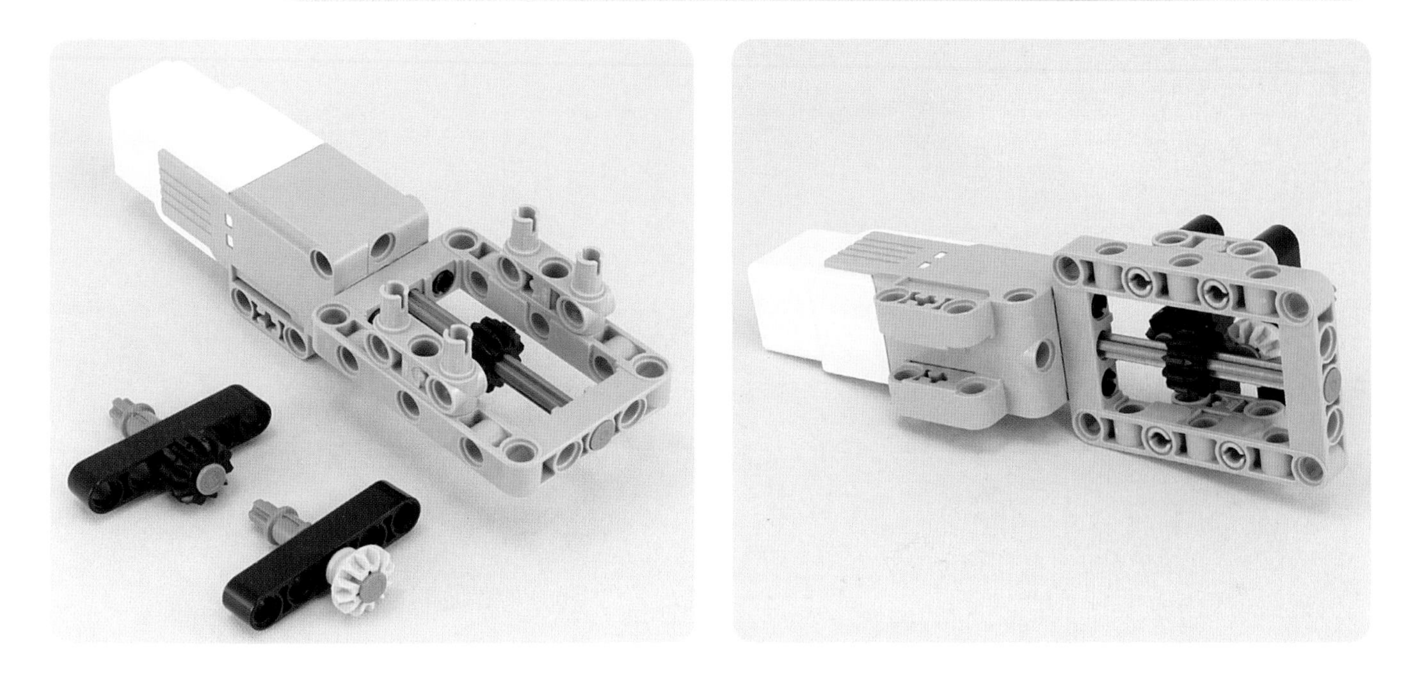

Using worm drives

#45

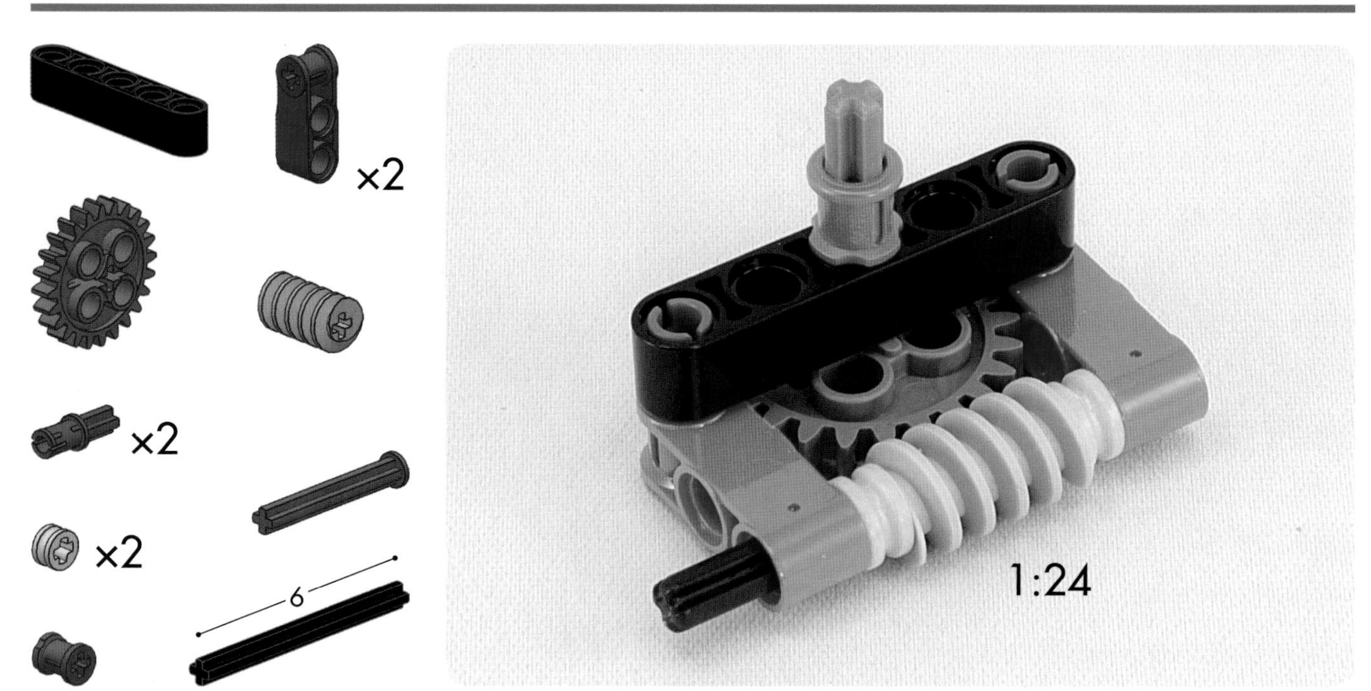

×2
×2
×2
6
1:24

#46

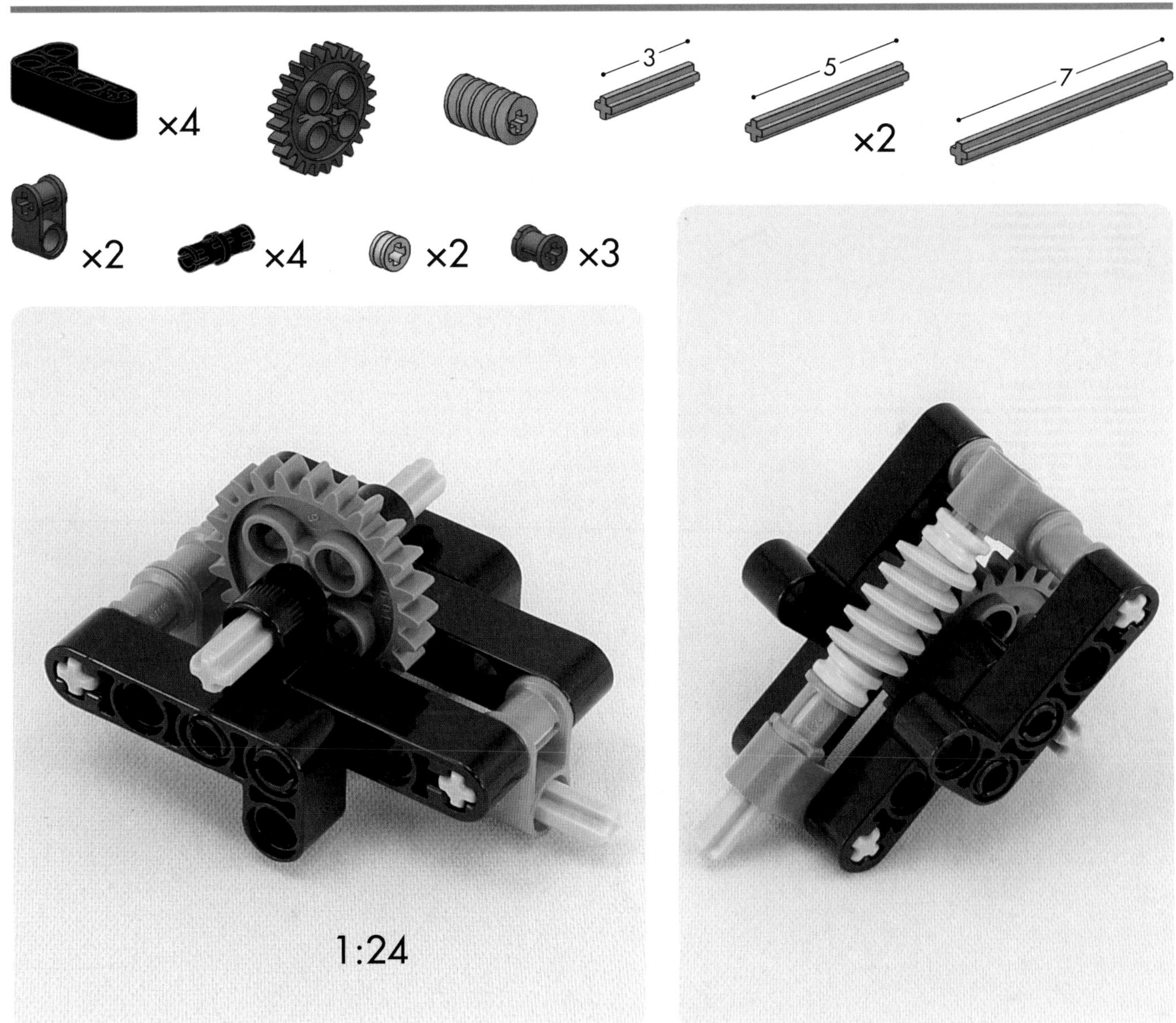
×4
3
5
×2
7
×2
×4
×2
×3
1:24

#47

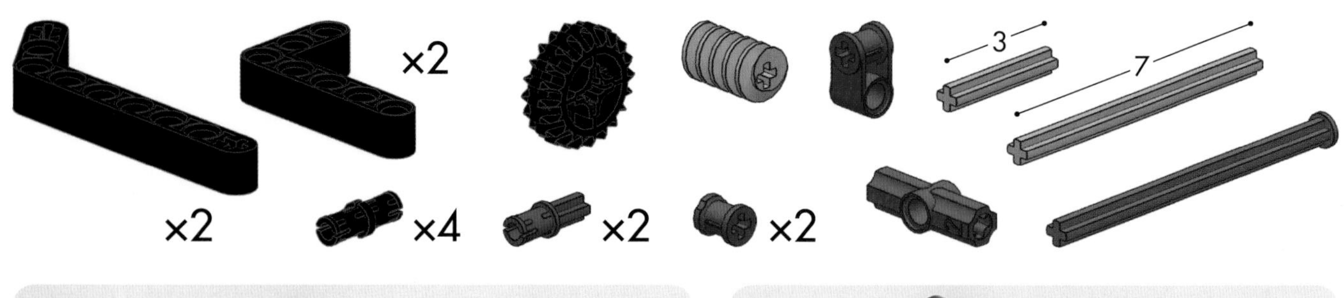
×2
×2
×4
×2
×2
3
7

1:20

#48

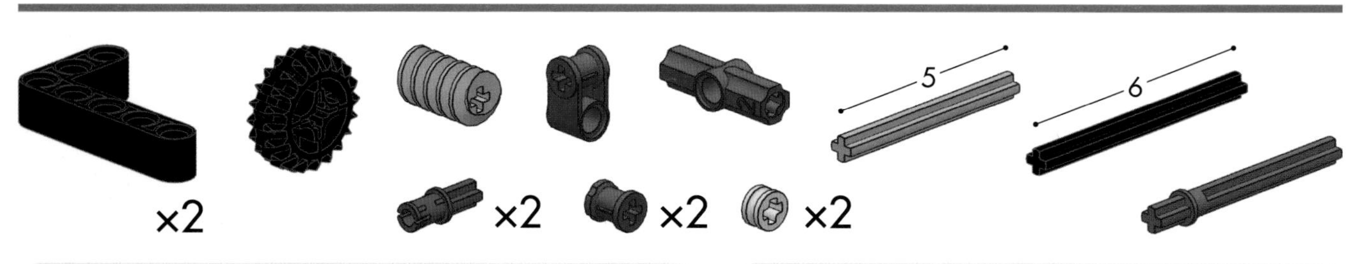
×2
×2
×2
×2
5
6

1:20

#49

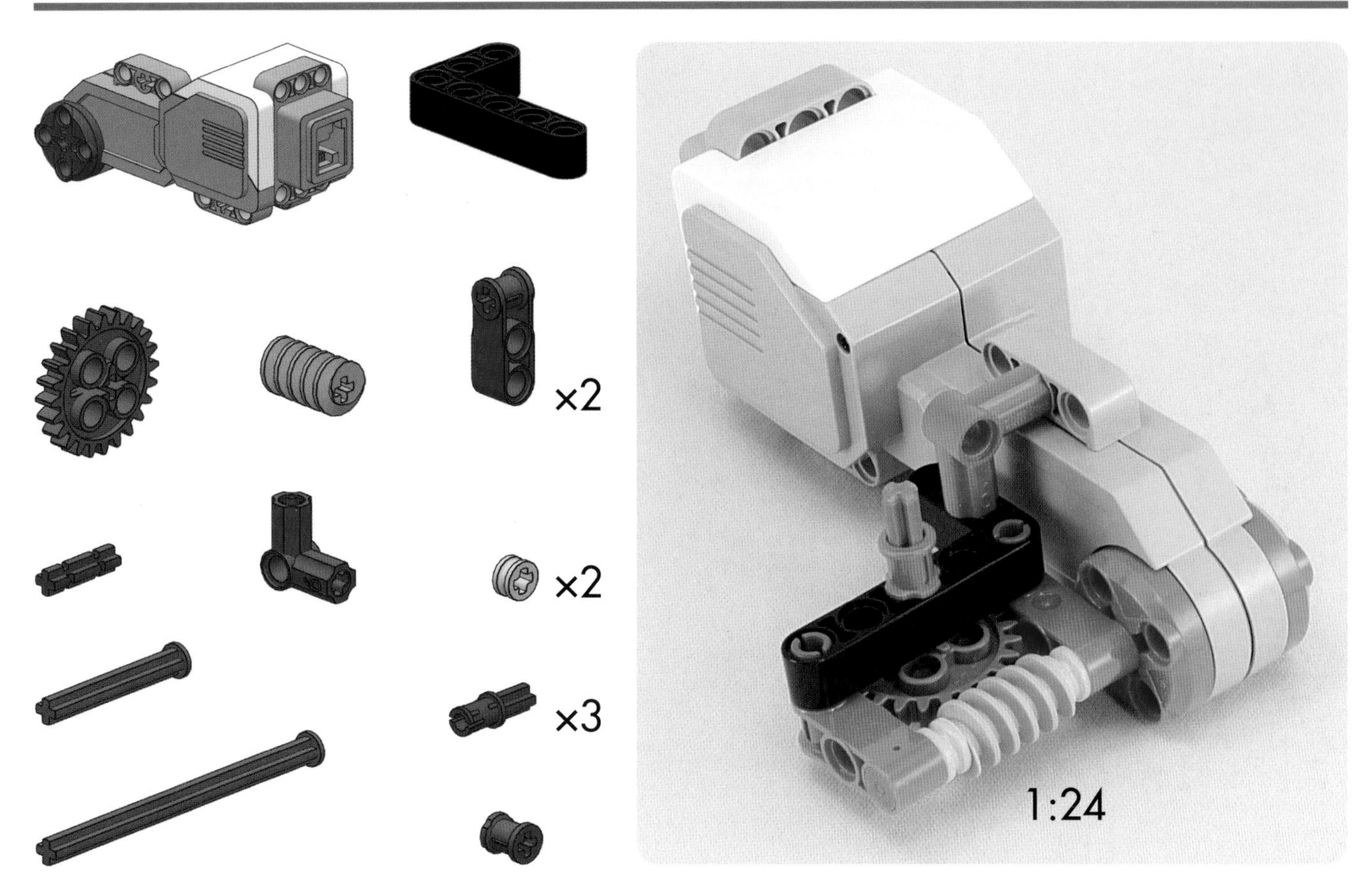
×2
×2
×3
1:24

#50

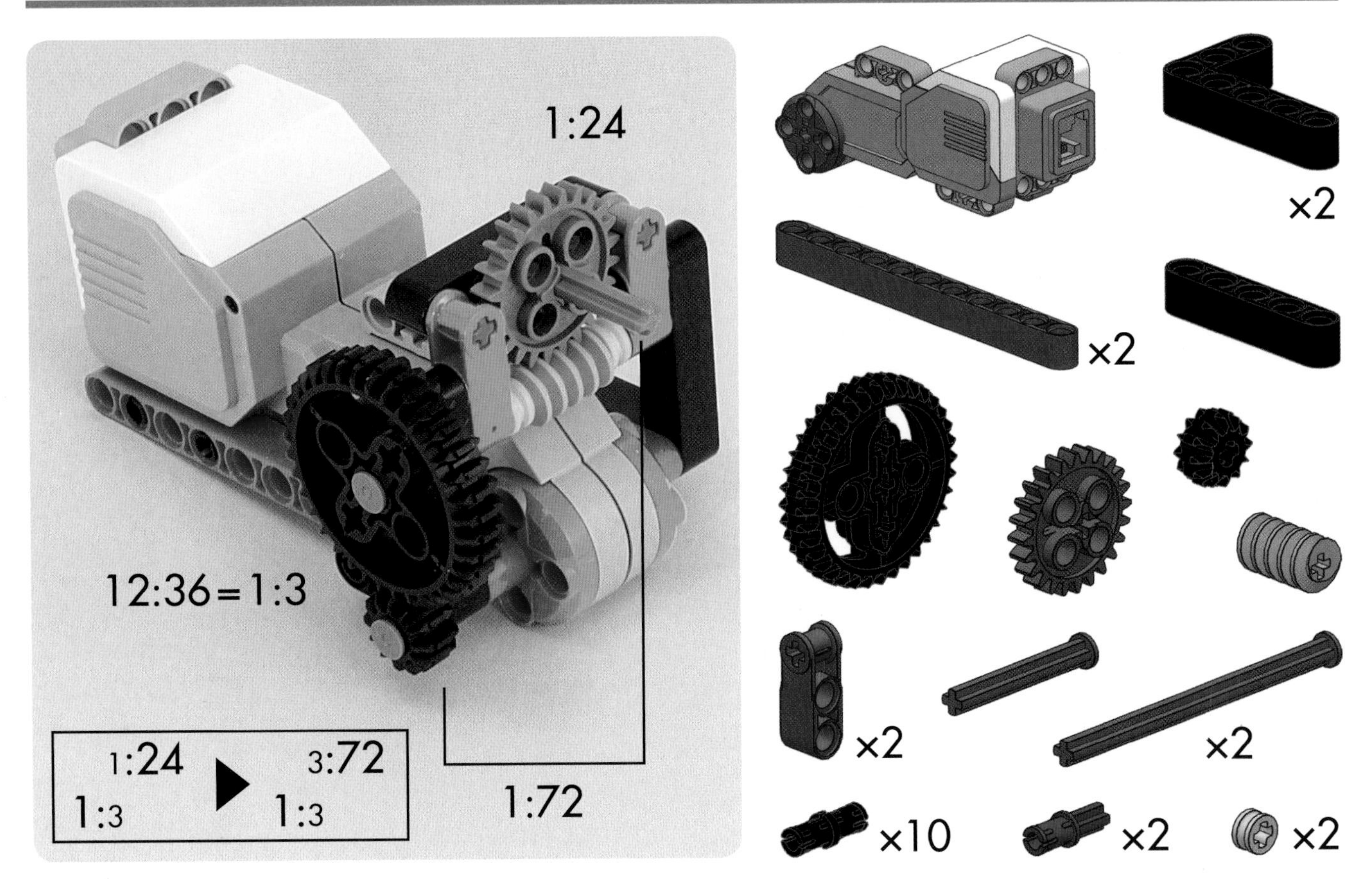
1:24
12:36=1:3
1:24
3:72
1:3
1:3
1:72
×2
×2
×2
×2
×2
×10
×2
×2

#51

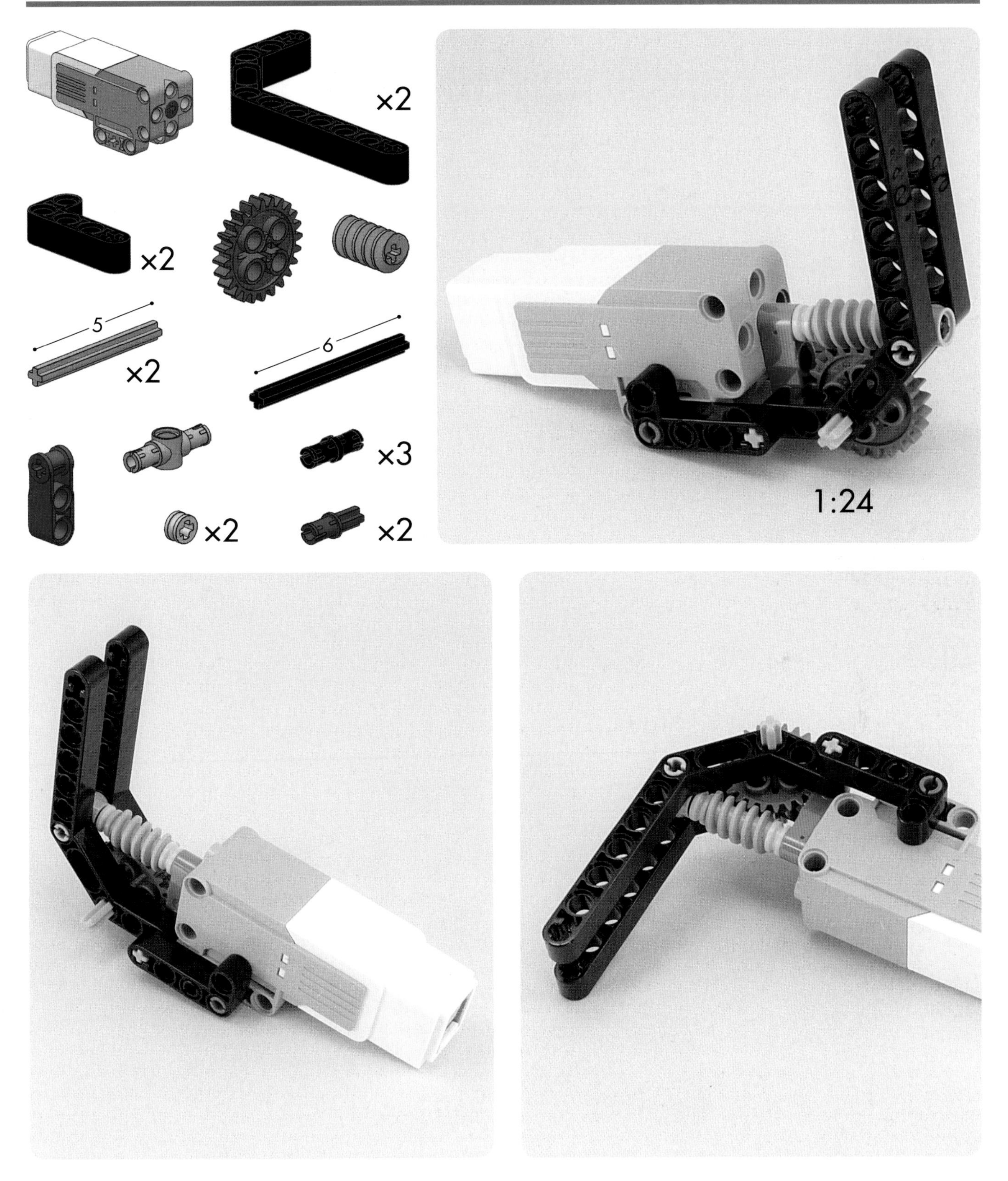
×2
×2
5
×2
6
×3
×2
×2
1:24

Swinging mechanisms

#52

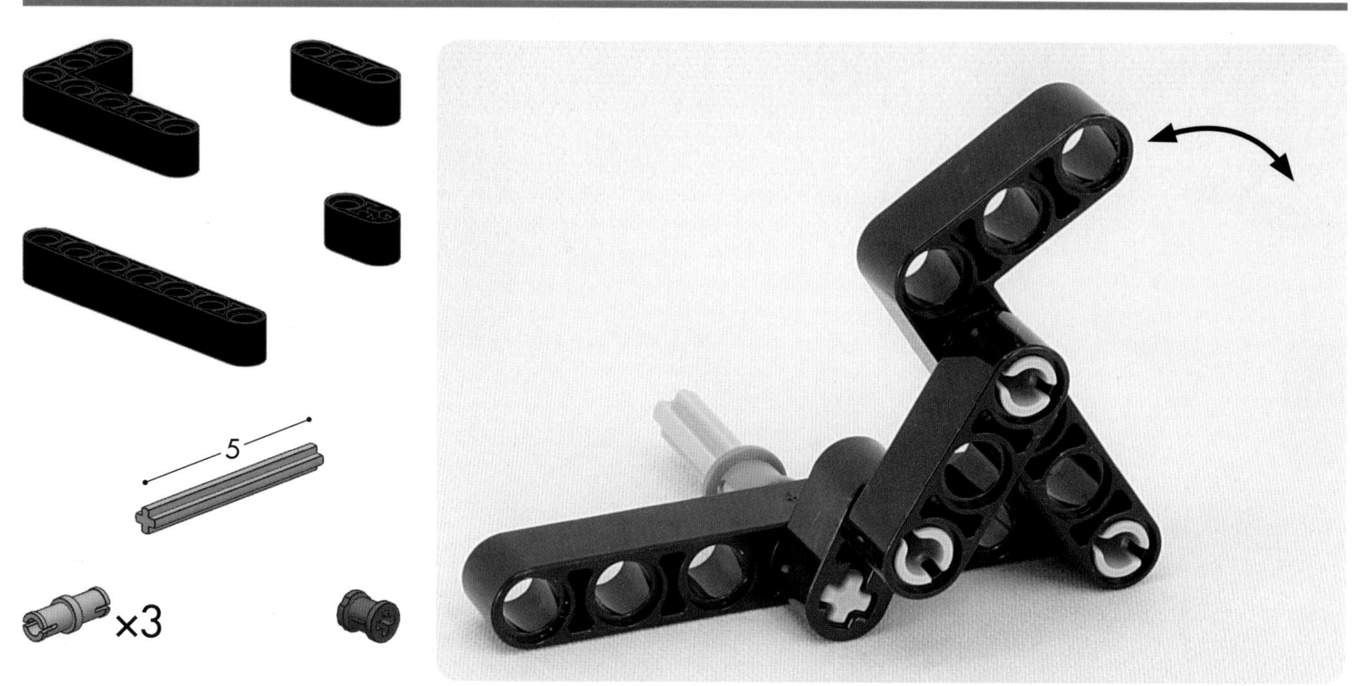
5
×3

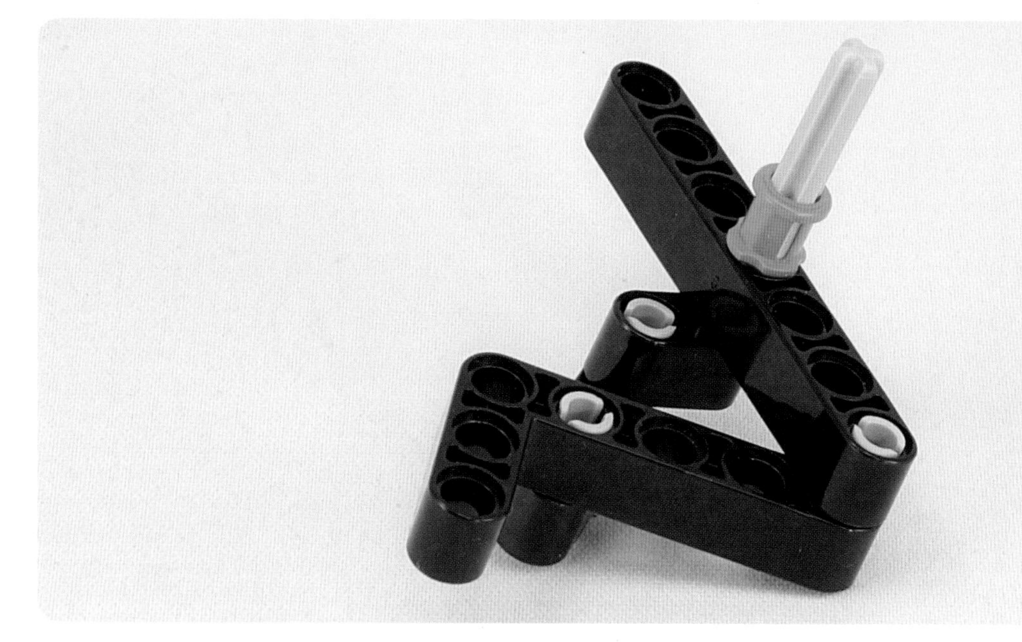

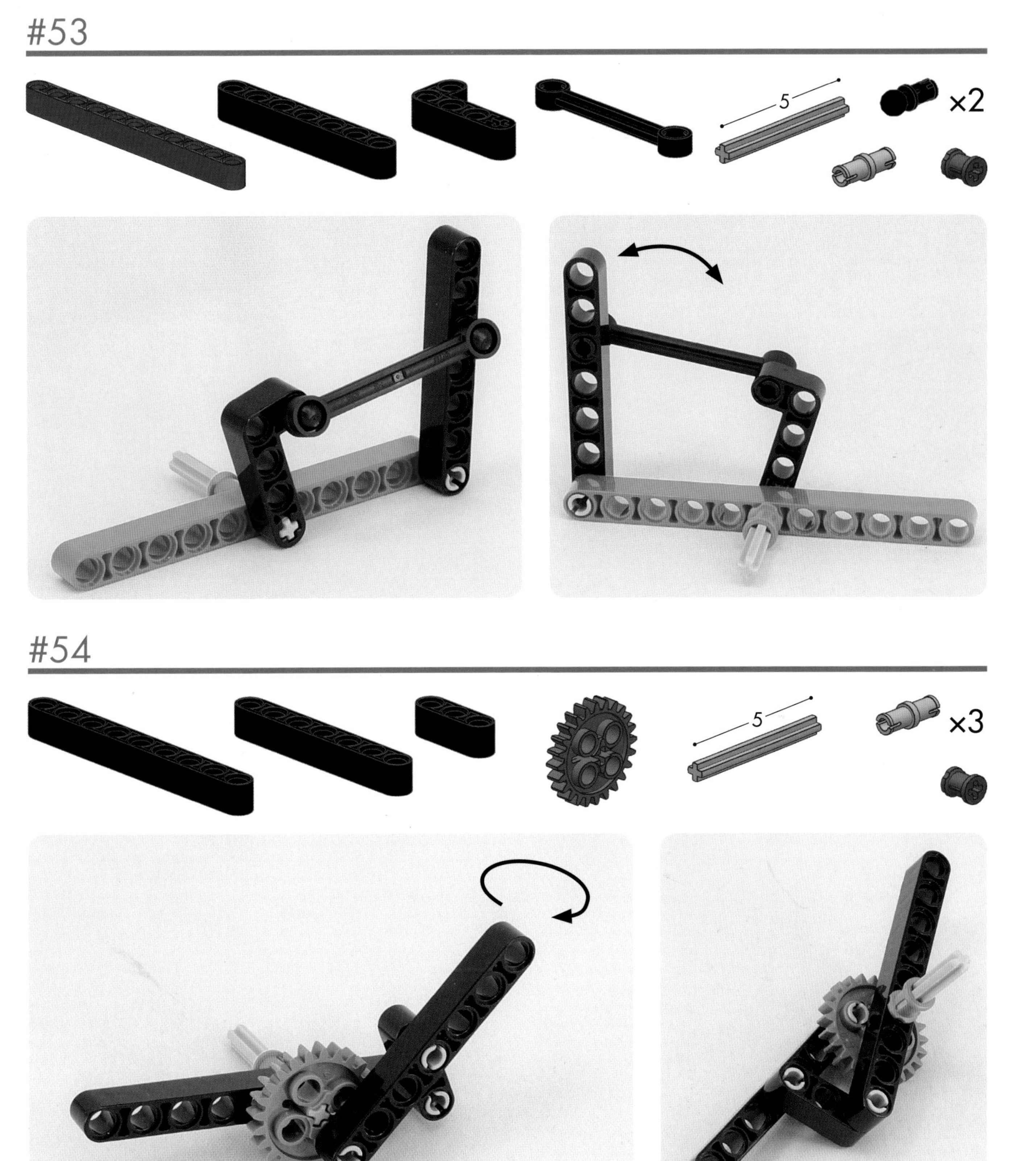
#53
5
×2
#54
5
×3

#55

×3
×3

#56

#57

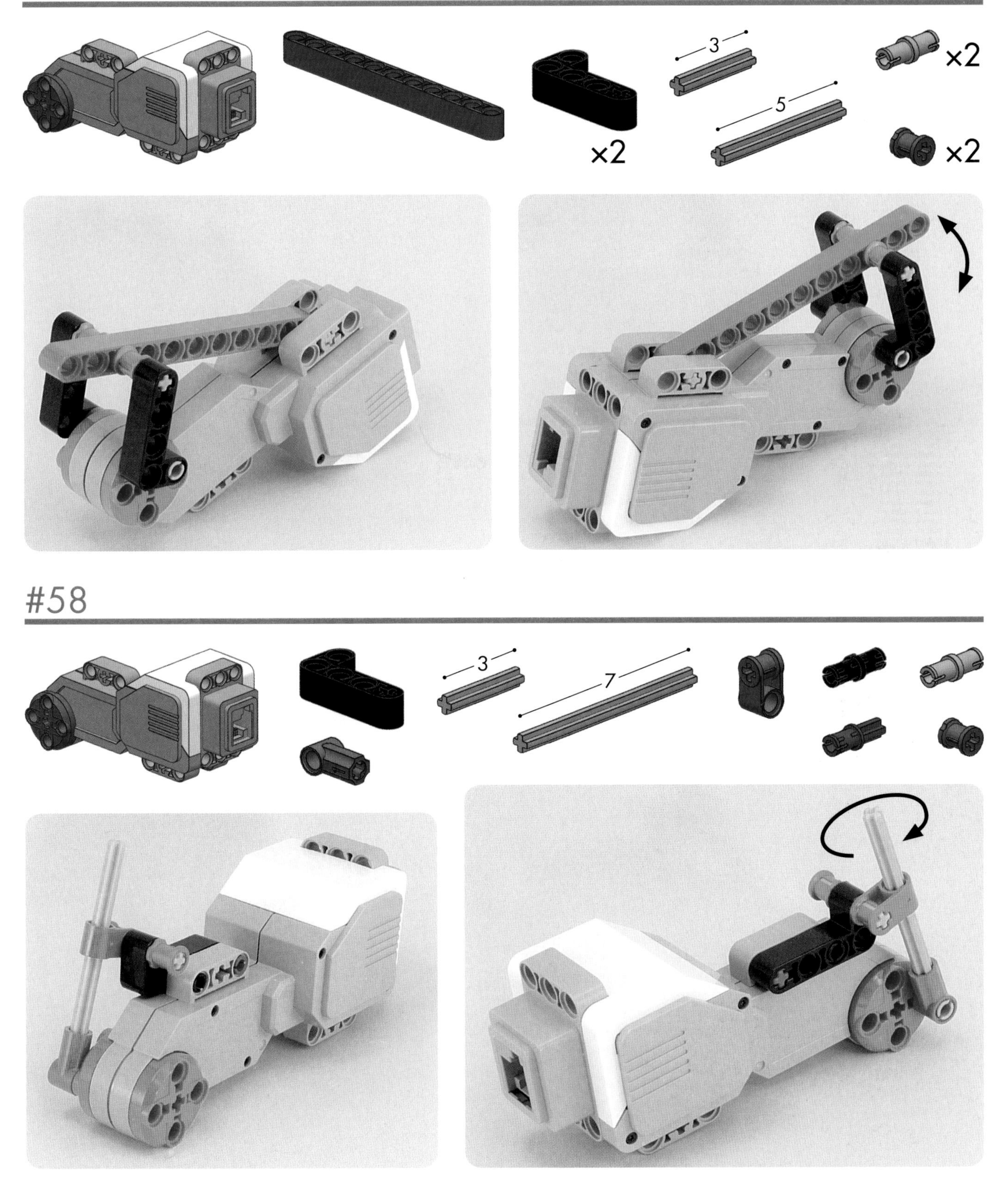
3
5
×2
×2
×2
#58
3
7

#59

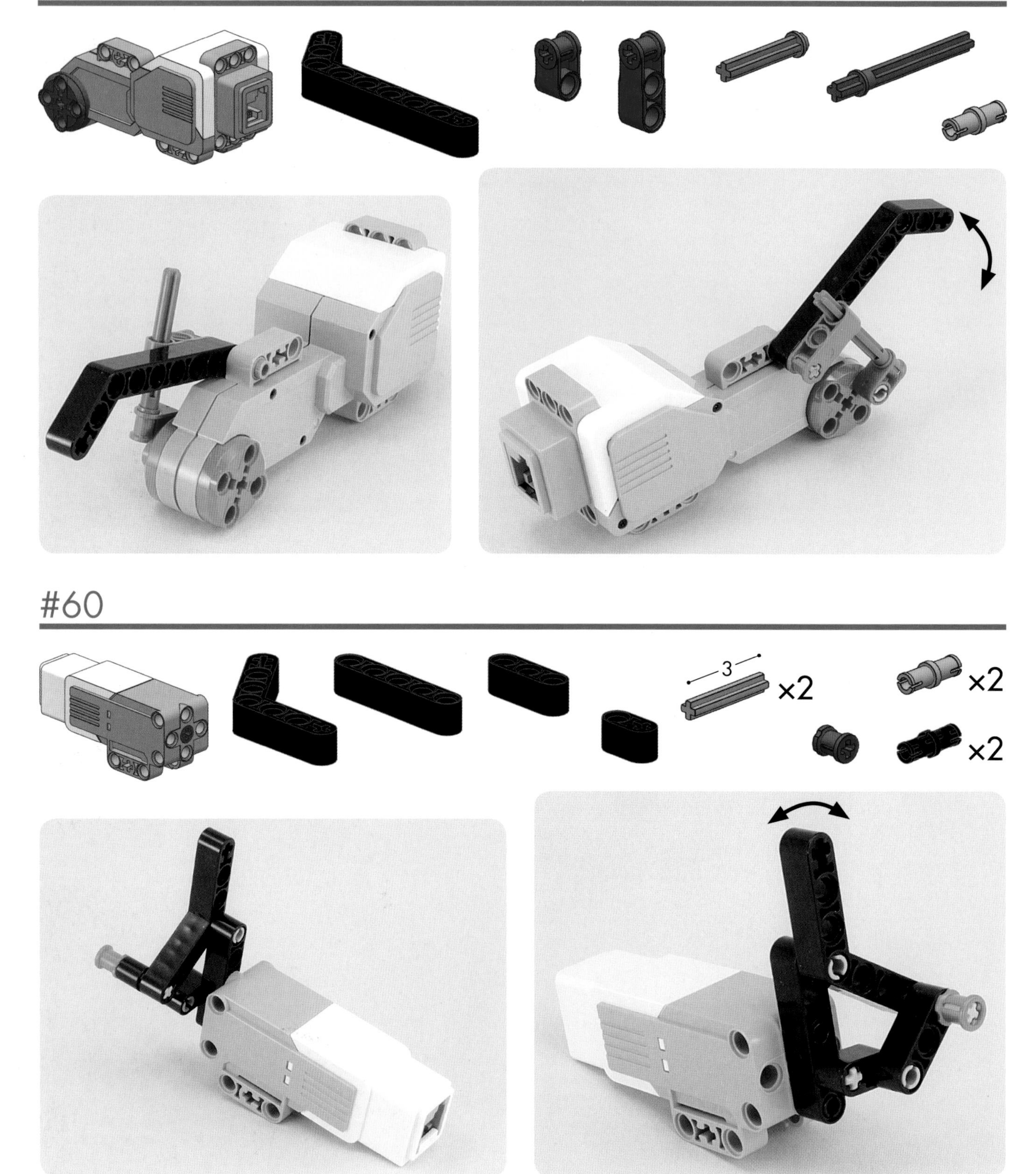
3
×2
×2
×2

#60

#61

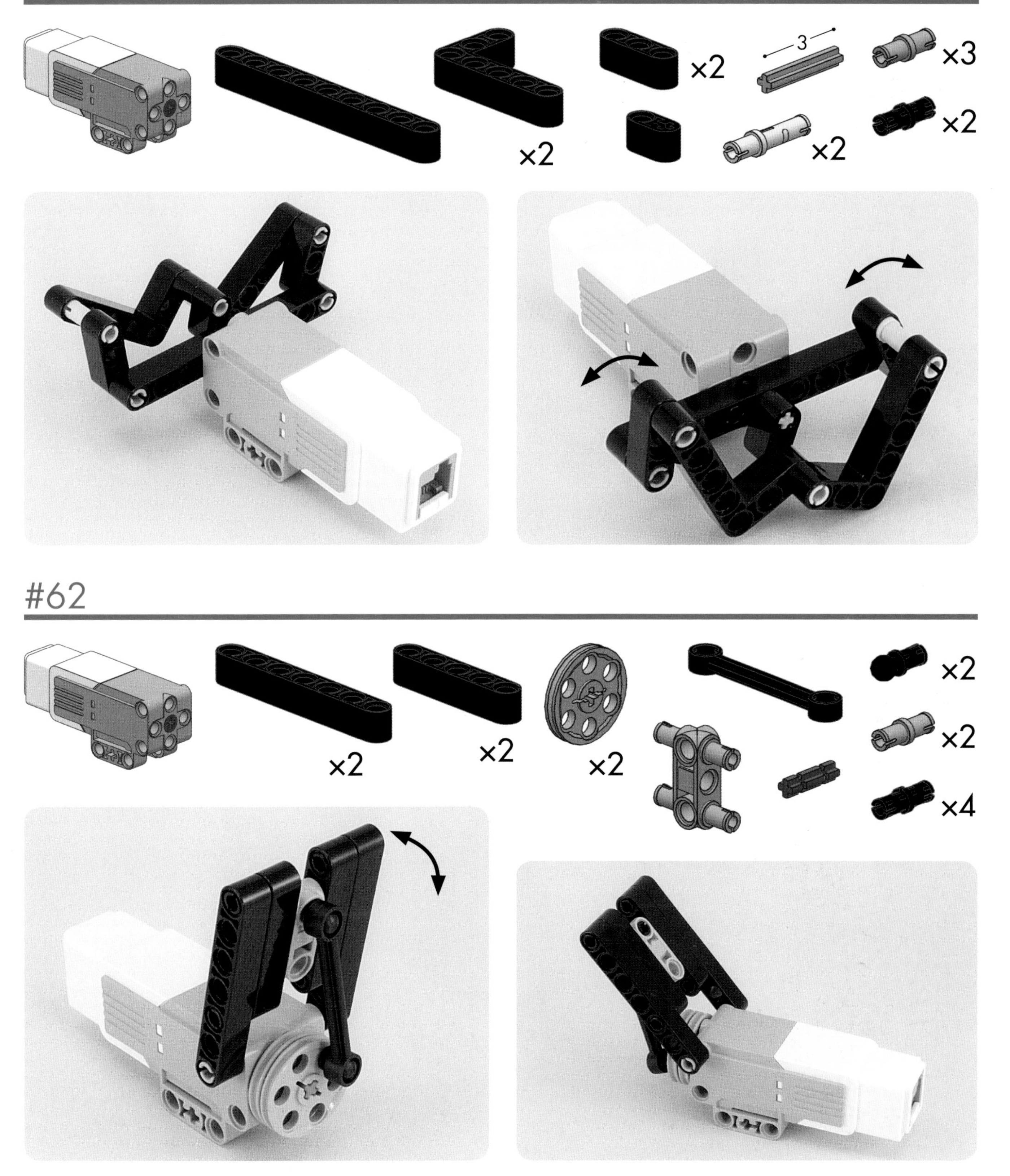
×2
3
×3
×2
×2
×2
#62
×2
×2
×2
×2
×2
×4

Reciprocating mechanisms

#63

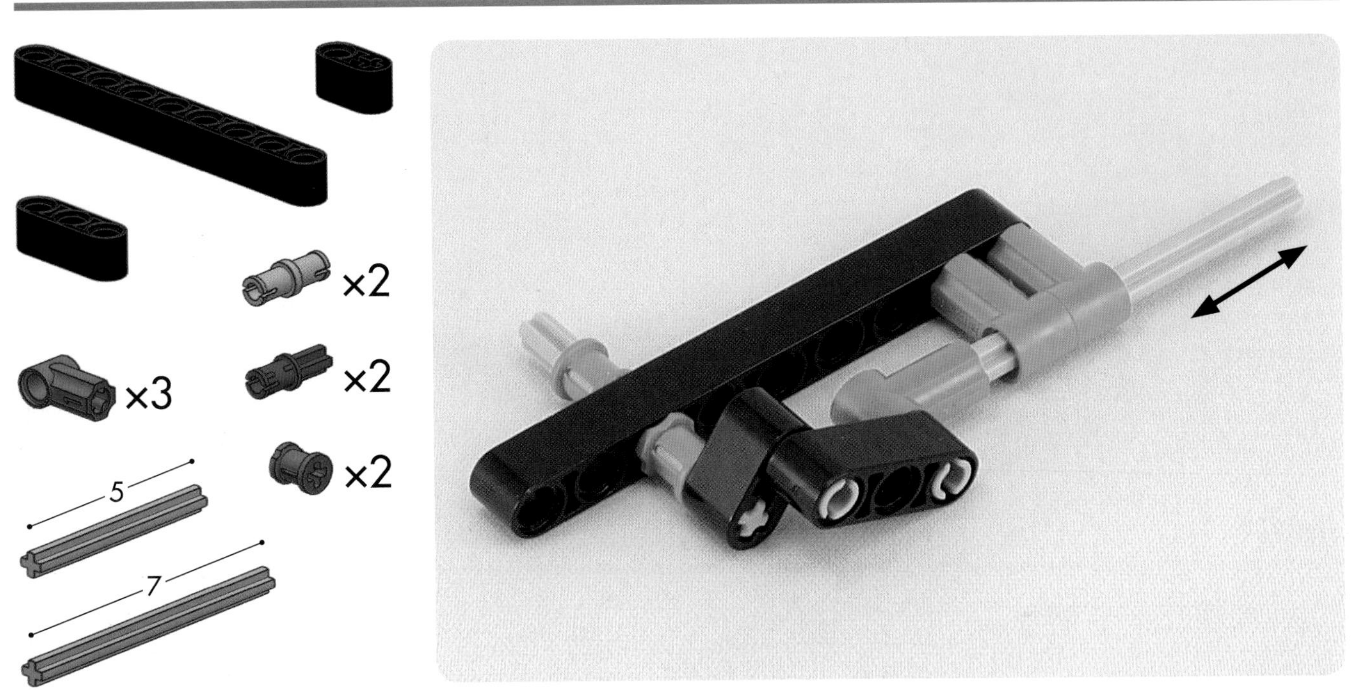
×2
×3
×2
×2
5
7

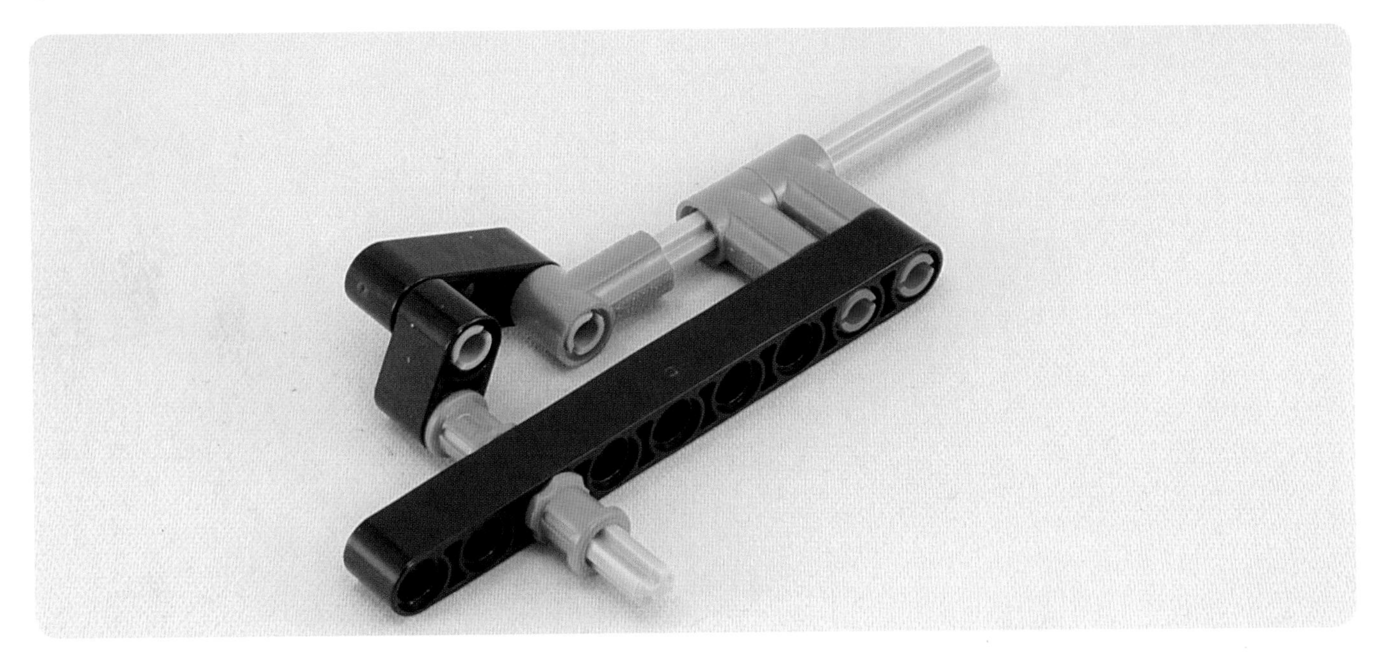

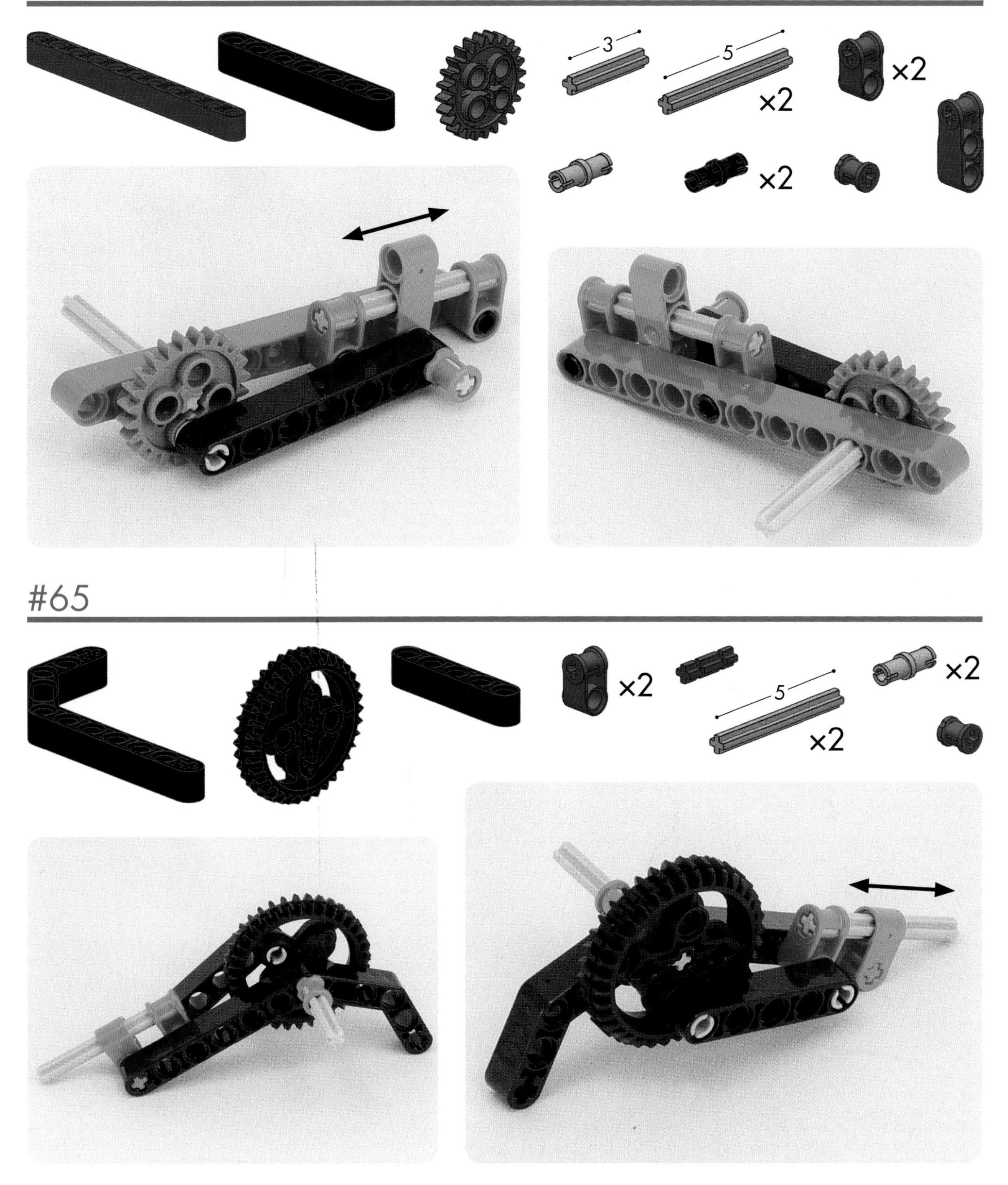
3
5
×2
×2
×2
#65
×2
5
×2
×2

#66

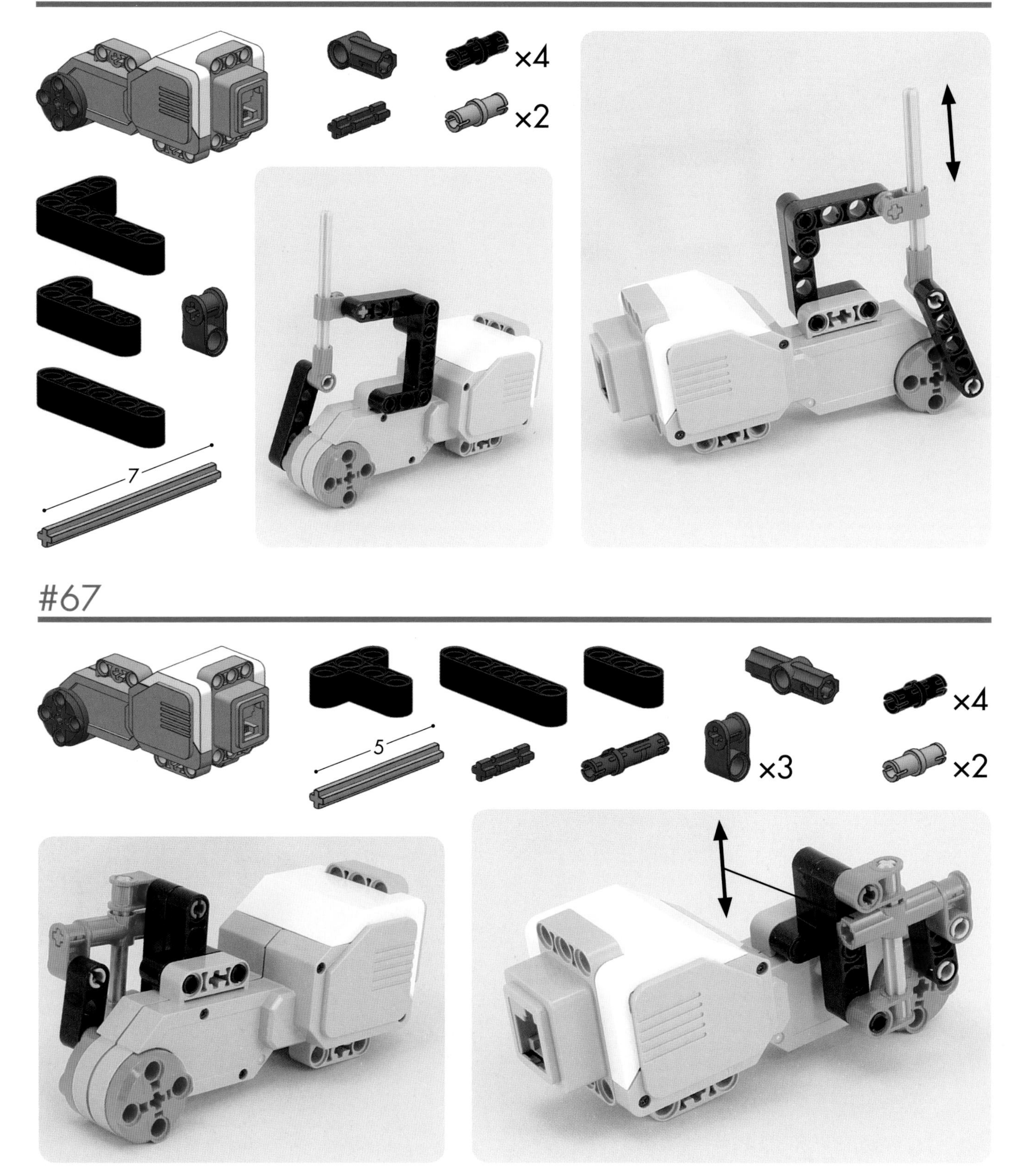
×4
×2
7
#67
5
×3
×4
×2

#68

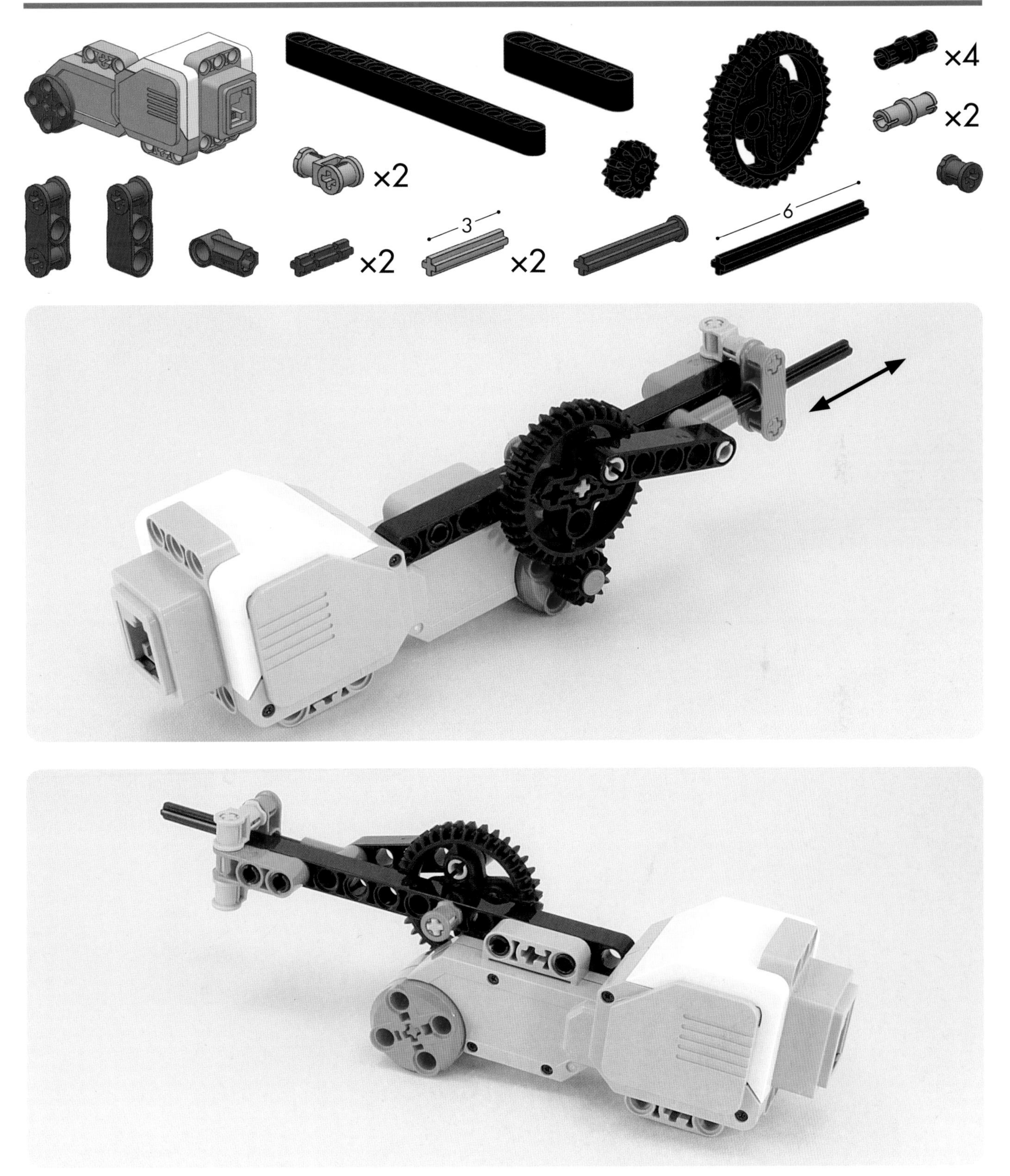

×4
×2
×2
×2
3
×2
6

#69

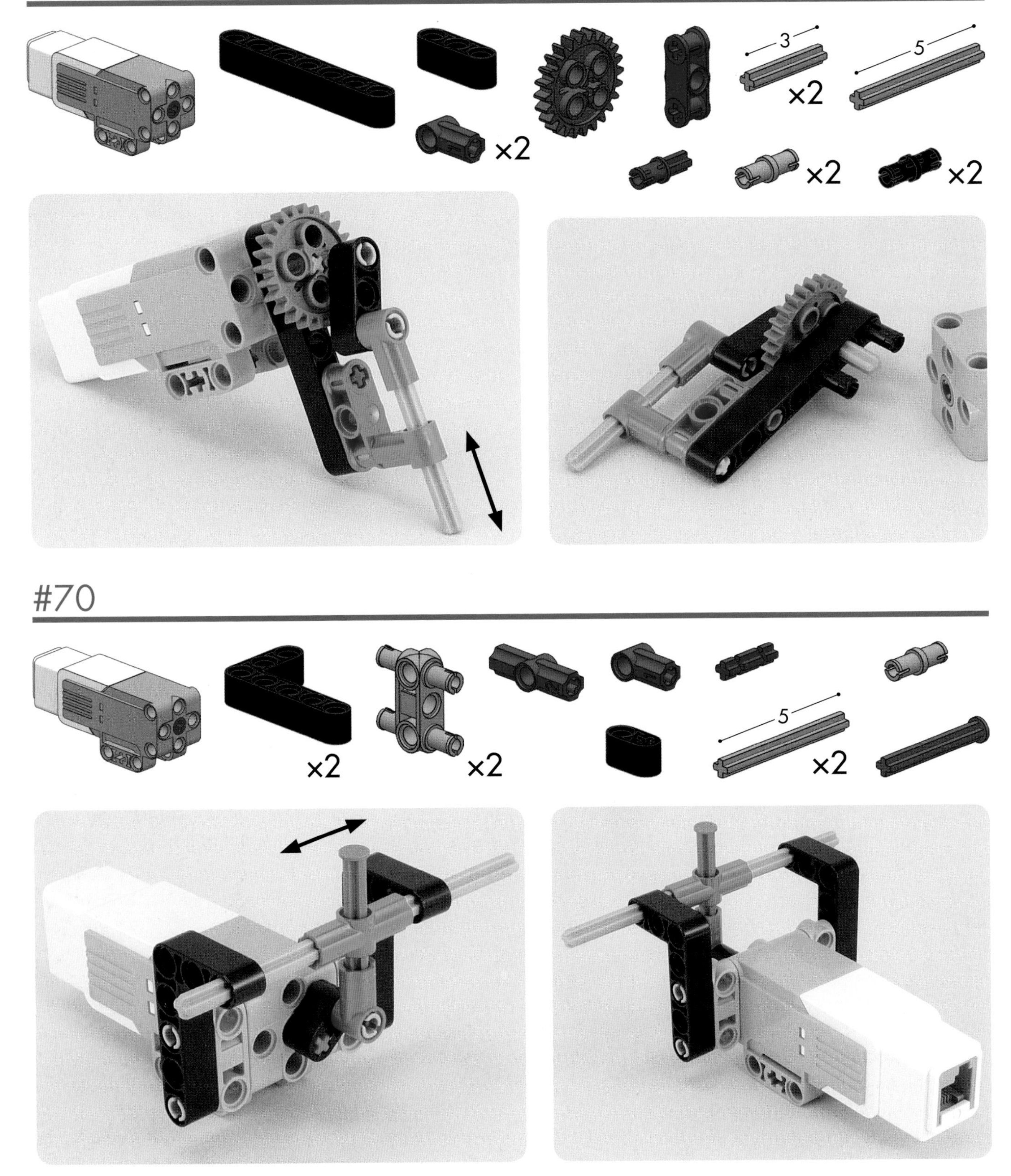
3
×2
5
×2
×2
×2
#70
×2
×2
5
×2

#71

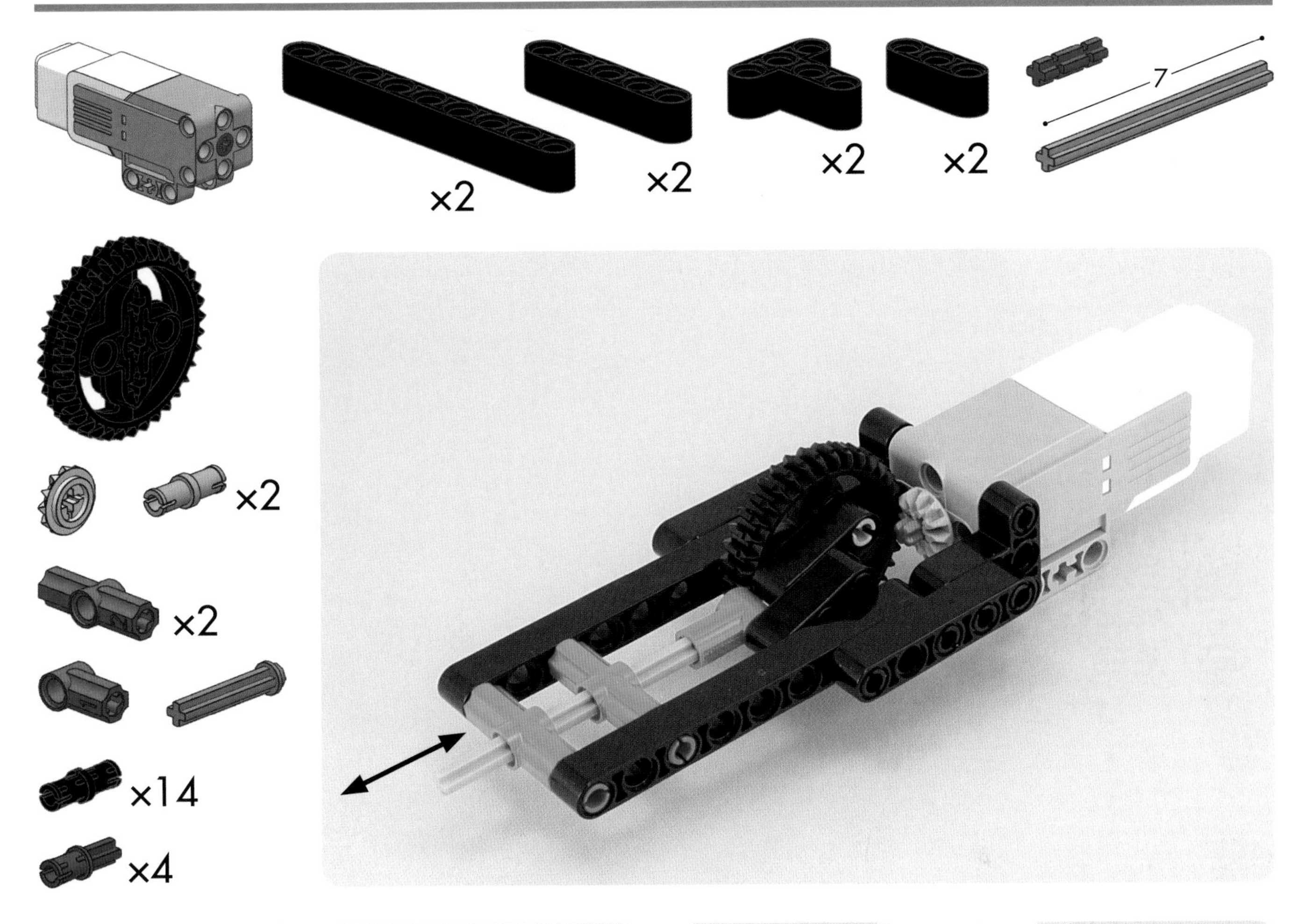
7
×2
×2
×2
×2
×2
×2
×14
×4

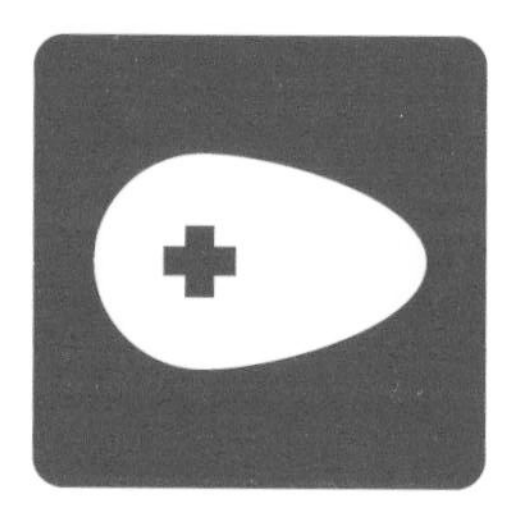

Cam mechanisms

#72

×2
5

#73

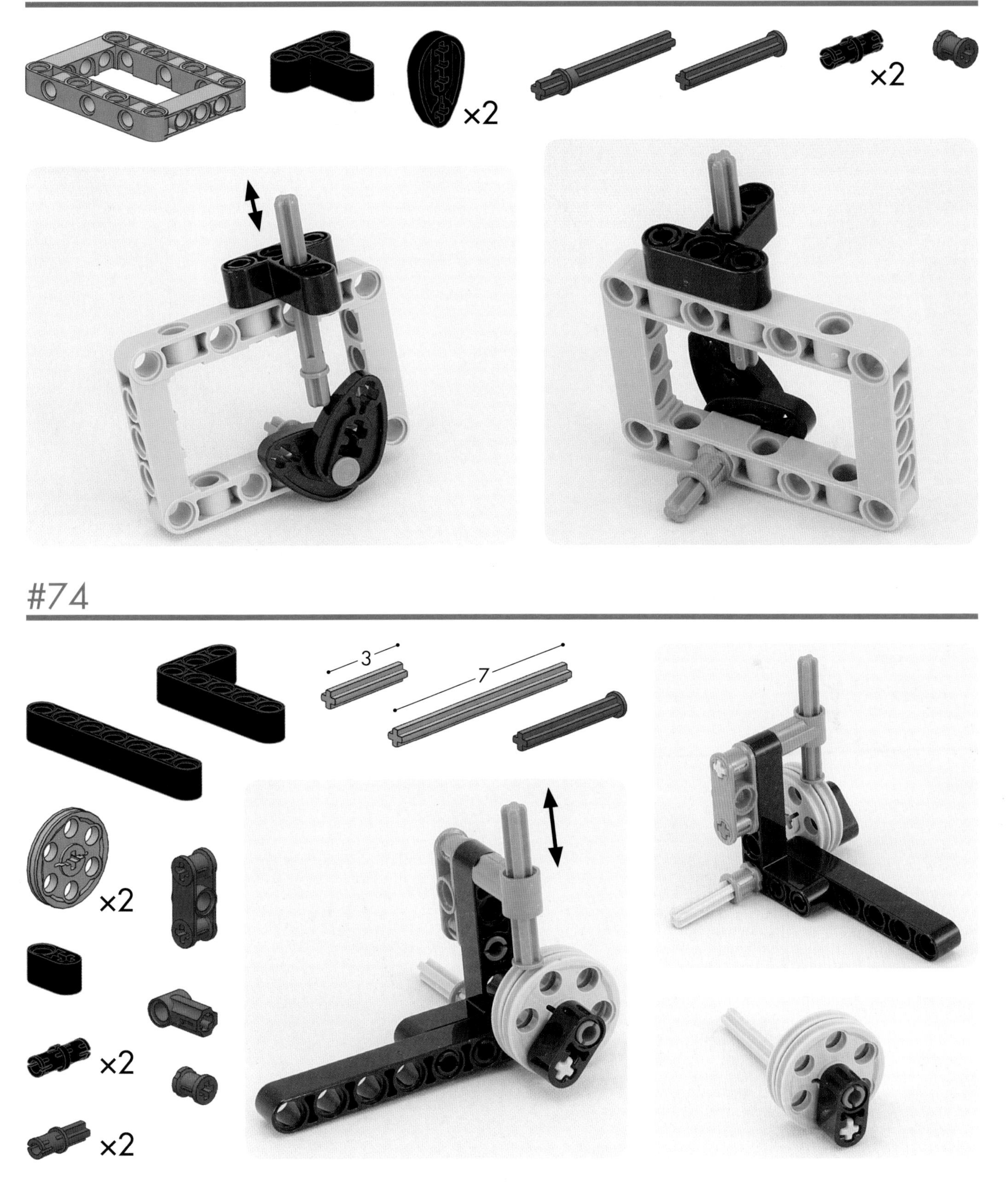
×2
×2
#74
3
7
×2
×2
×2

#75

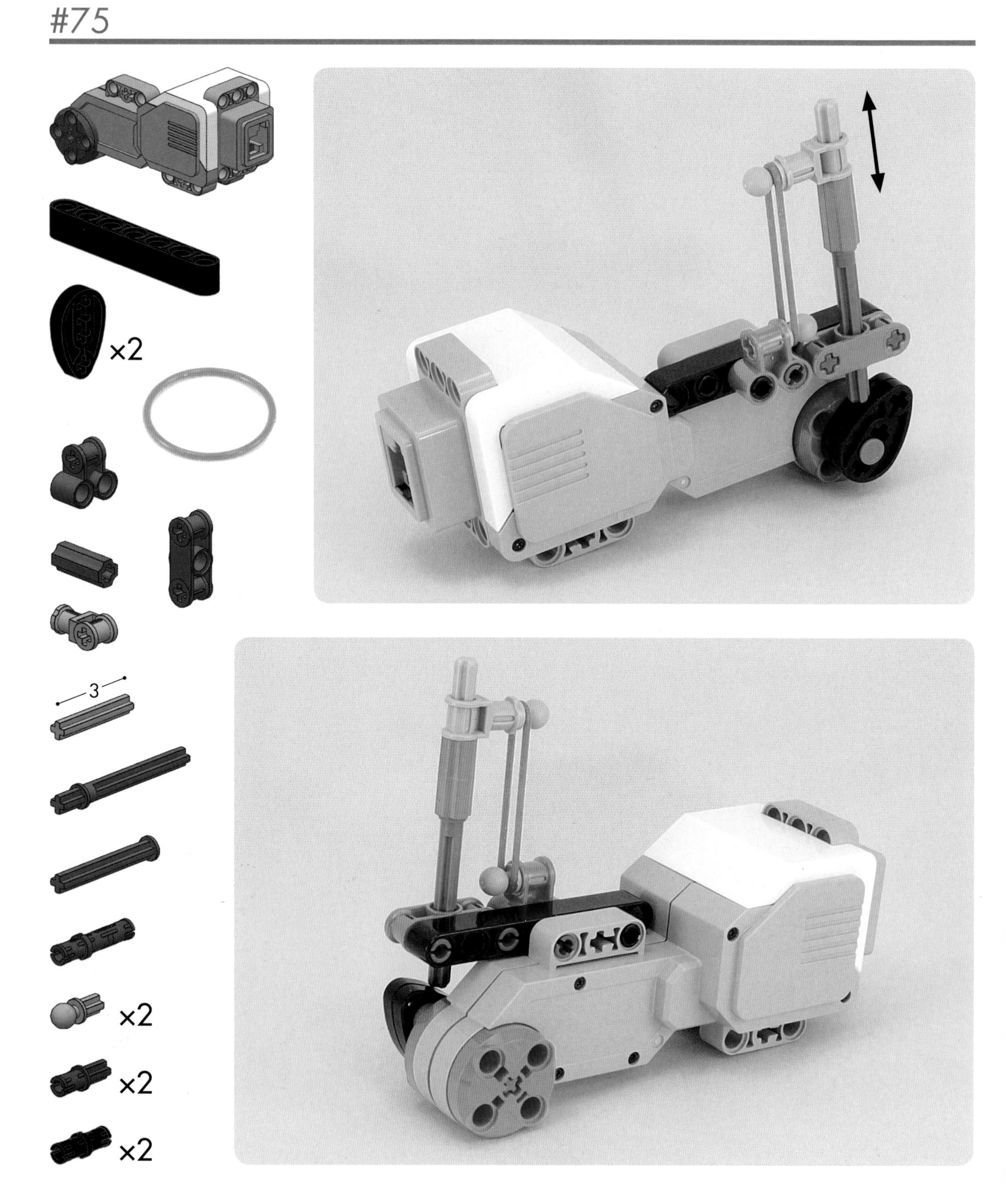
×2
3
×2
×2
×2

#76

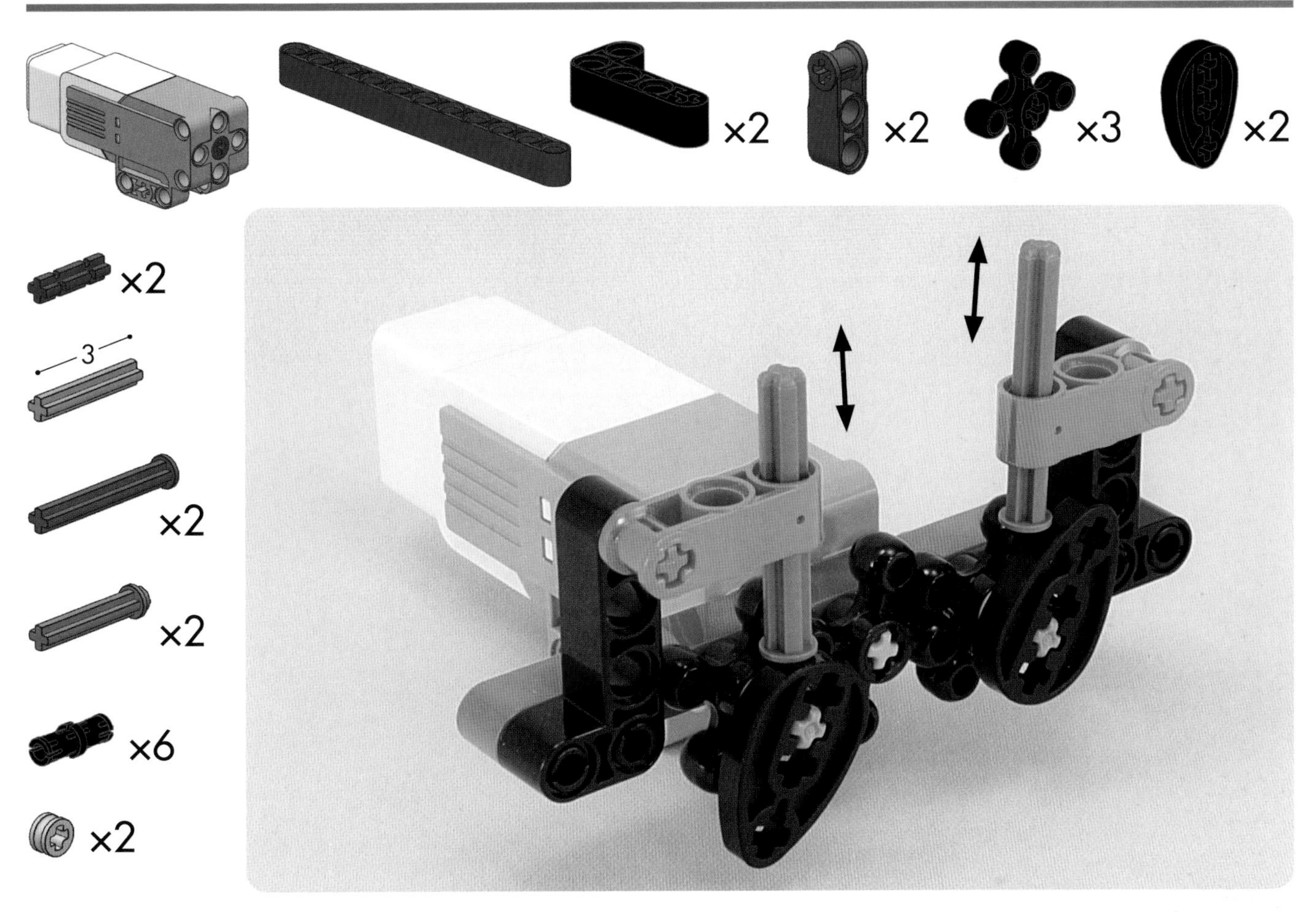
×2
×2
×3
×2
×2
3
×2
×2
×6
×2

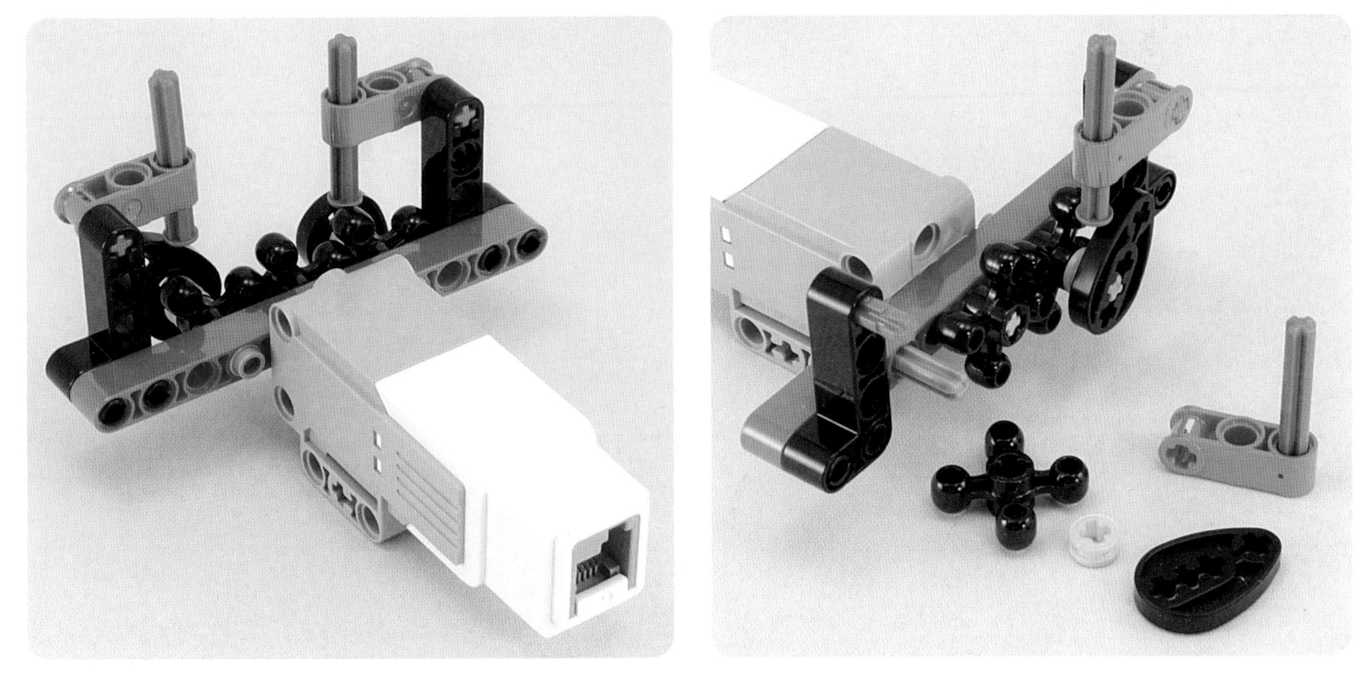

Intermittent motion

#77

×2
5
×2
×2
×2

#78

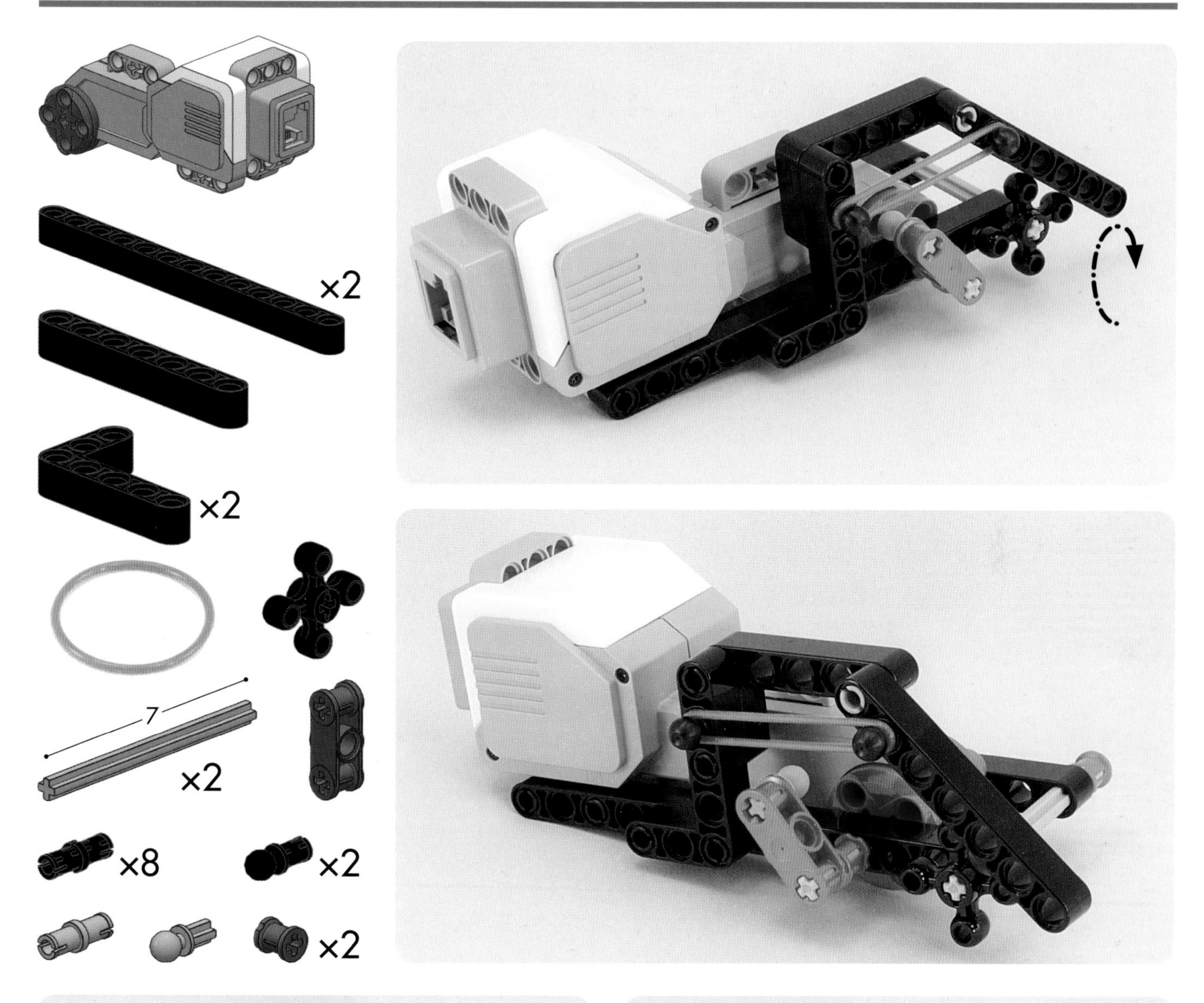
×2
×2
7
×2
×8
×2
×2

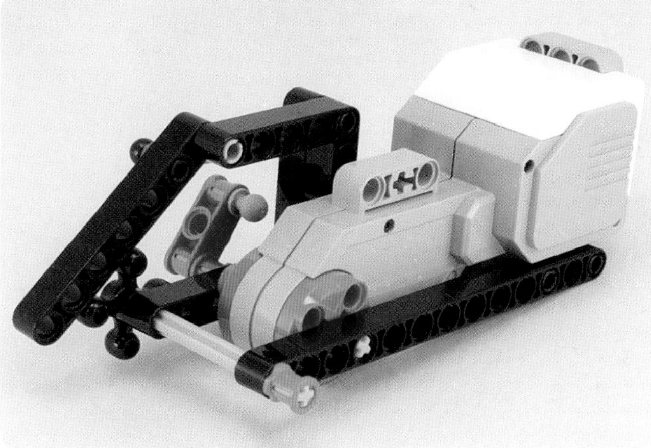

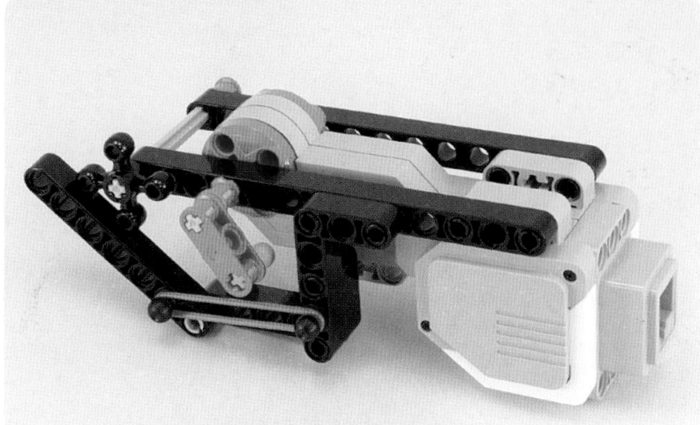

#79

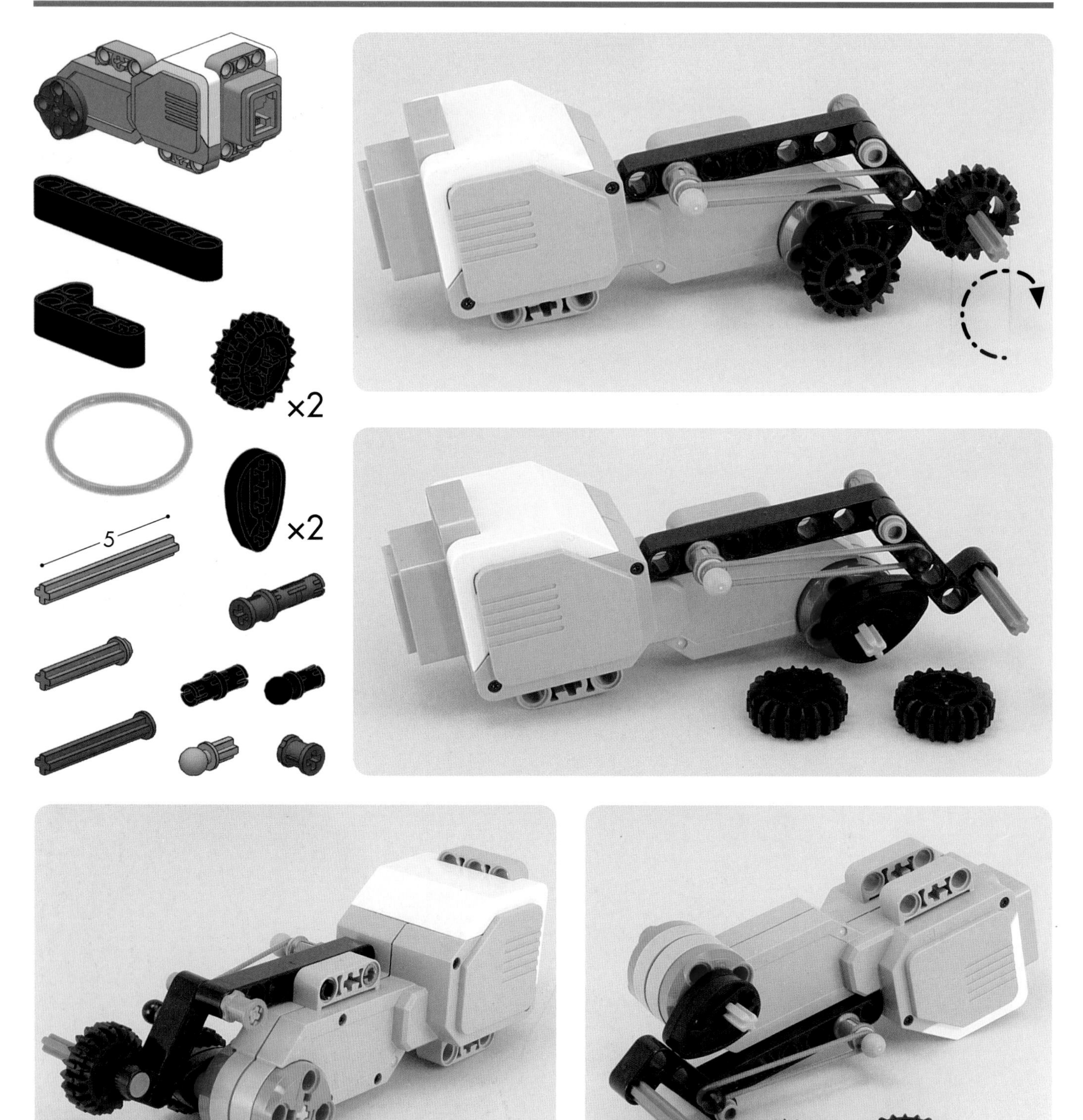
×2
×2
5

#80

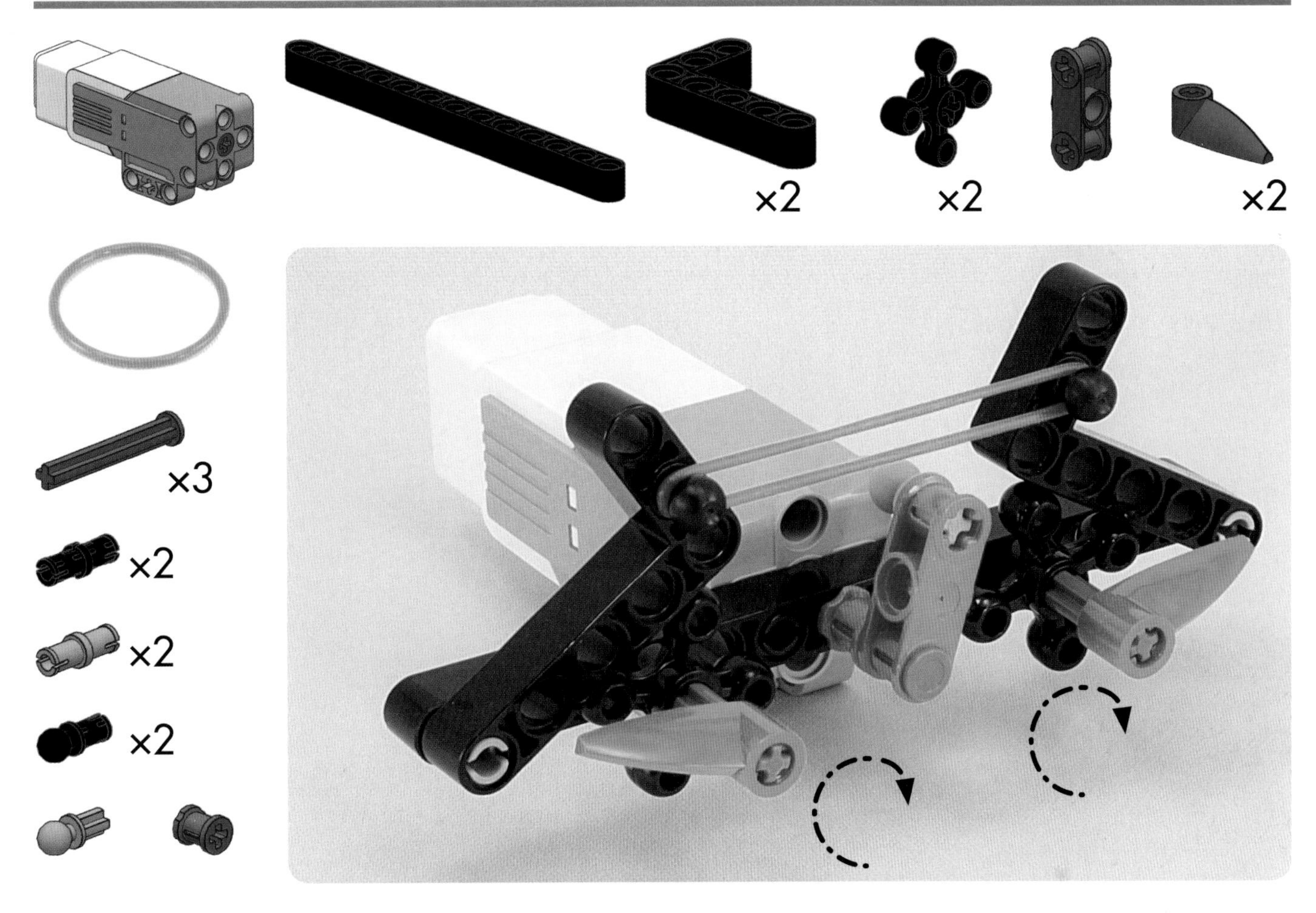
×2
×2
×2
×3
×2
×2
×2

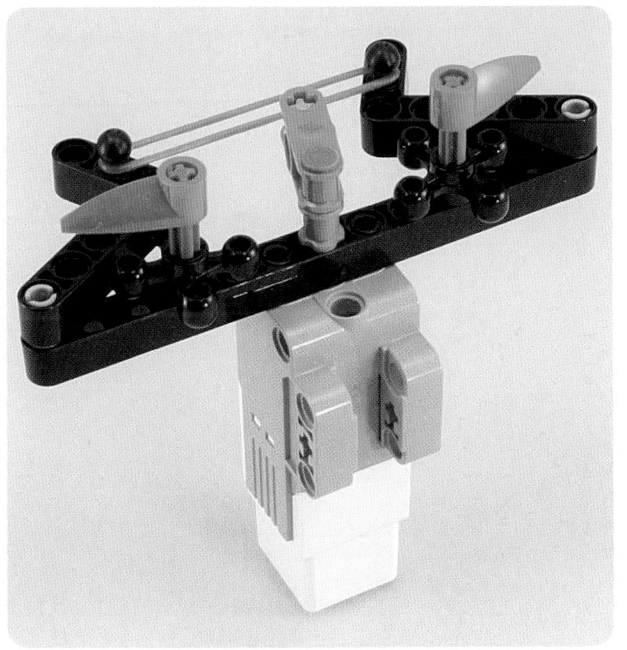

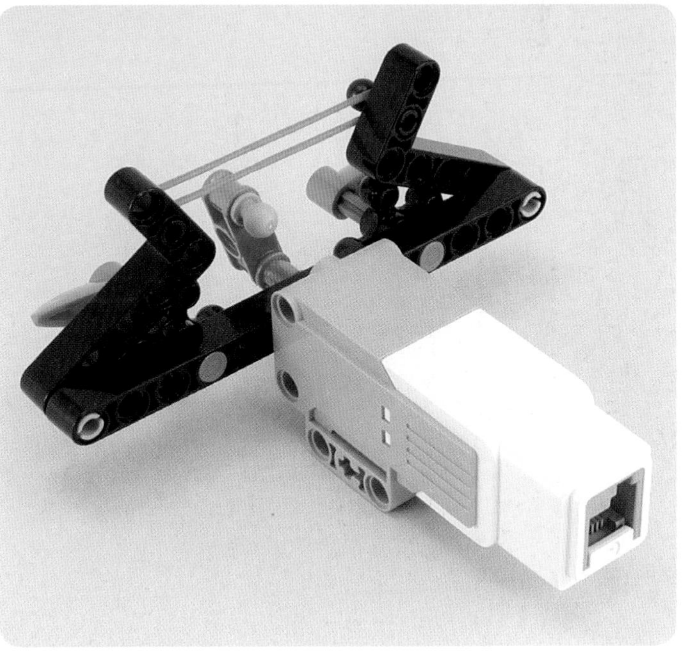

Transmitting rotation with rubber bands

#81

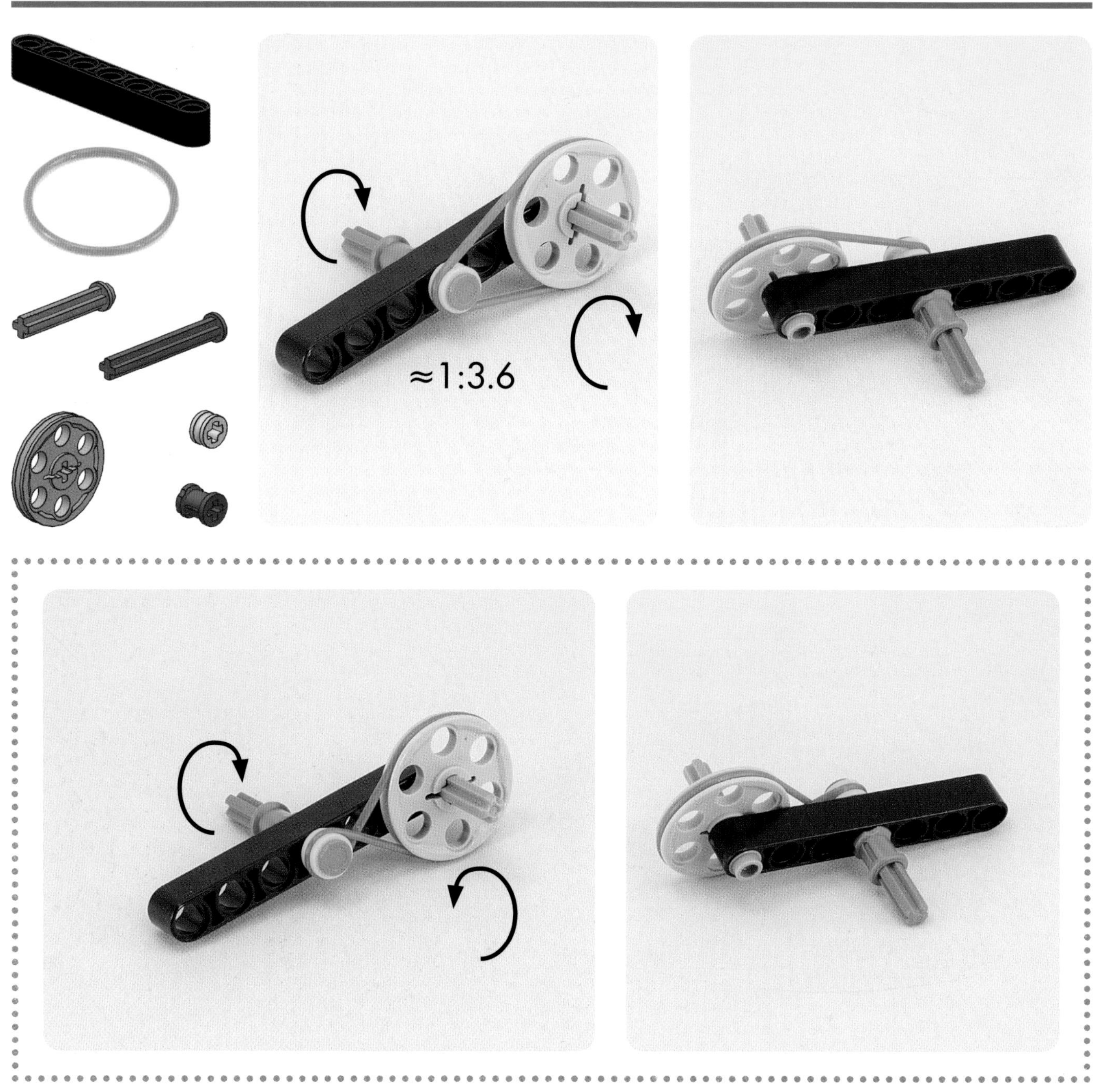
≈1:3.6

#82

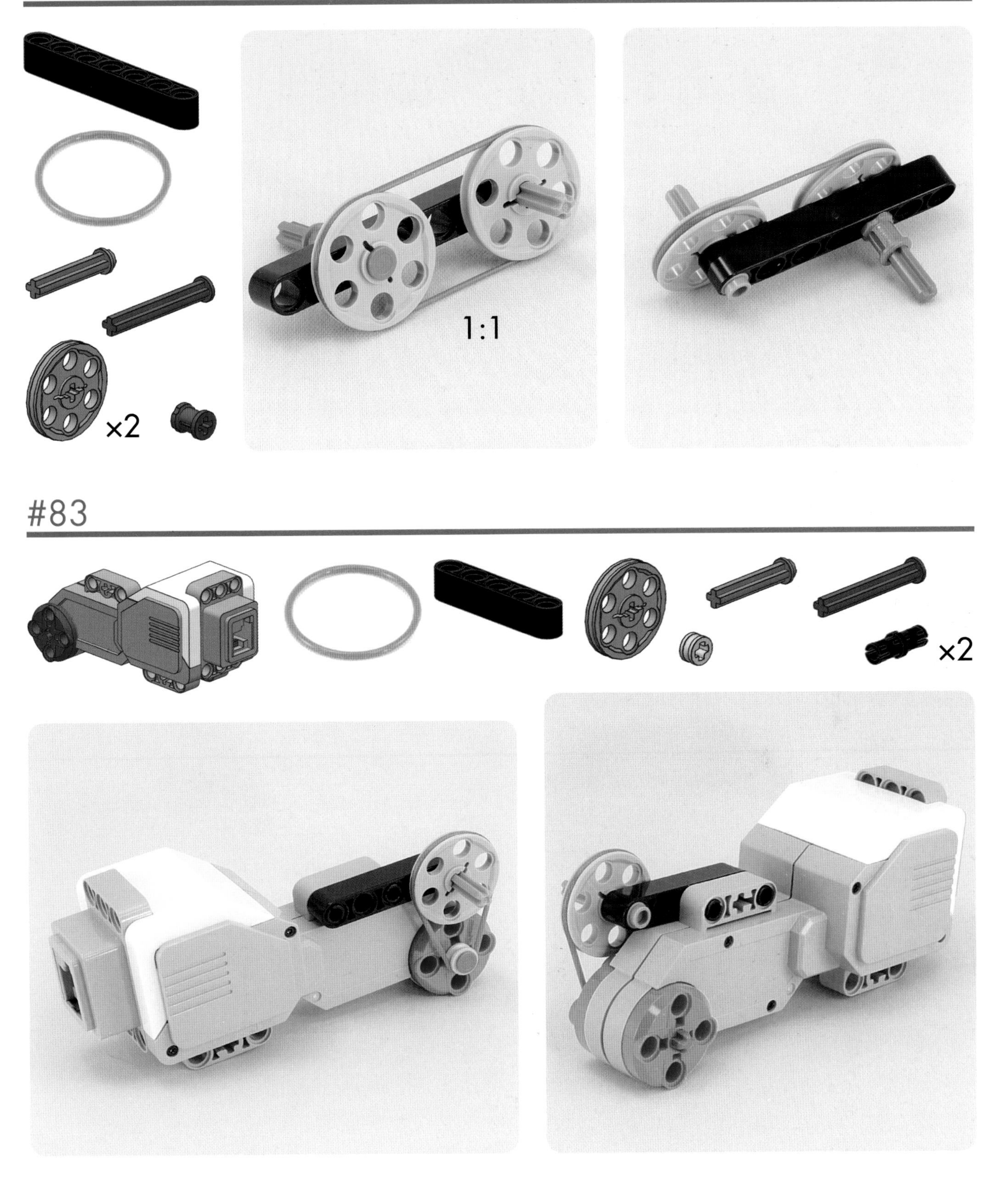
×2
1:1
#83
×2

#84

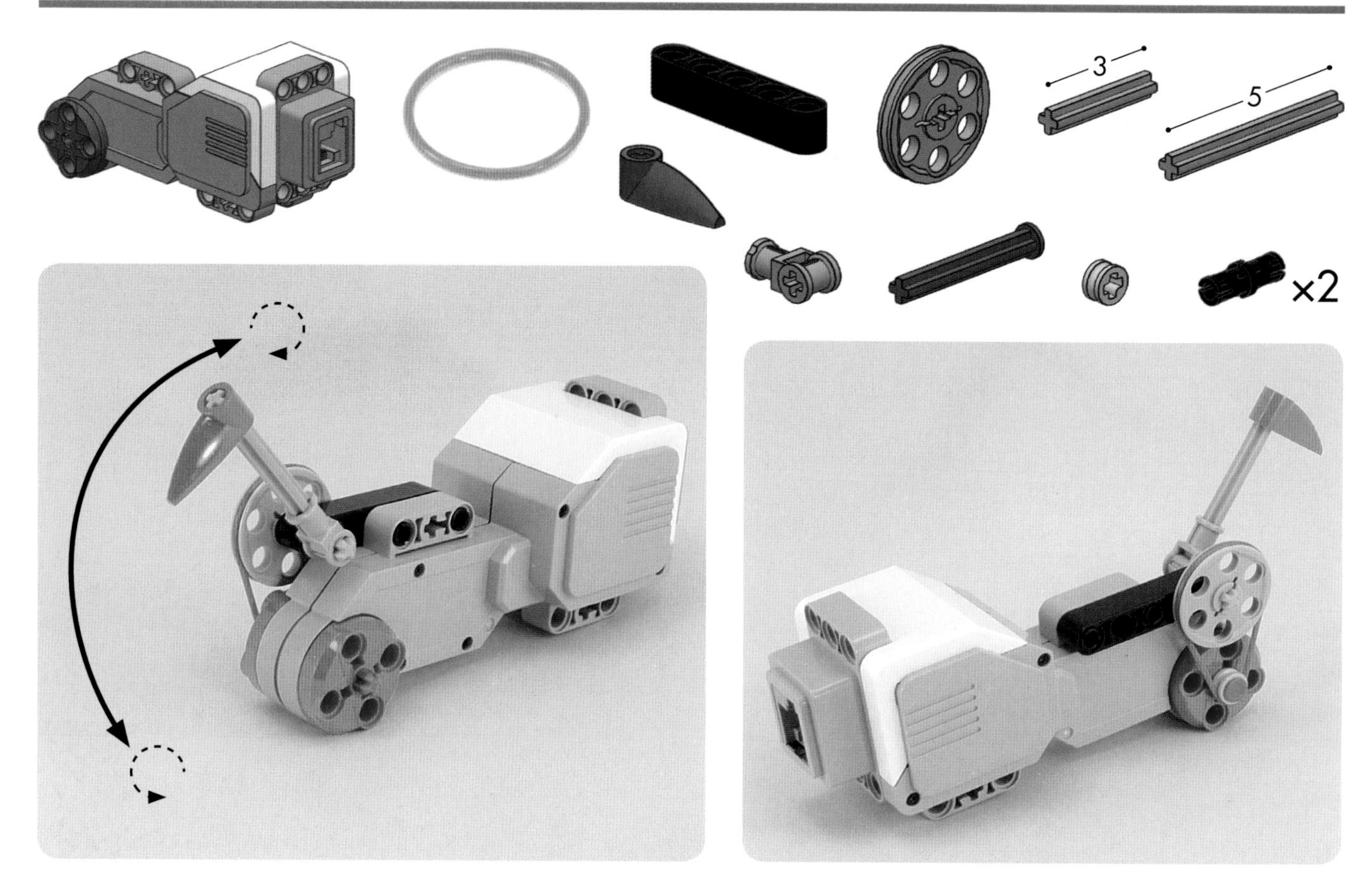
3
5
×2

#85

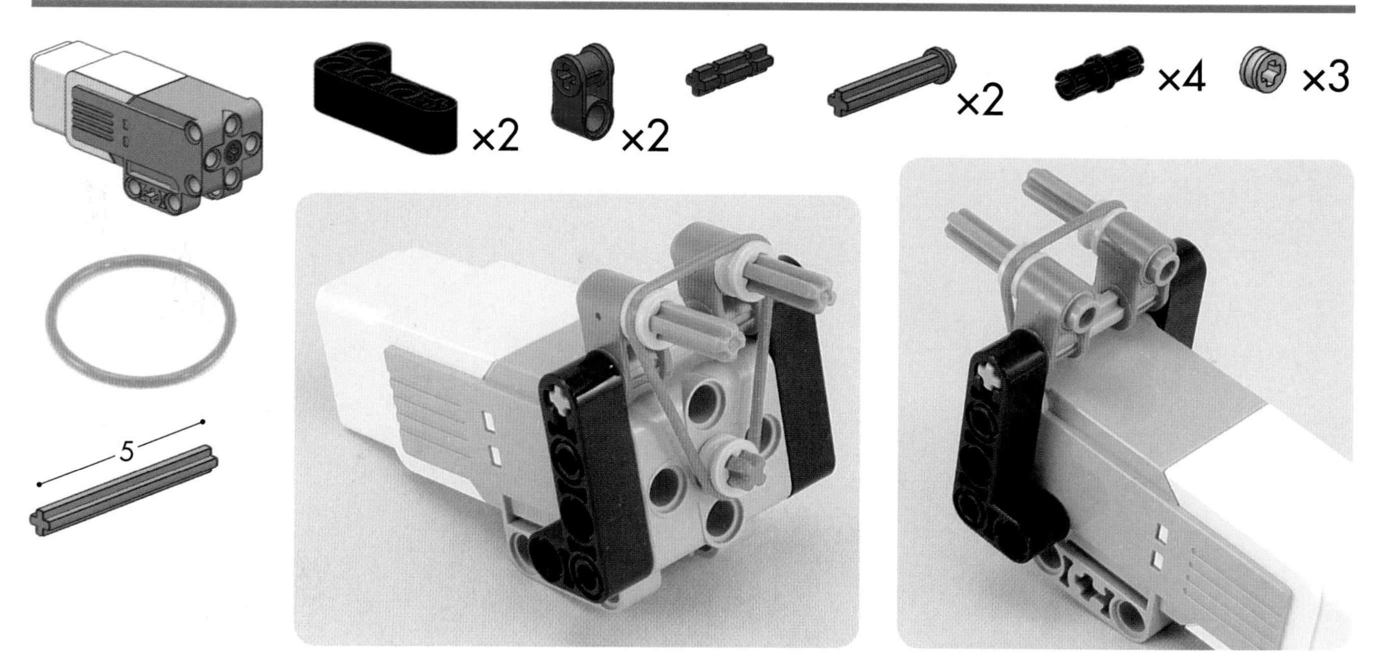
×2
×2
×2
×4
×3
5

#86

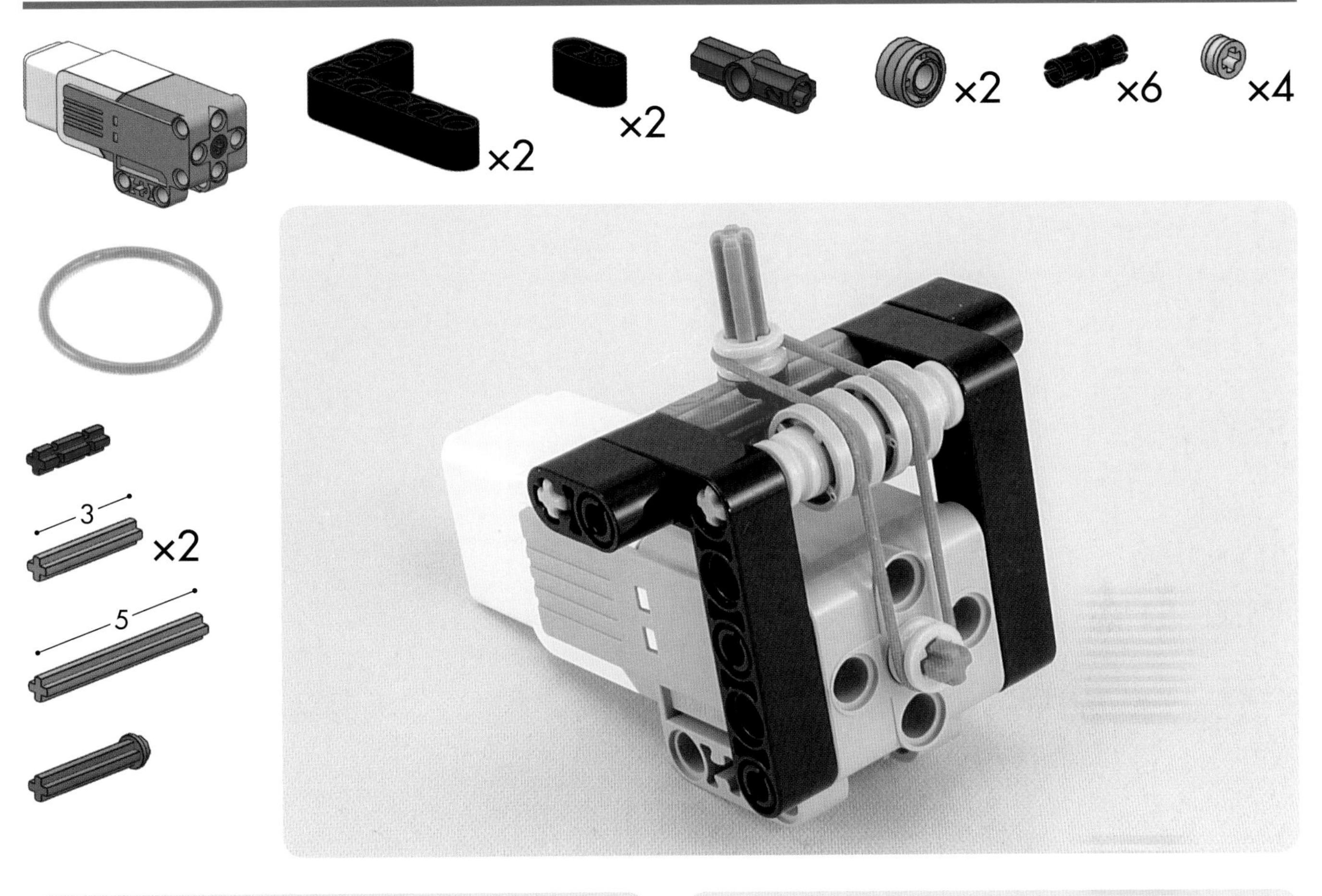

×2
×2
×2
×6
×4
3
×2
5

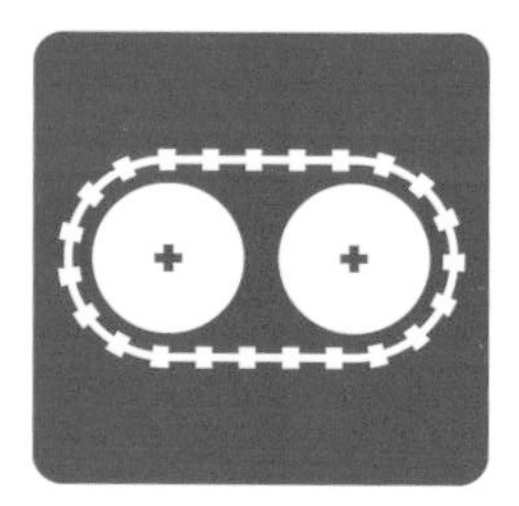

Transmitting rotation with caterpillar treads

#87

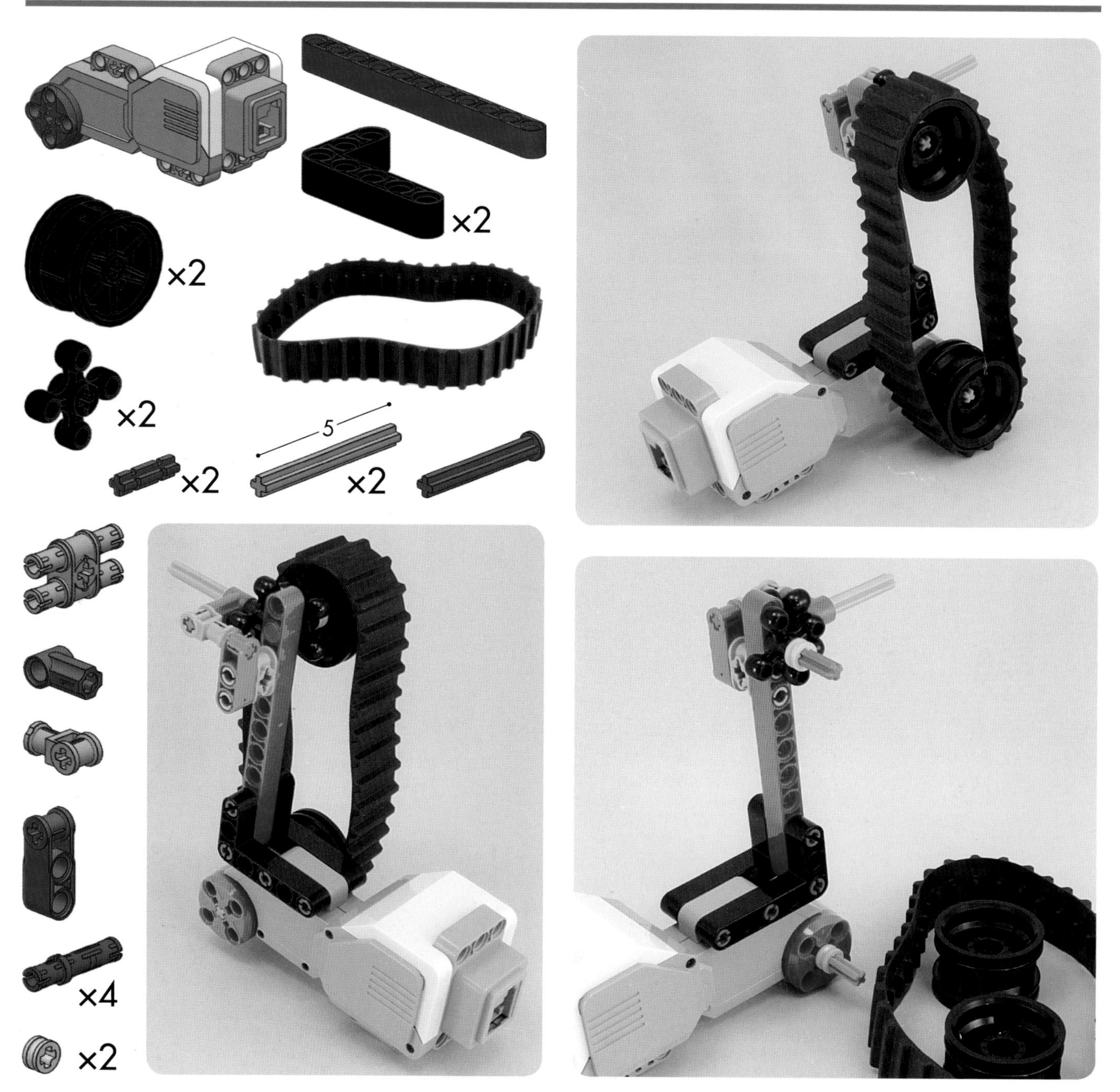

×2
×2
×2
×2
5
×2
×4
×2

#88

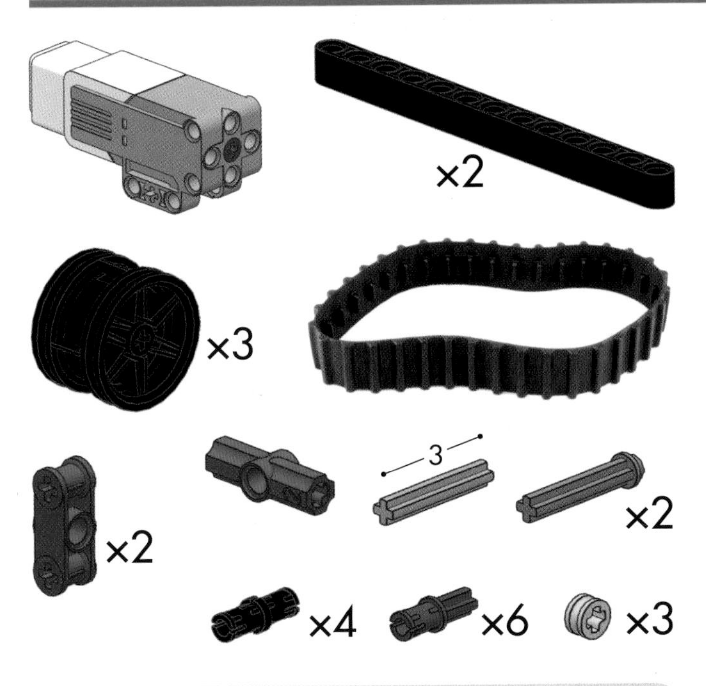
×2
×3
3
×2
×2
×4
×6
×3

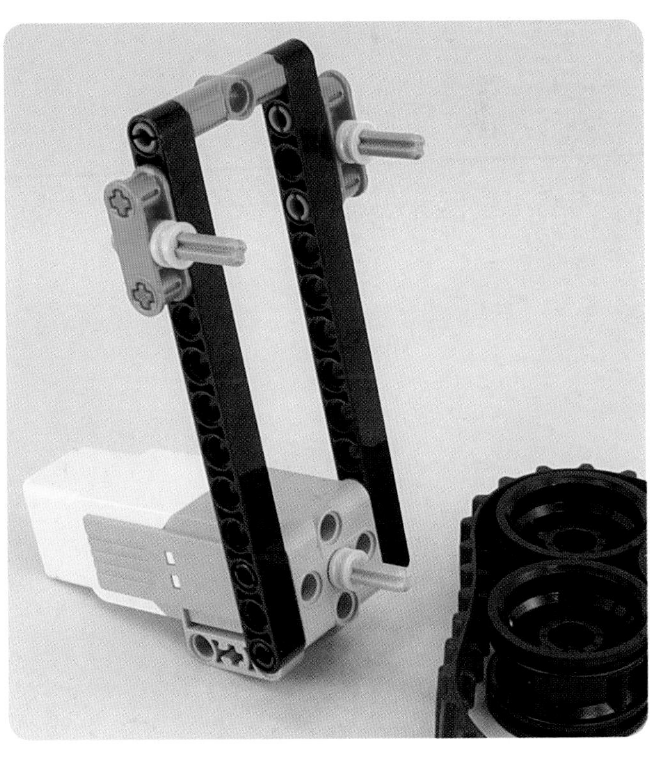

Transmitting rotation over a long distance

#89

×2
3
×4
×2
×4
×2

#90

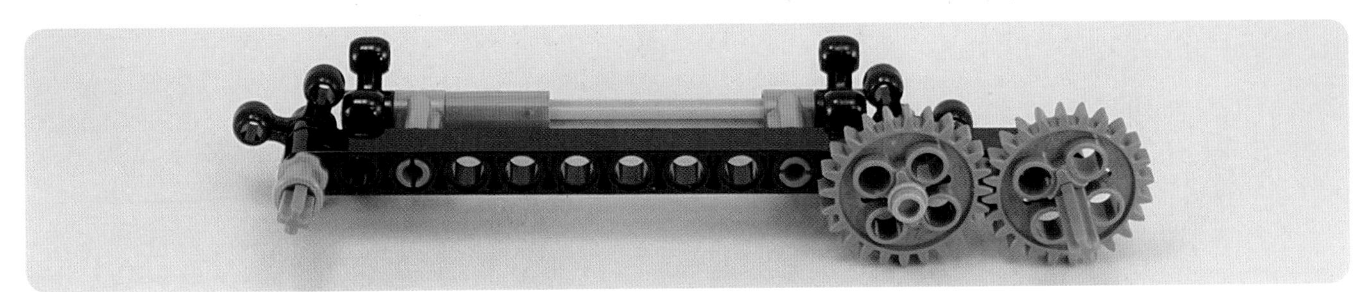

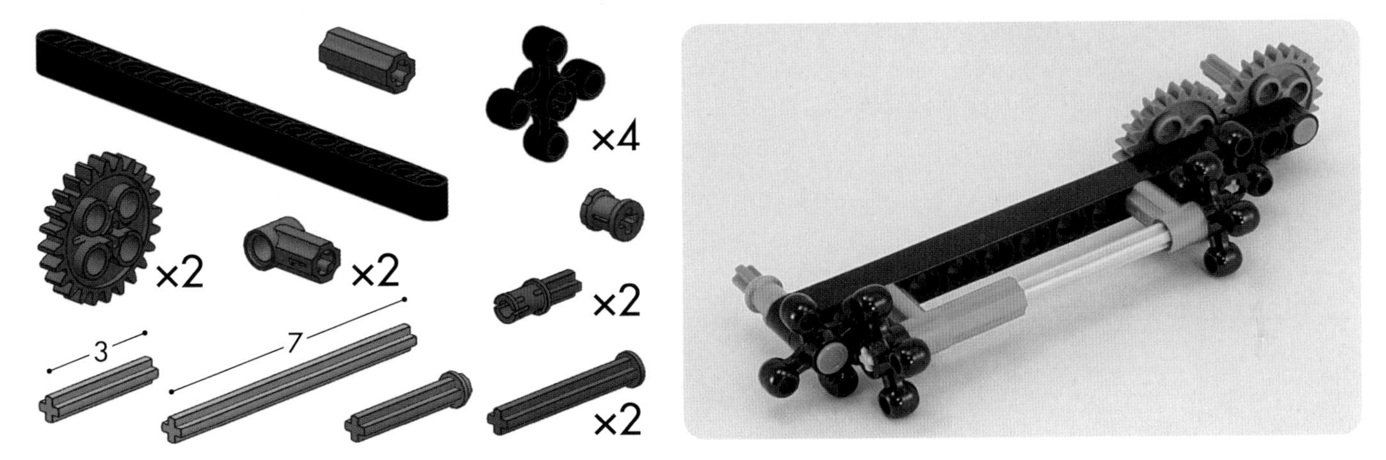
×4
×2
×2
×2
3
7
×2

#91

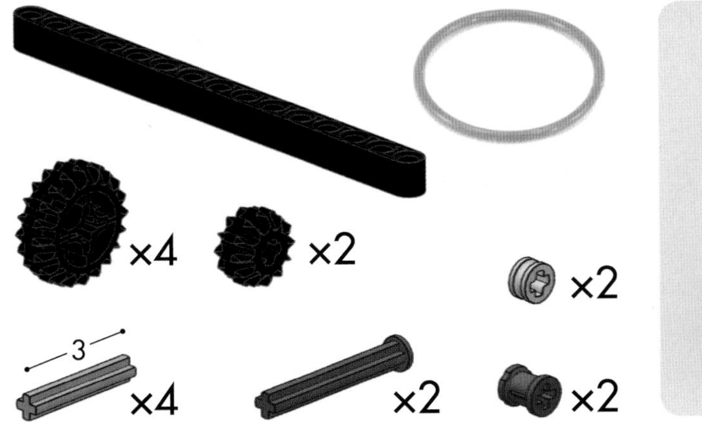
×4
×2
×2
3
×4
×2
×2

Off-center axes of rotation

#92

#93

×2
×3

#94

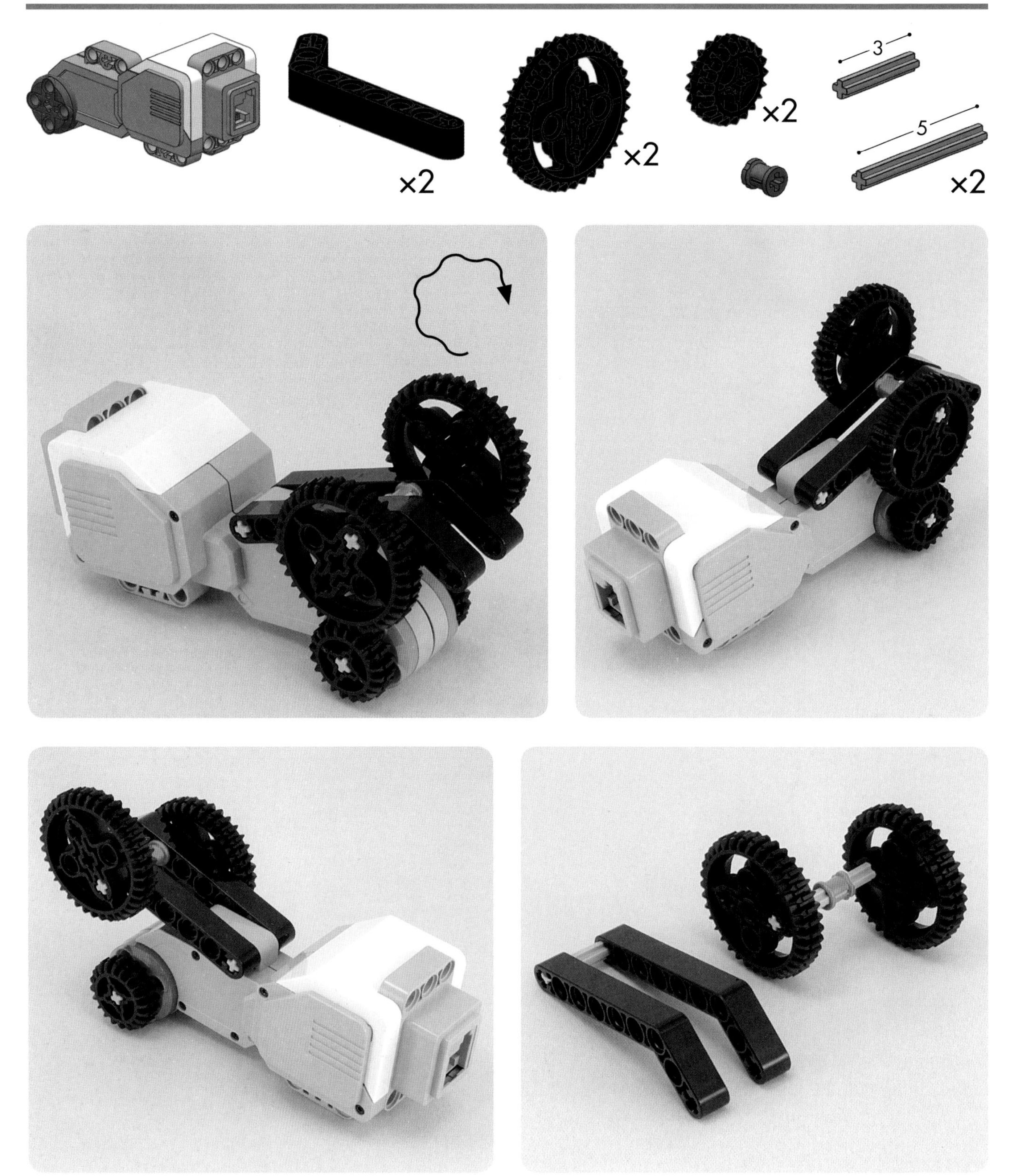
×2
×2
×2
3
5
×2

#95

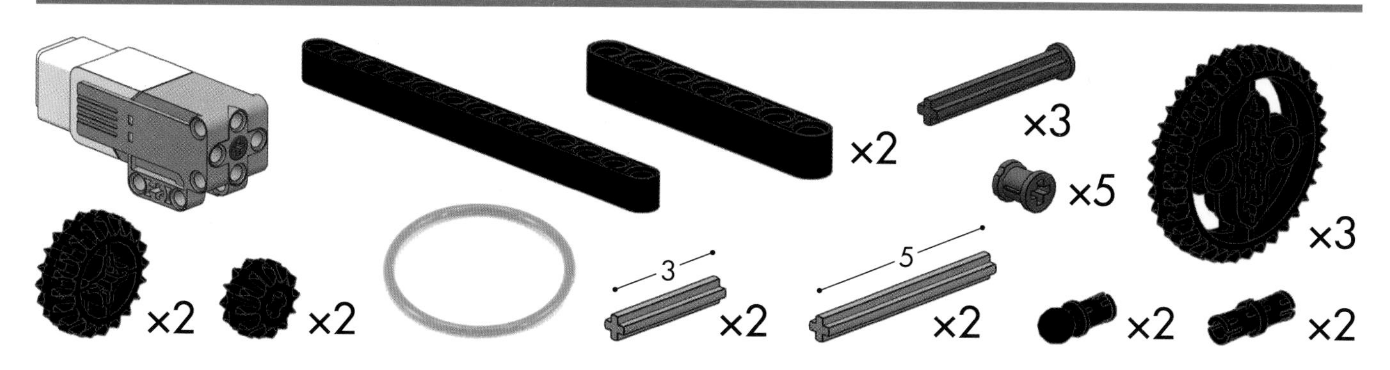
×2
×3
×5
×3
×2
×2
3
×2
5
×2
×2
×2

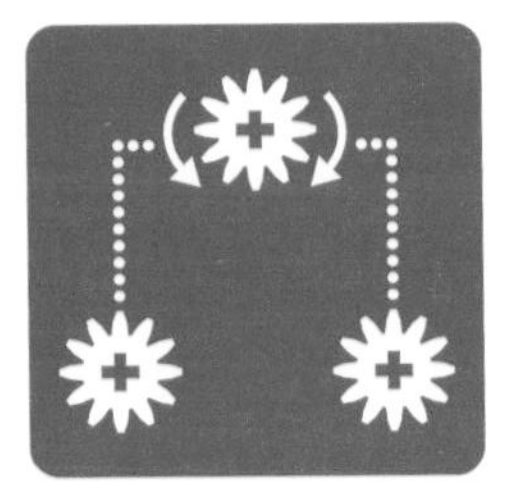

Changeover mechanisms using rotational direction

#96

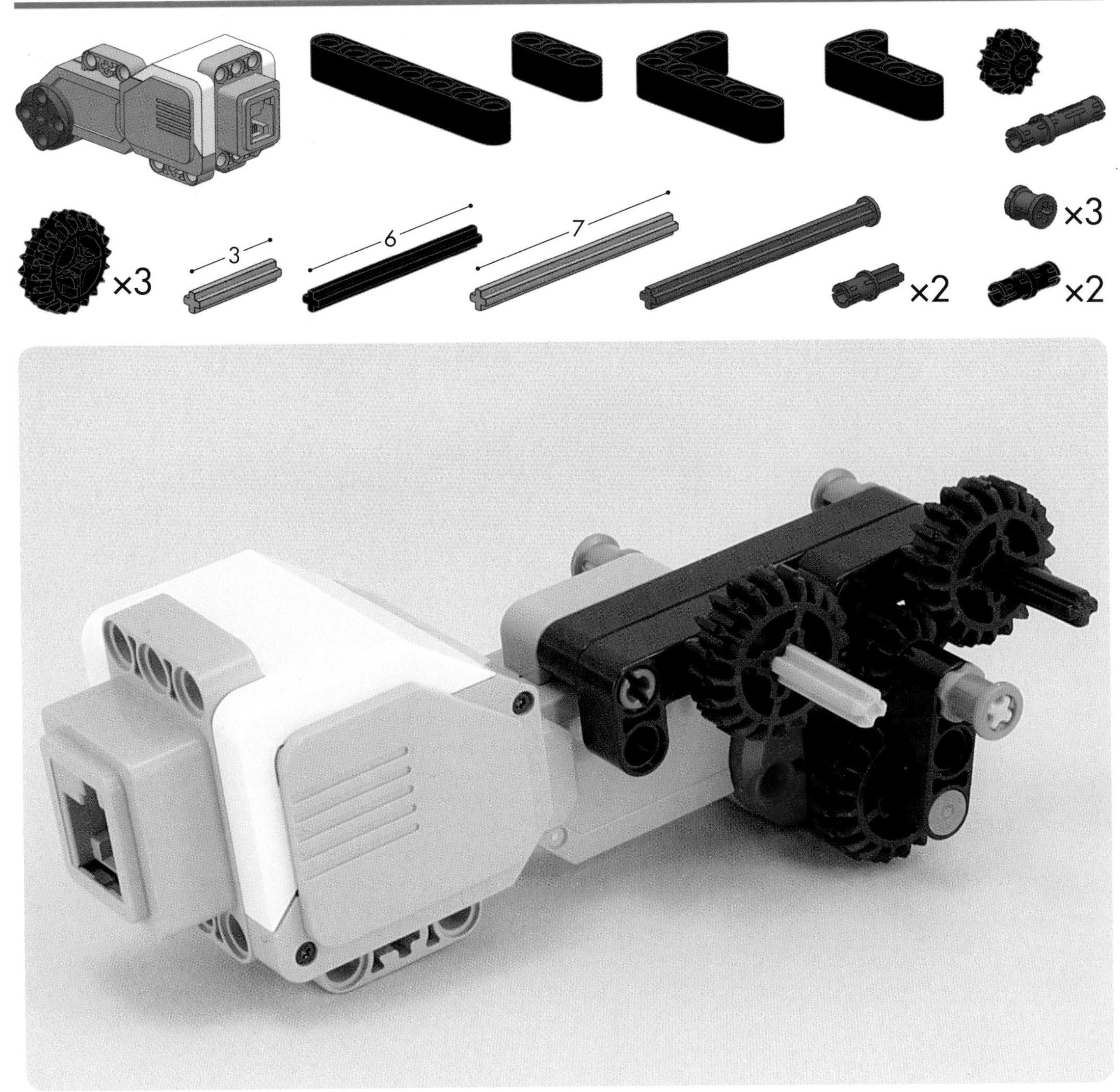

#97

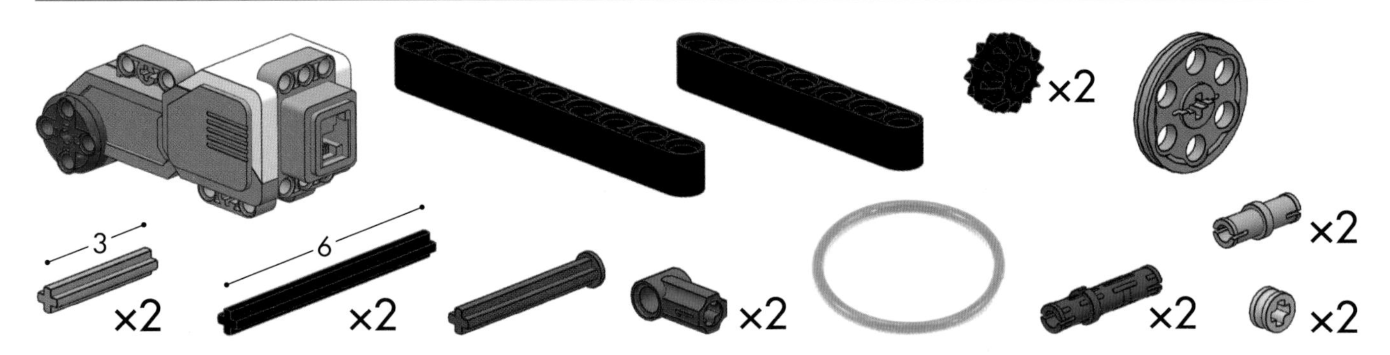
×2
3
×2
6
×2
×2
×2
×2
×2

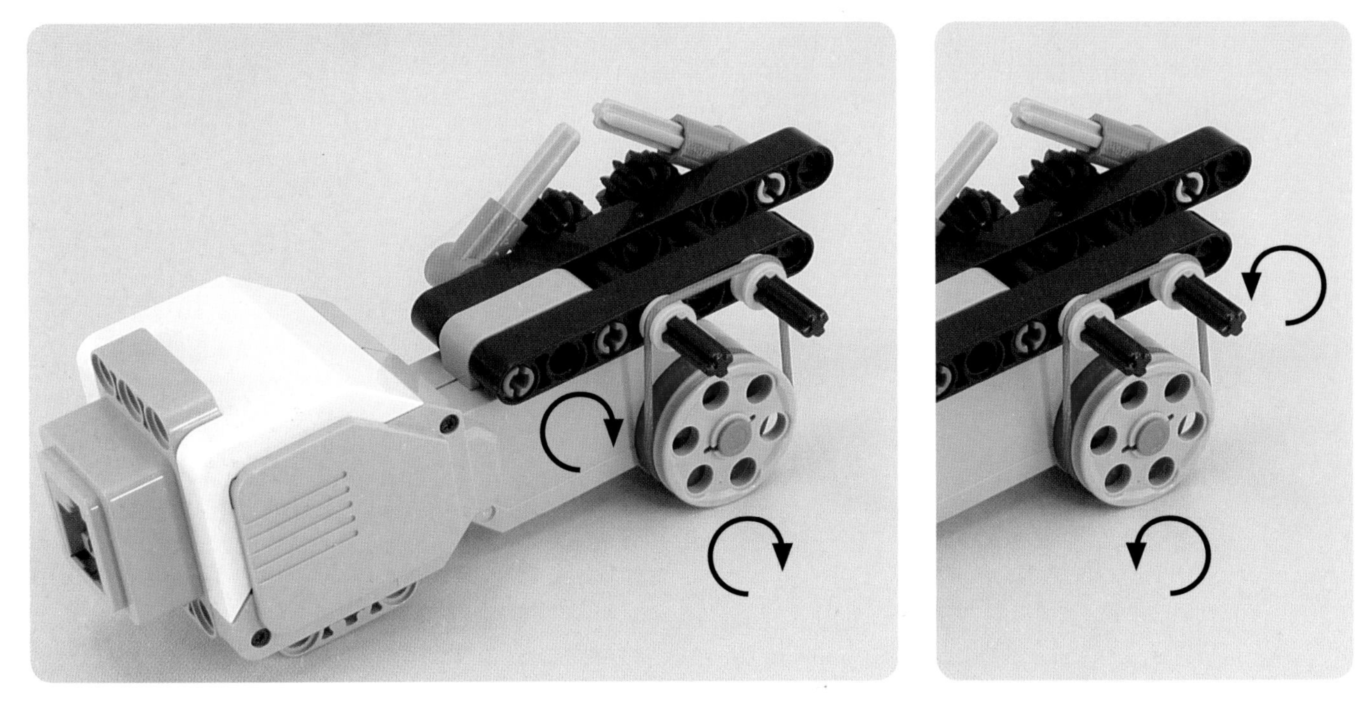

#98

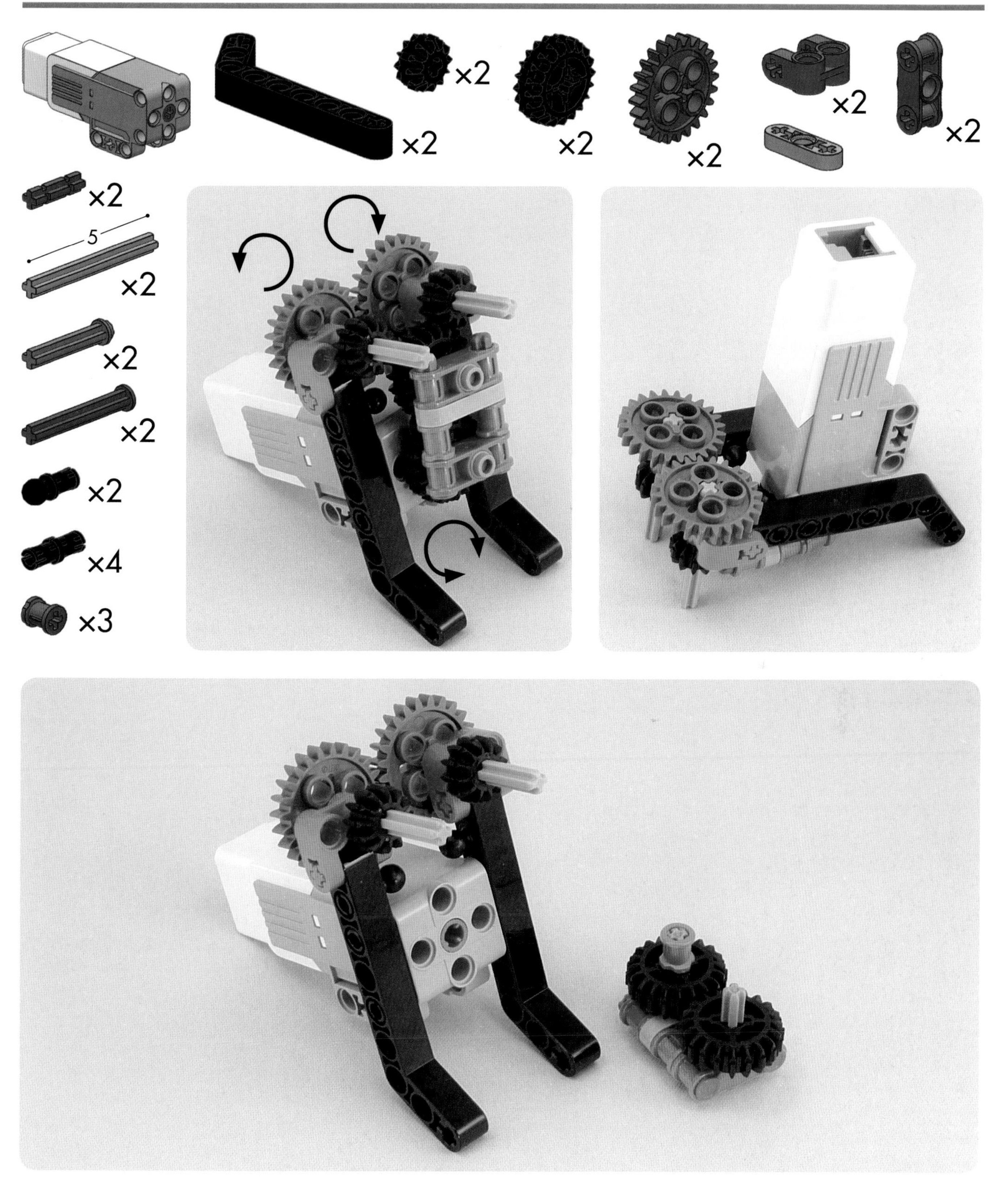

×2
×2
×2
×2
×2
×2
×2
5
×2
×2
×2
×2
×4
×3

#99

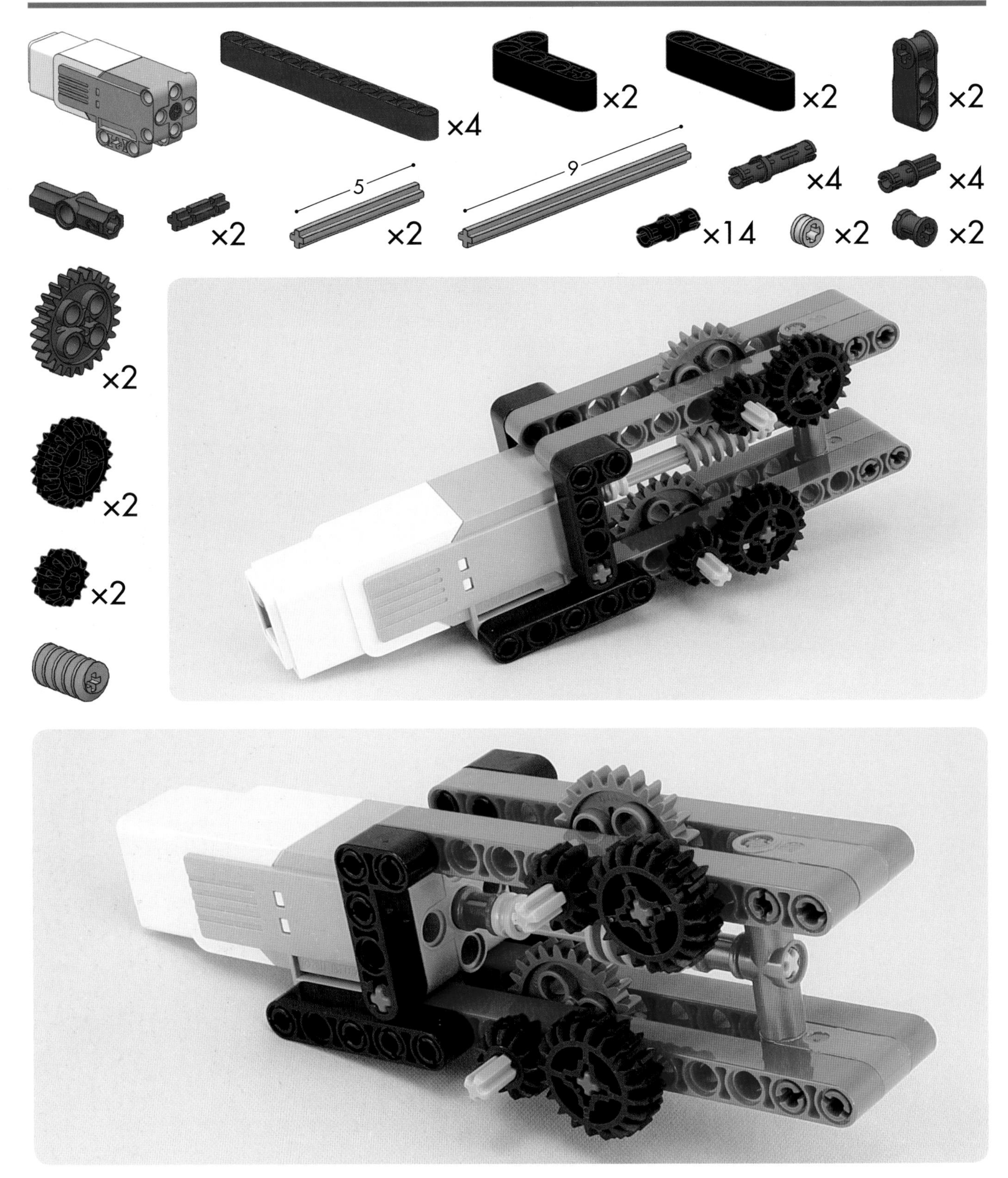
×4
×2
×2
×2
5
9
×2
×2
×2
×4
×4
×14
×2
×2
×2
×2
×2

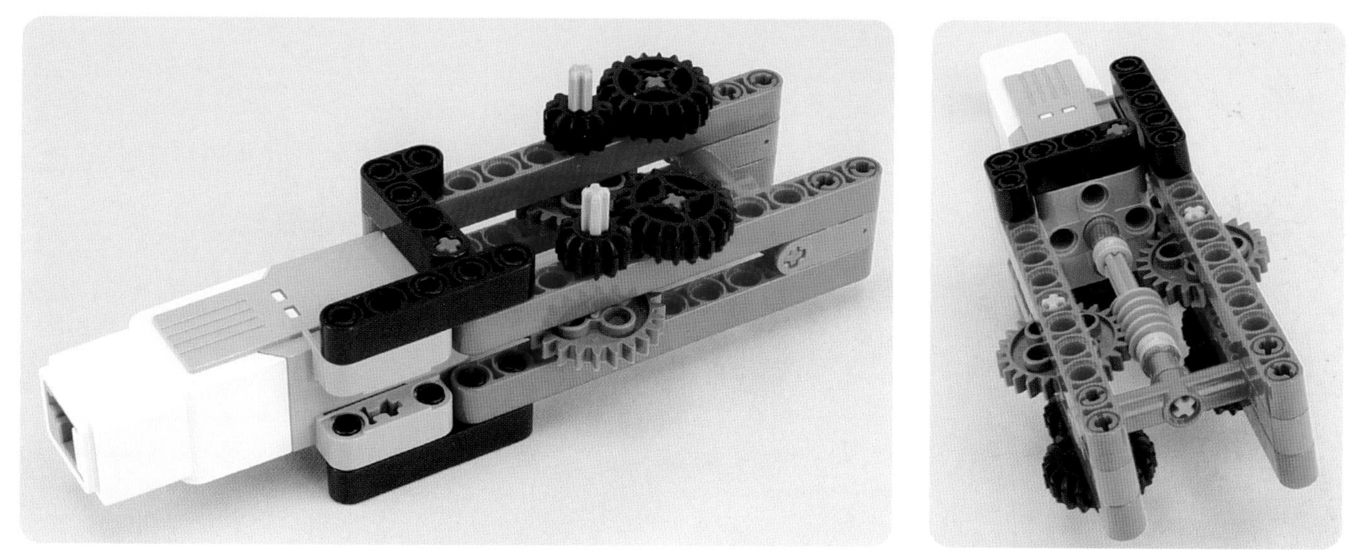

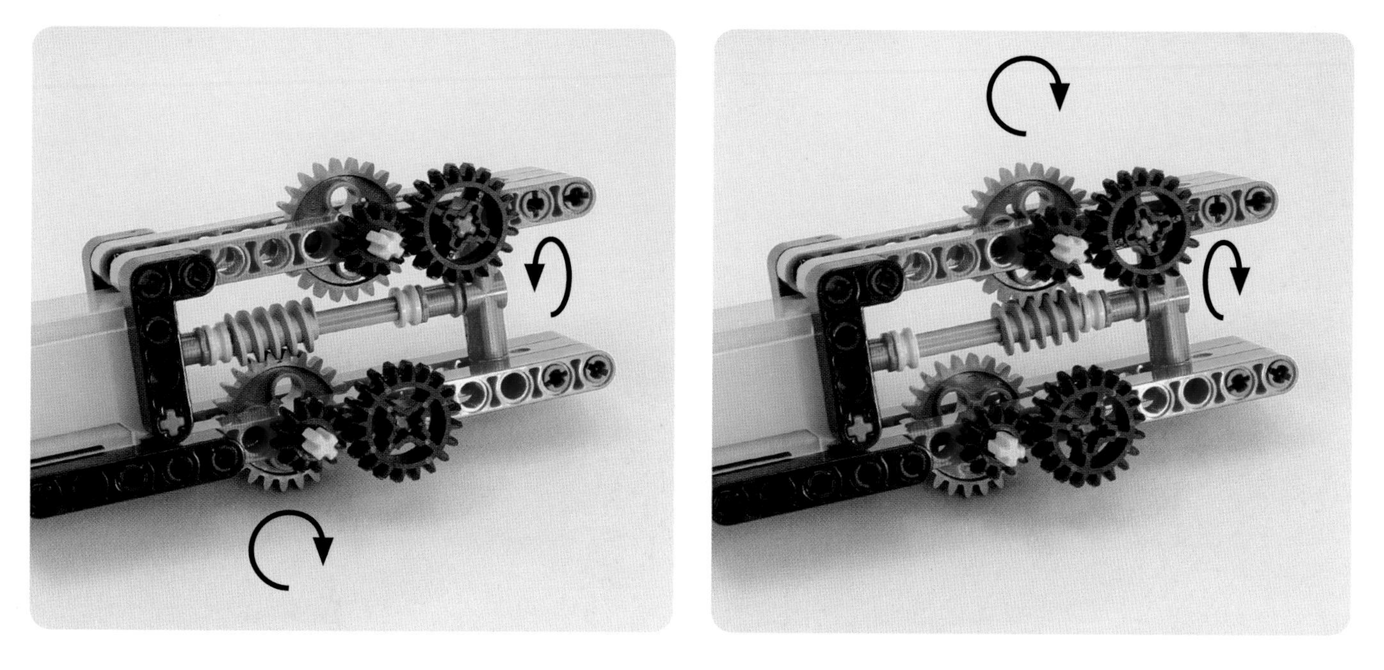

Universal joints

#100

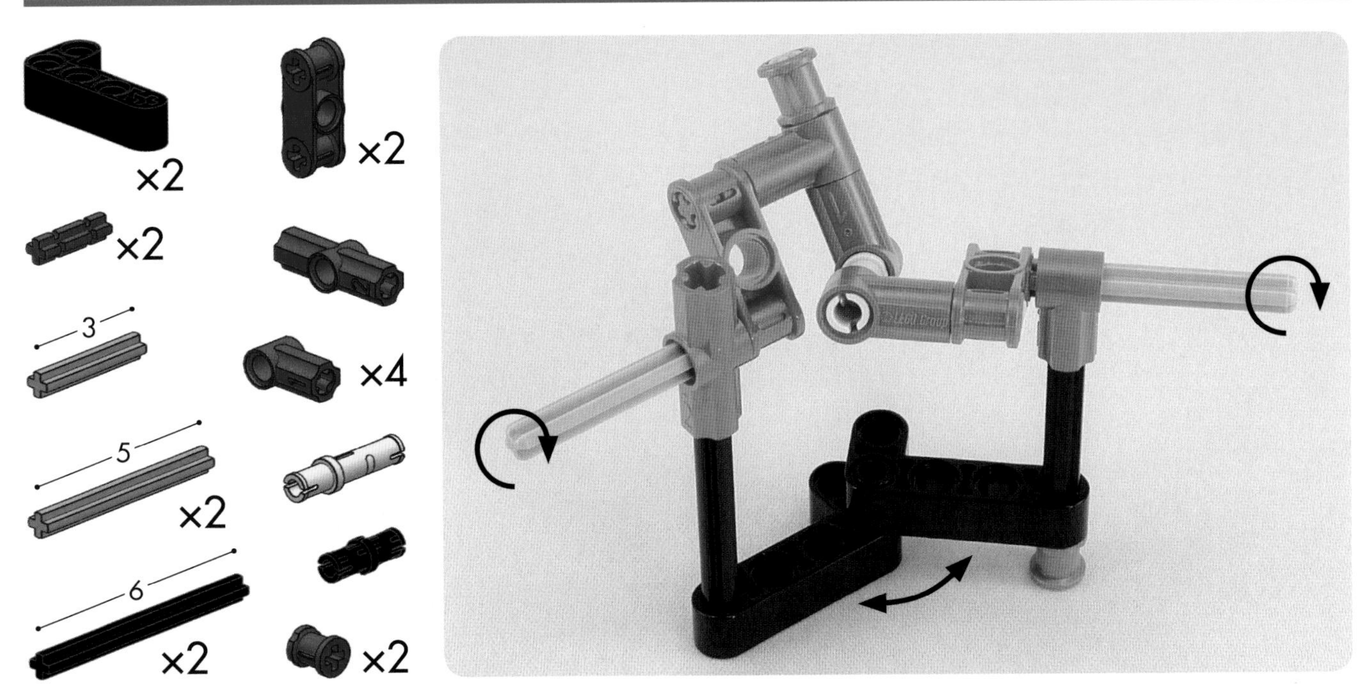

×2
×2
×2
3
×4
5
×2
6
×2
×2

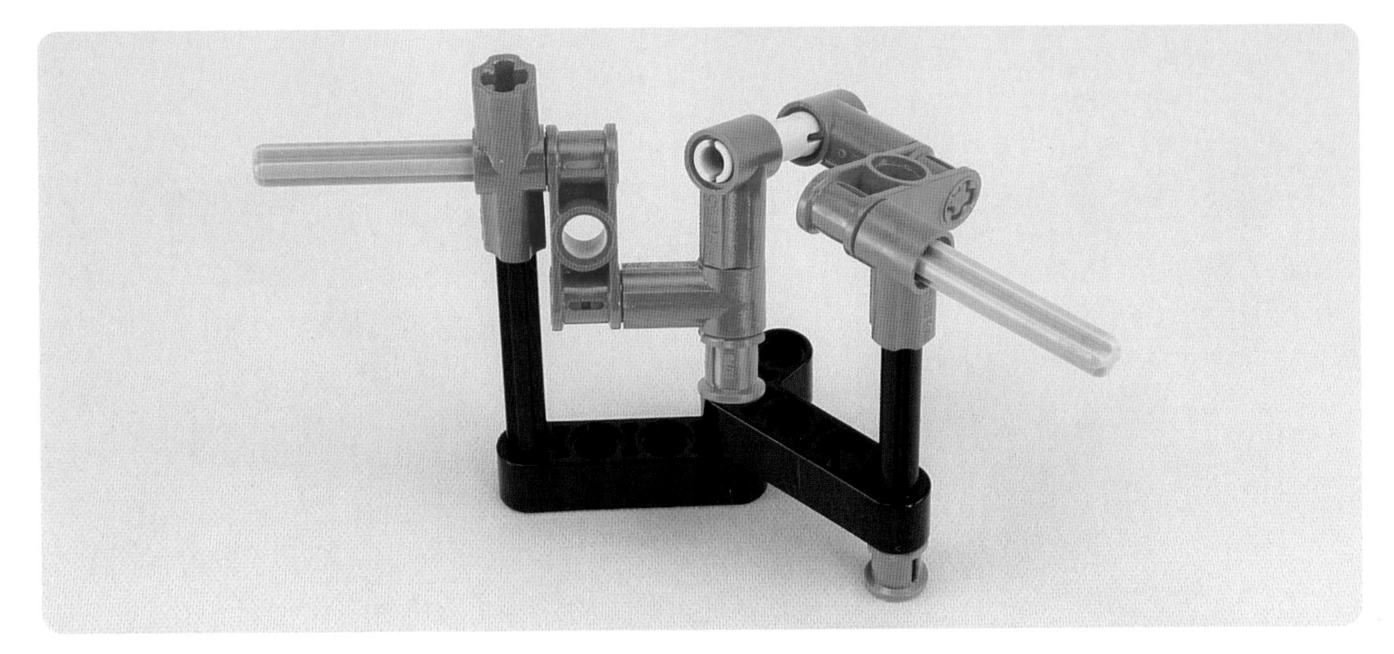

#101

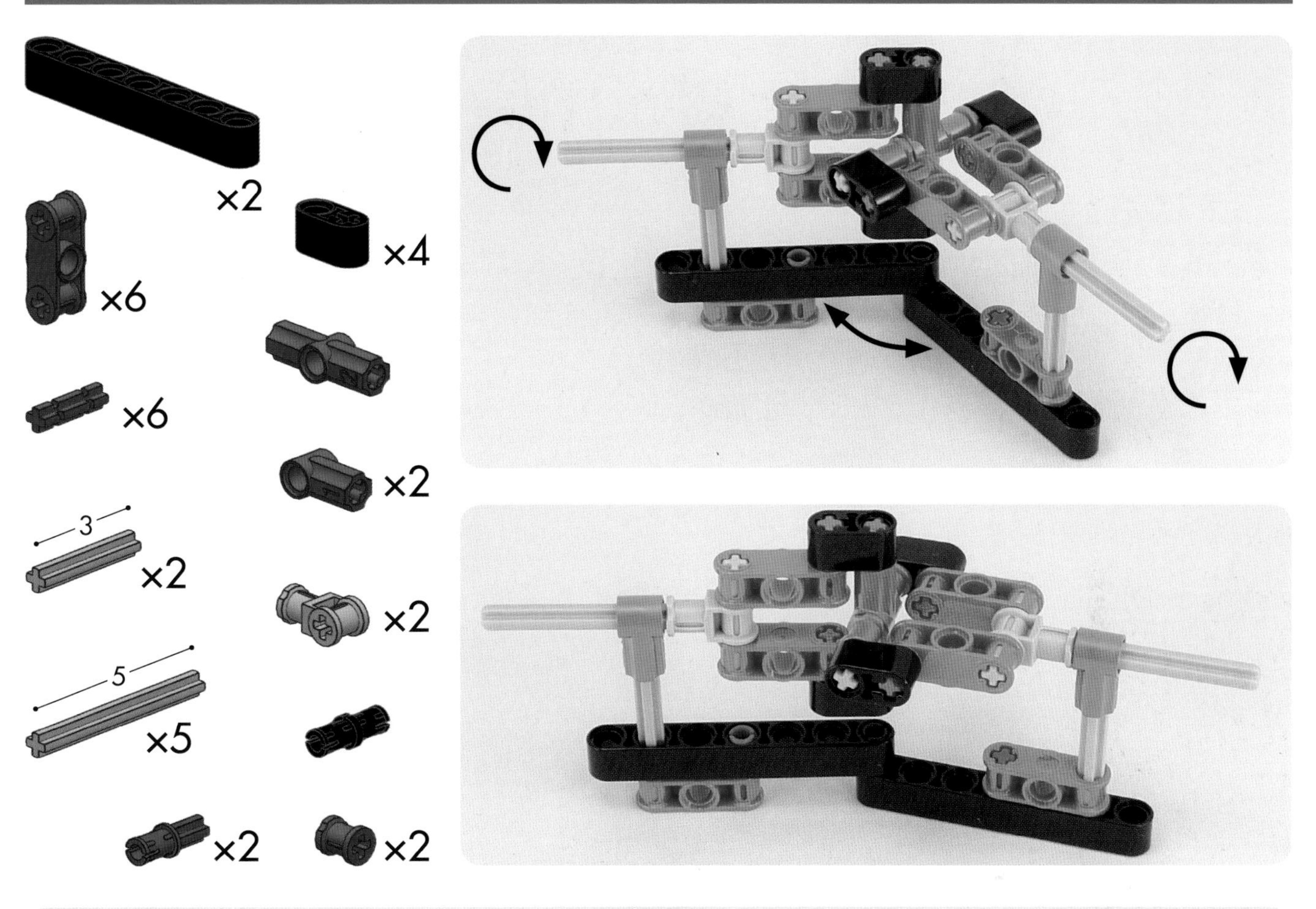
×2
×4
×6
×6
×2
3
×2
×2
5
×5
×2
×2

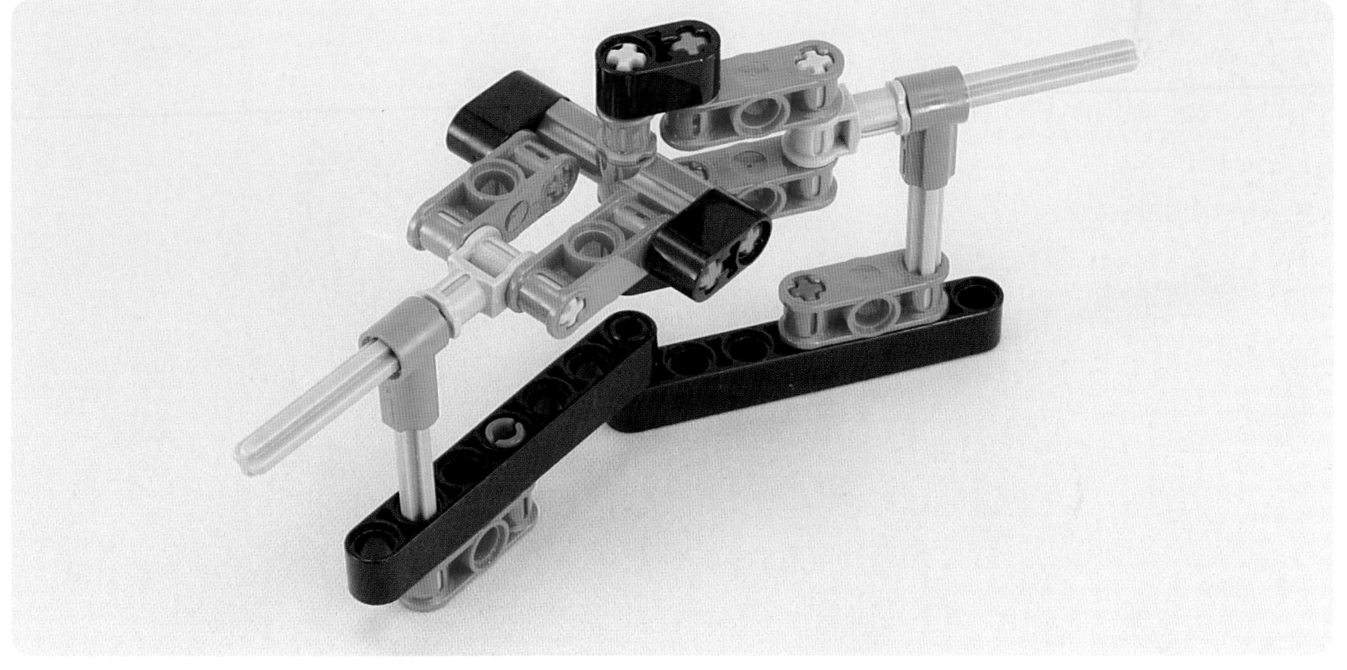

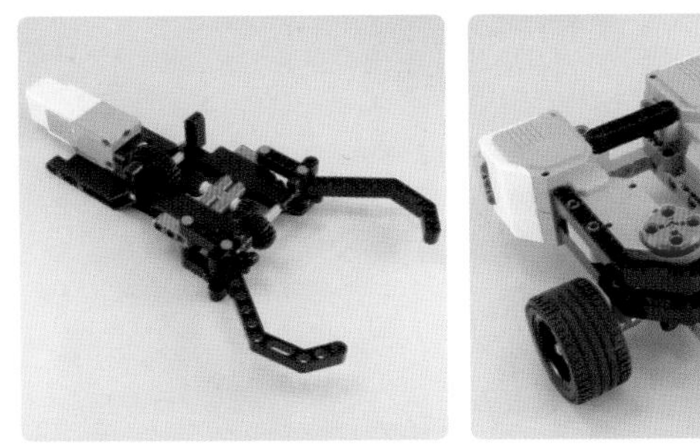
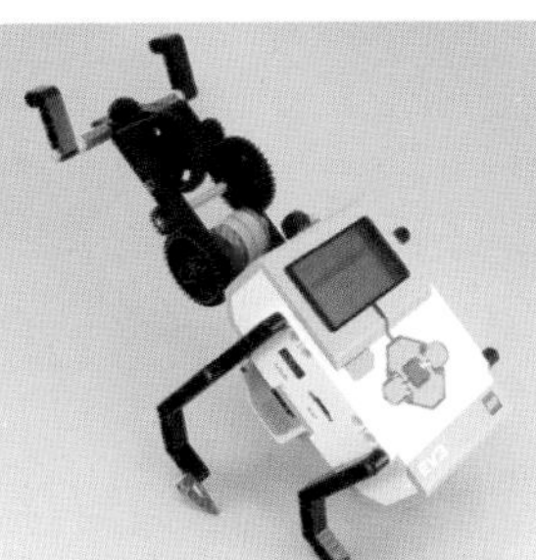

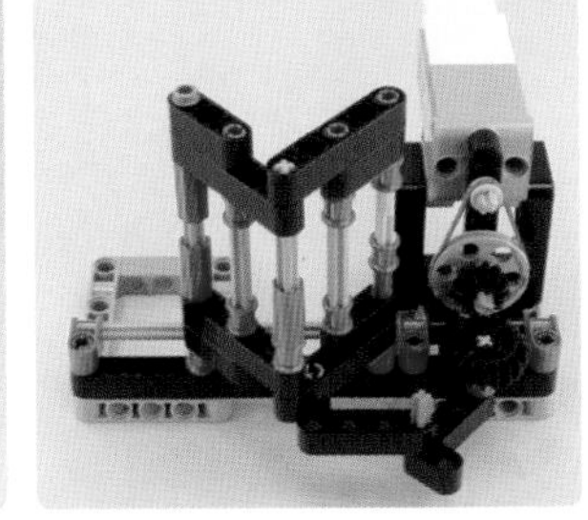
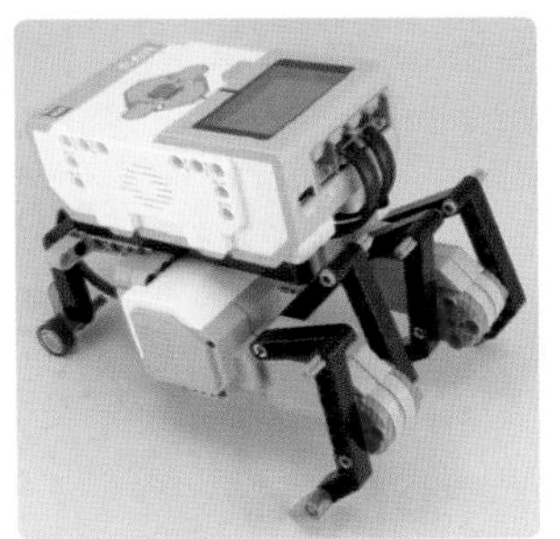

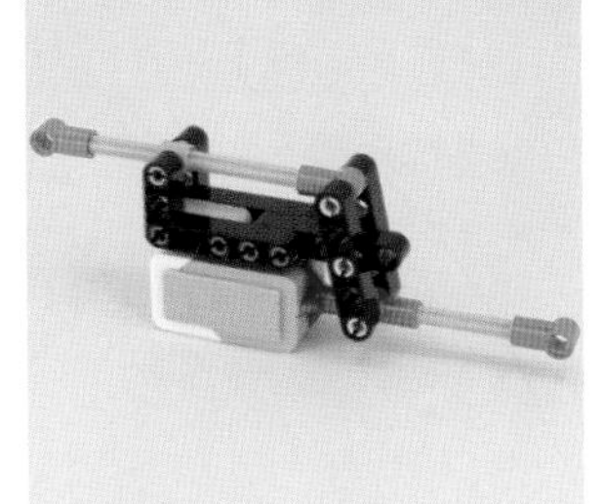

PART 2

Vehicles

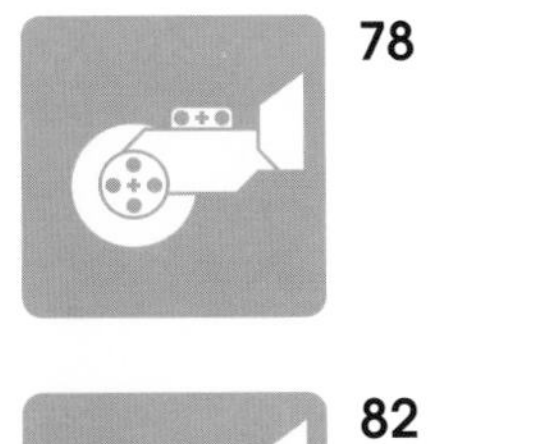
78

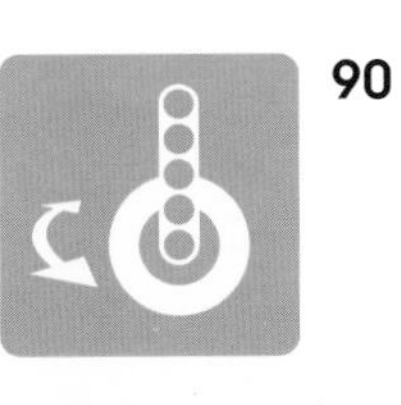
90

100

82

94

104

Driving wheels with a motor

#102

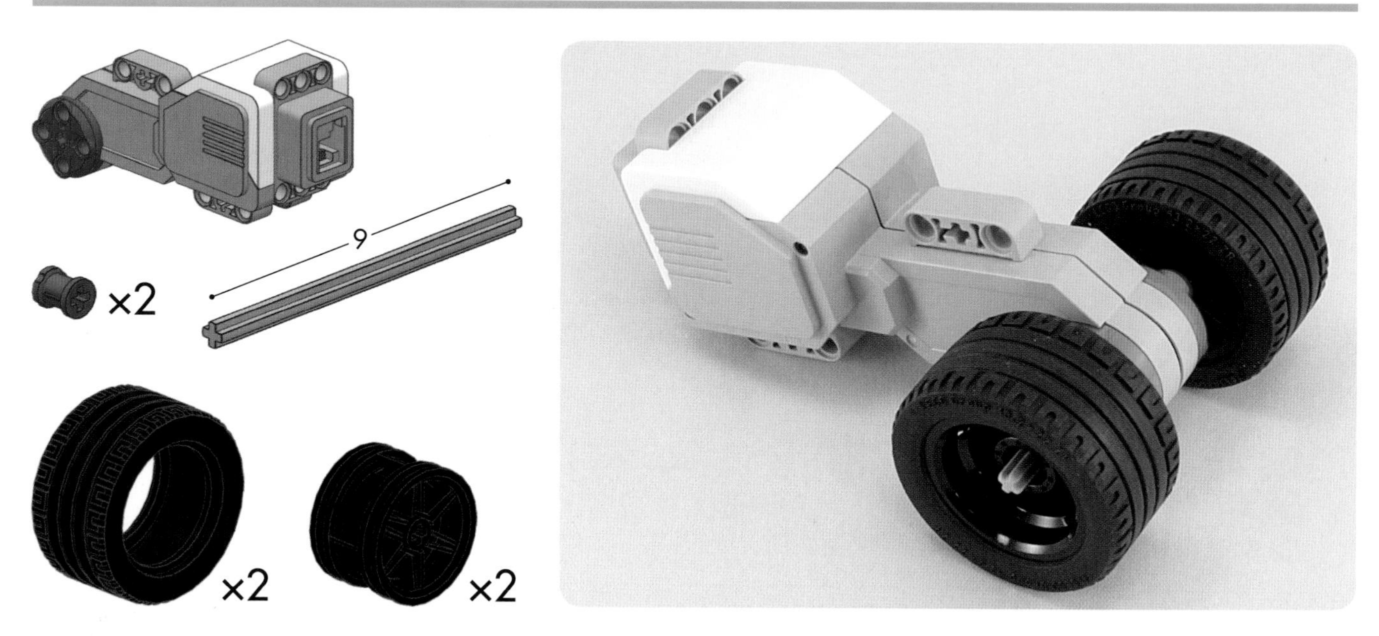
9
×2
×2
×2

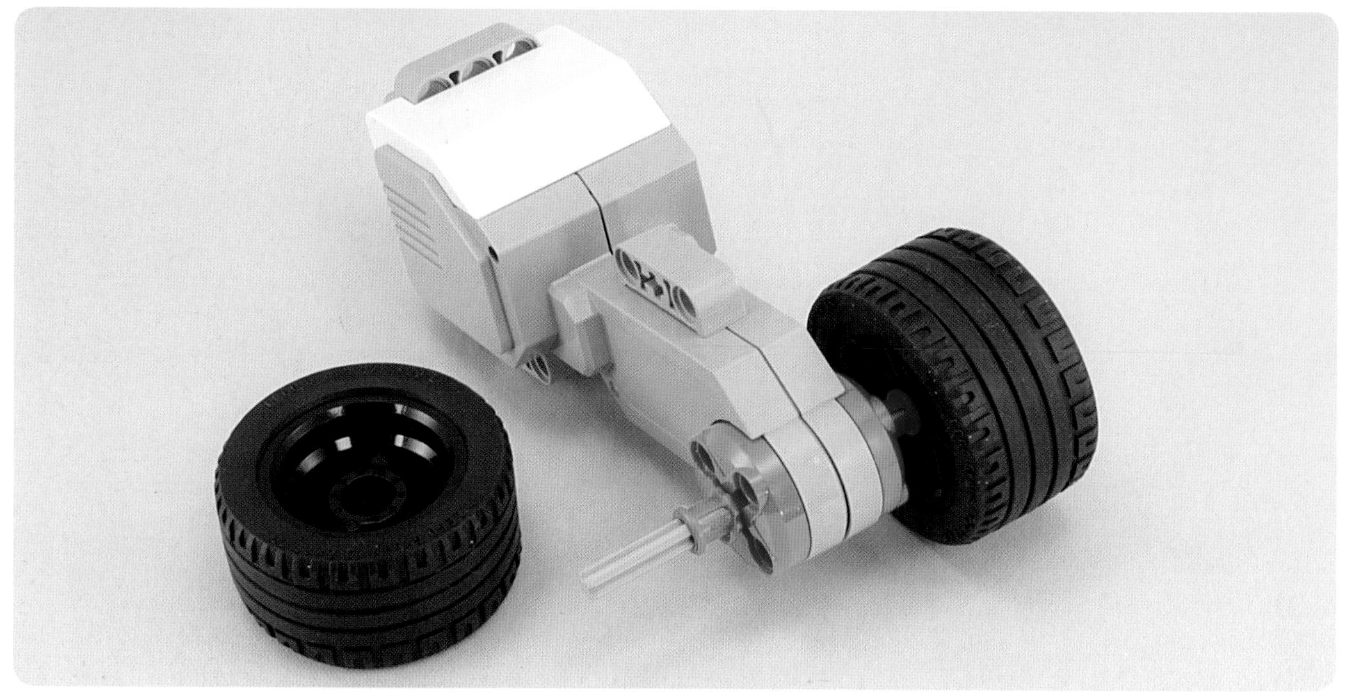

#103

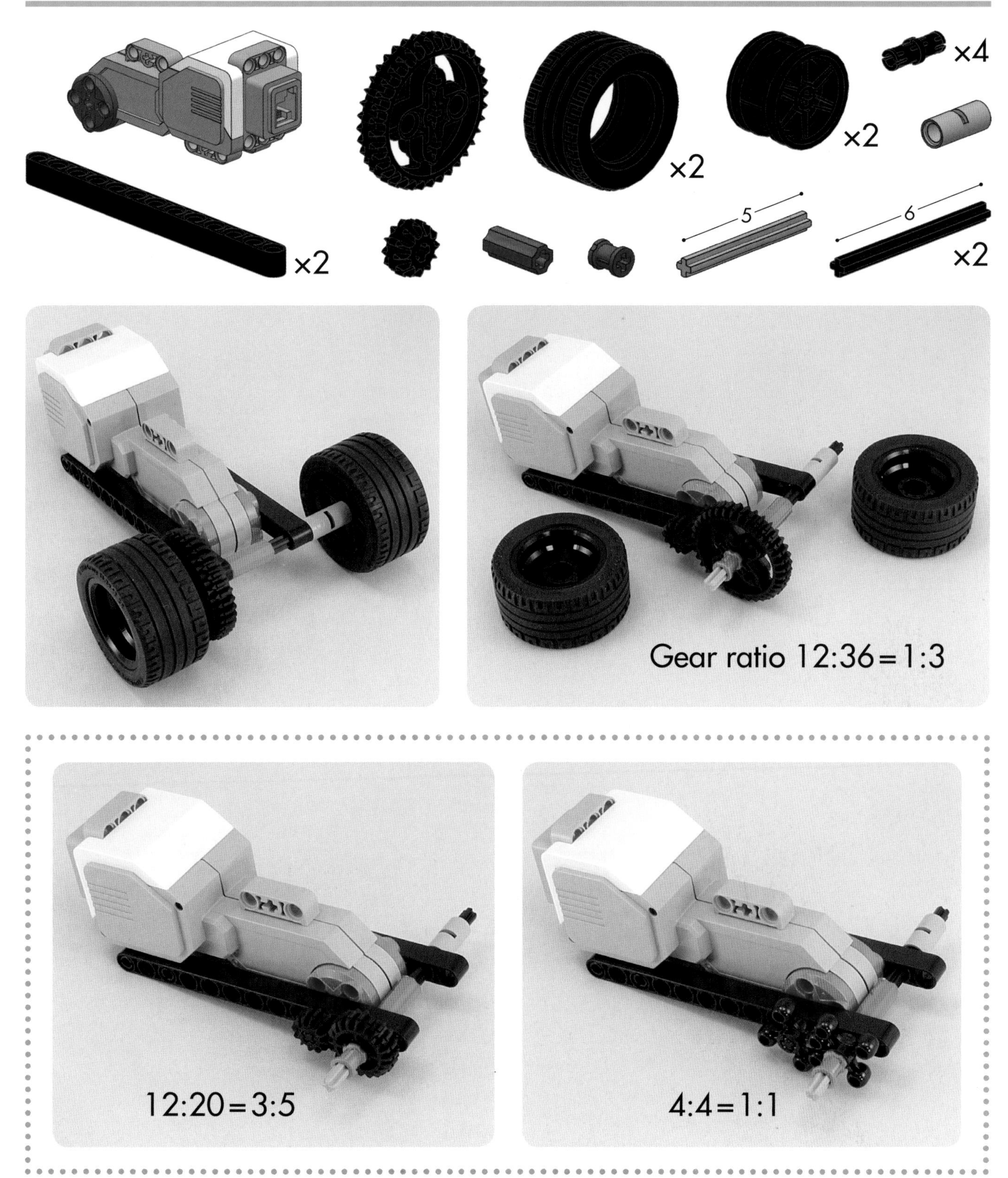
×4
×2
×2
×2
5
6
×2
Gear ratio 12:36 = 1:3
12:20 = 3:5
4:4 = 1:1

9
×2
×2
×2
×2
×3

#105

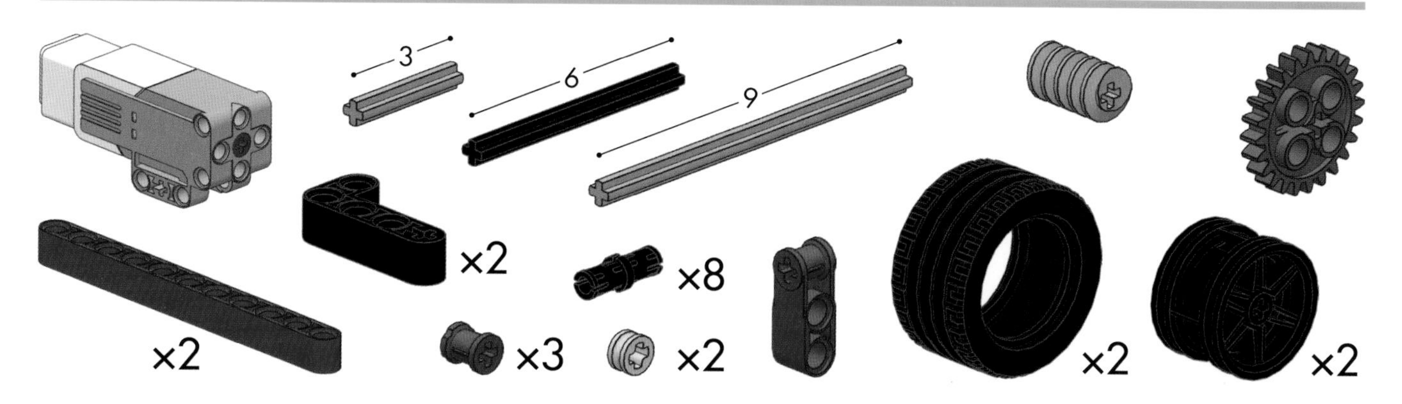
3
6
9
×2
×2
×8
×3
×2
×2
×2

1:24

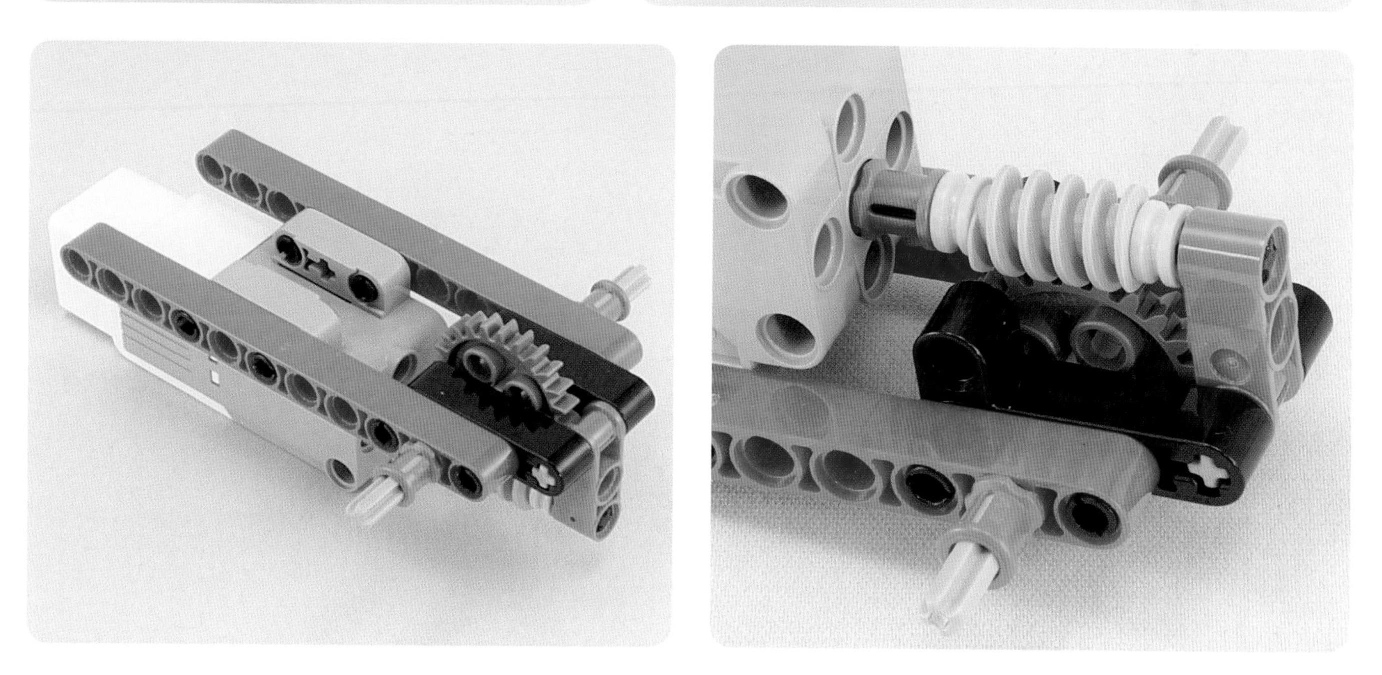

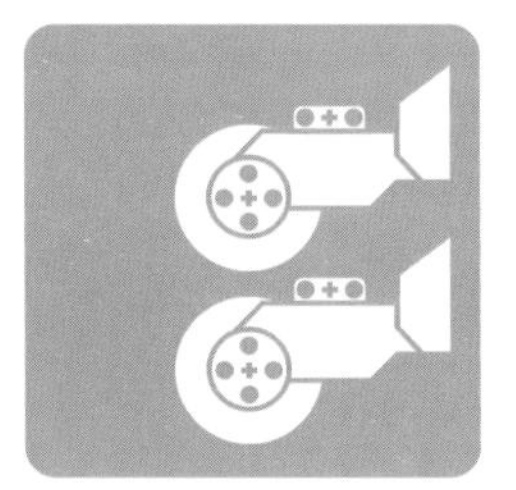

Driving wheels with two motors

#106

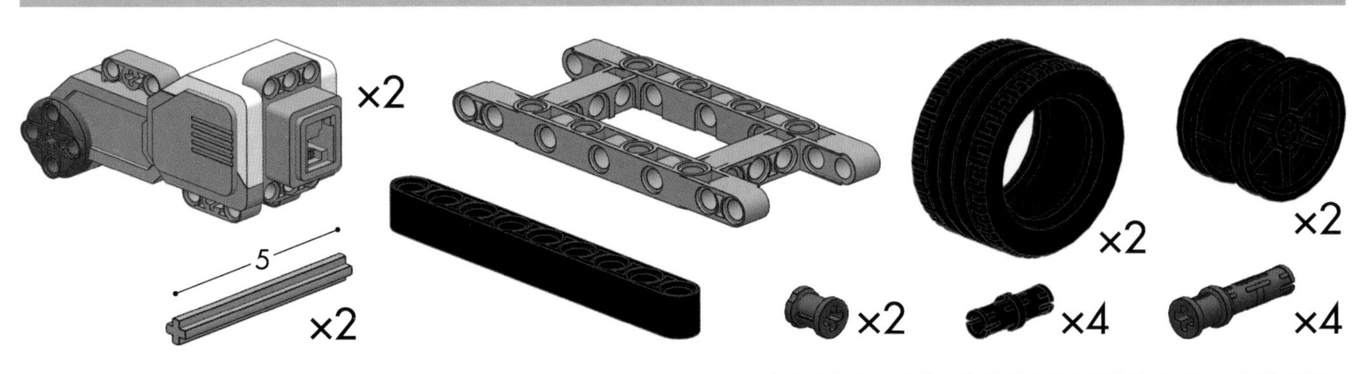
×2
5
×2
×2
×2
×2
×4
×4

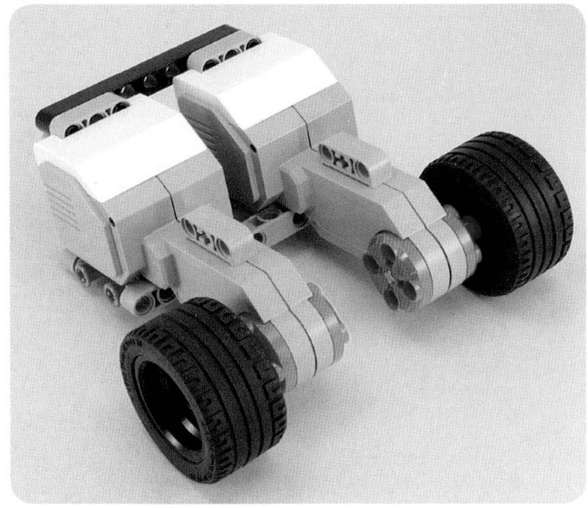

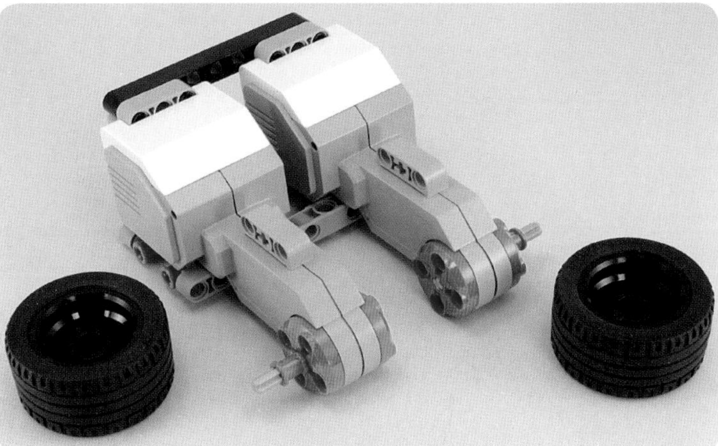

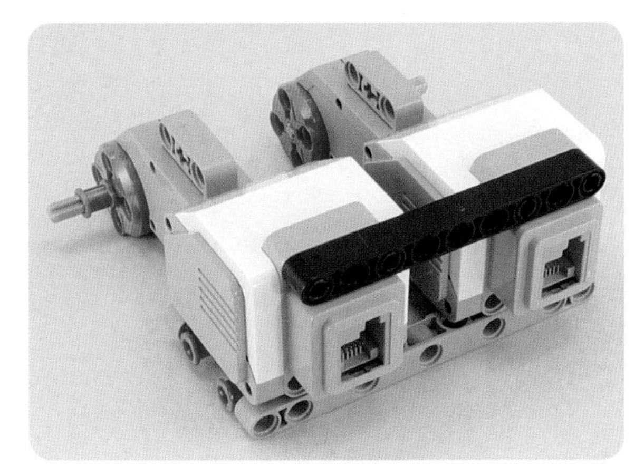

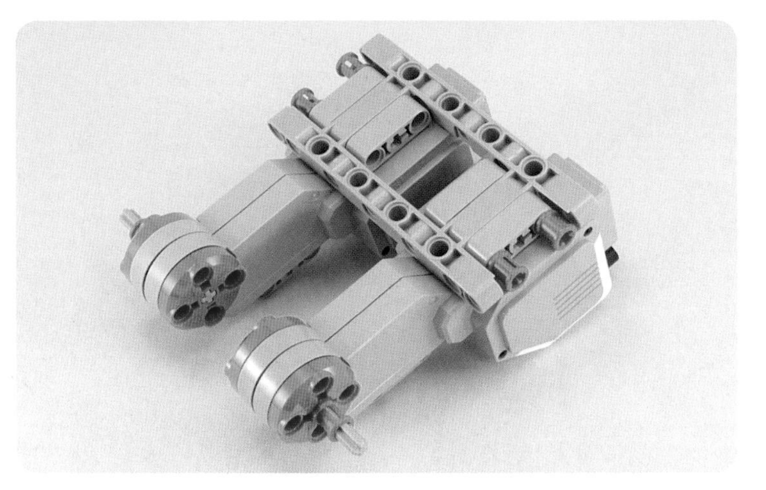

×2
×2
×2
×2
×2
×2
×2
×2
×2
3
5
×3
×2
×2
×6
×4
12:20 = 3:5

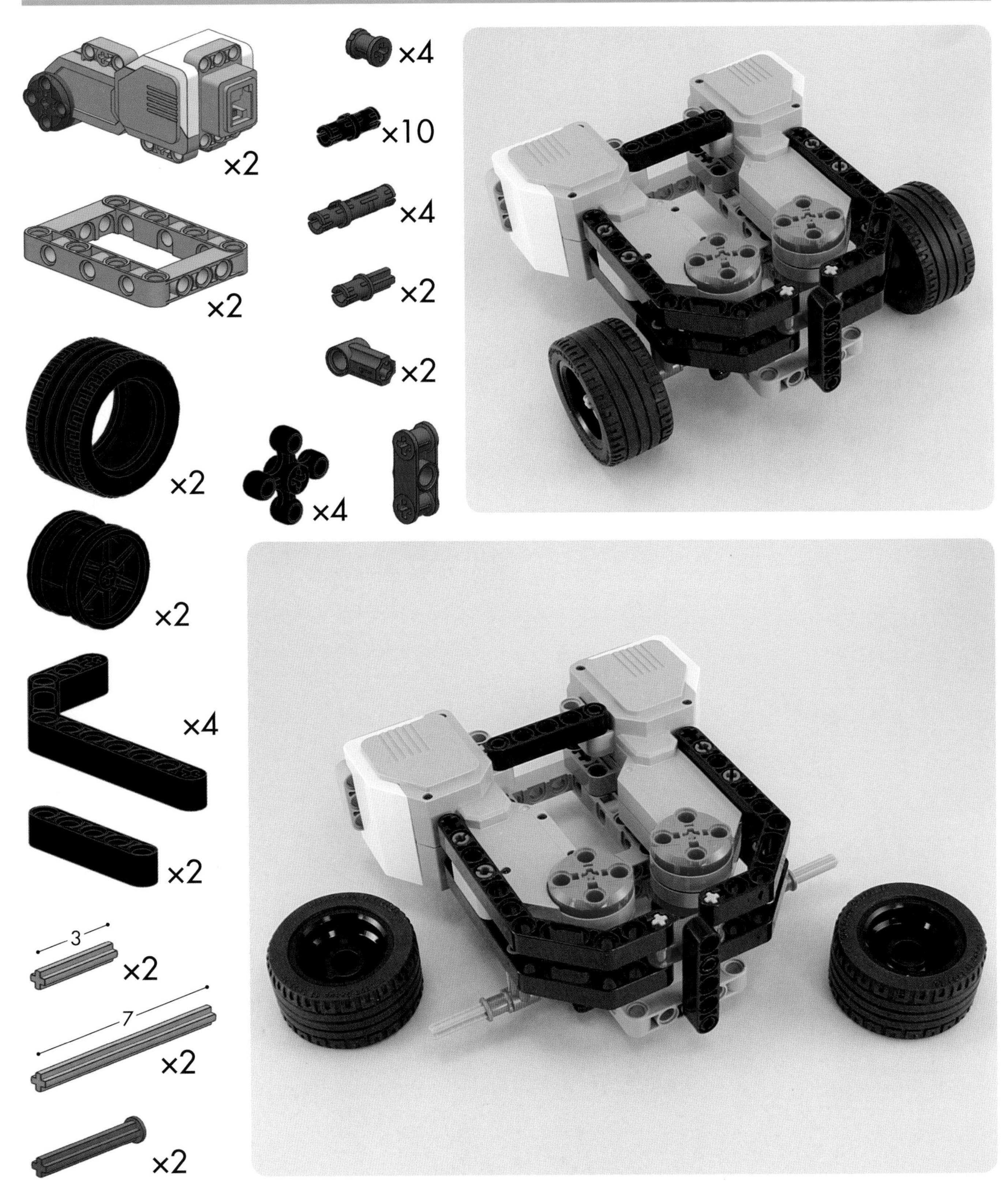
×2
×4
×10
×2
×4
×2
×2
×2
×4
×2
×4
×2
3
×2
7
×2
×2

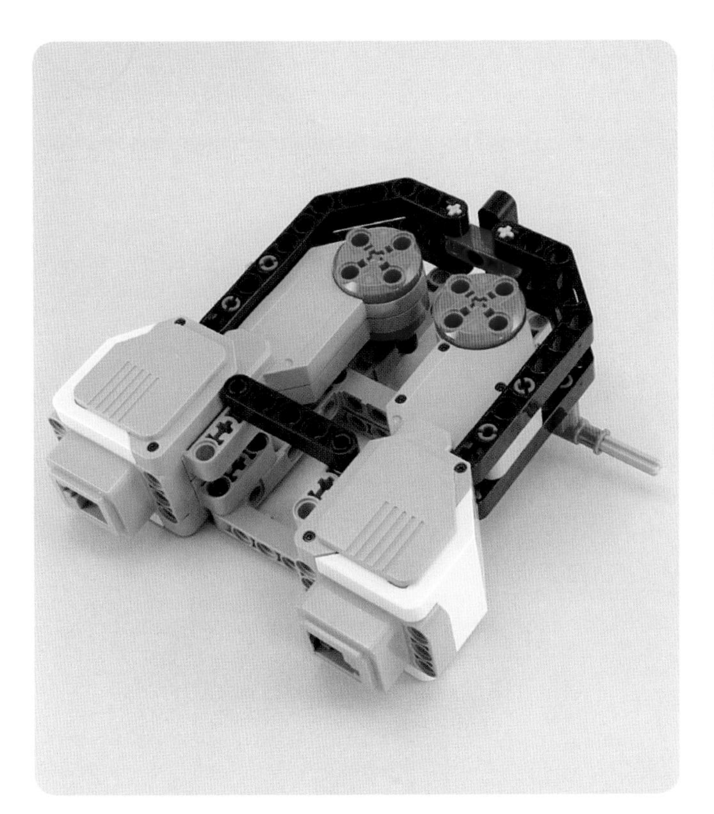

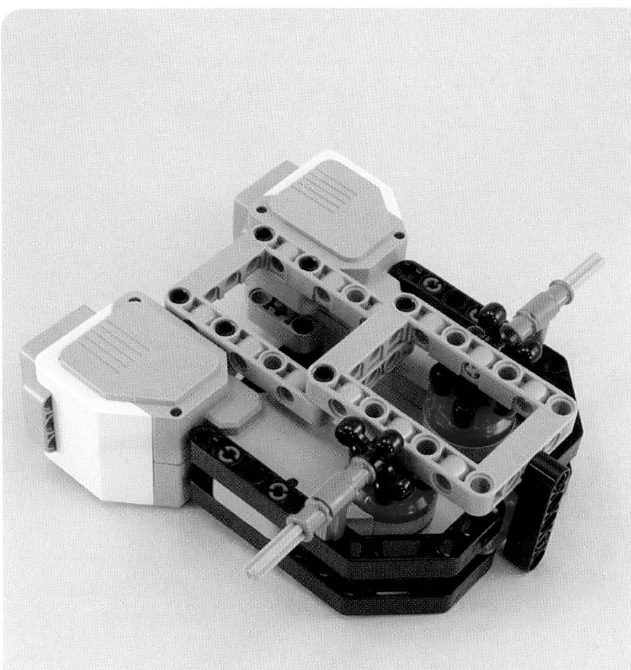

4:4=1:1

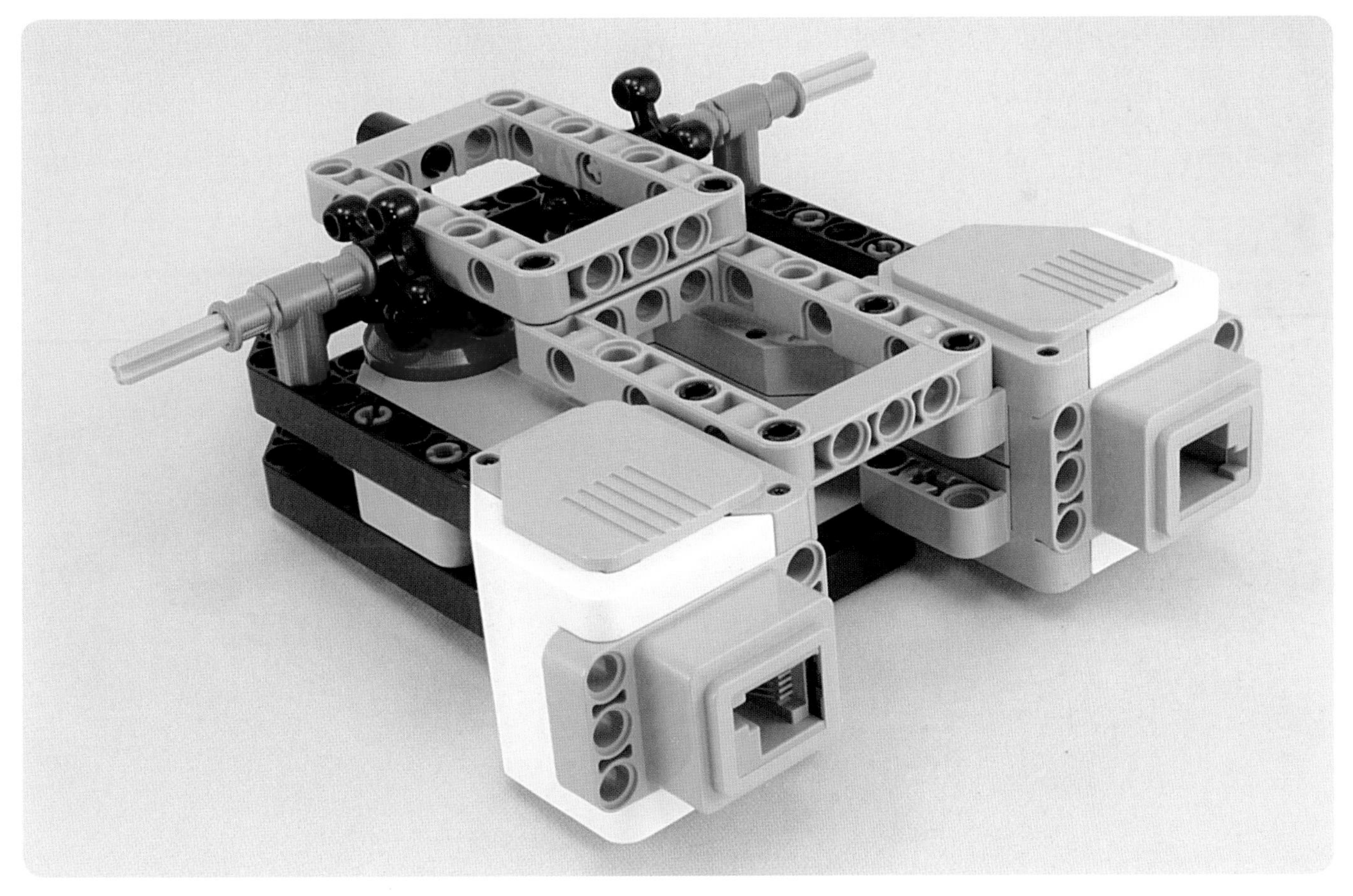

#109

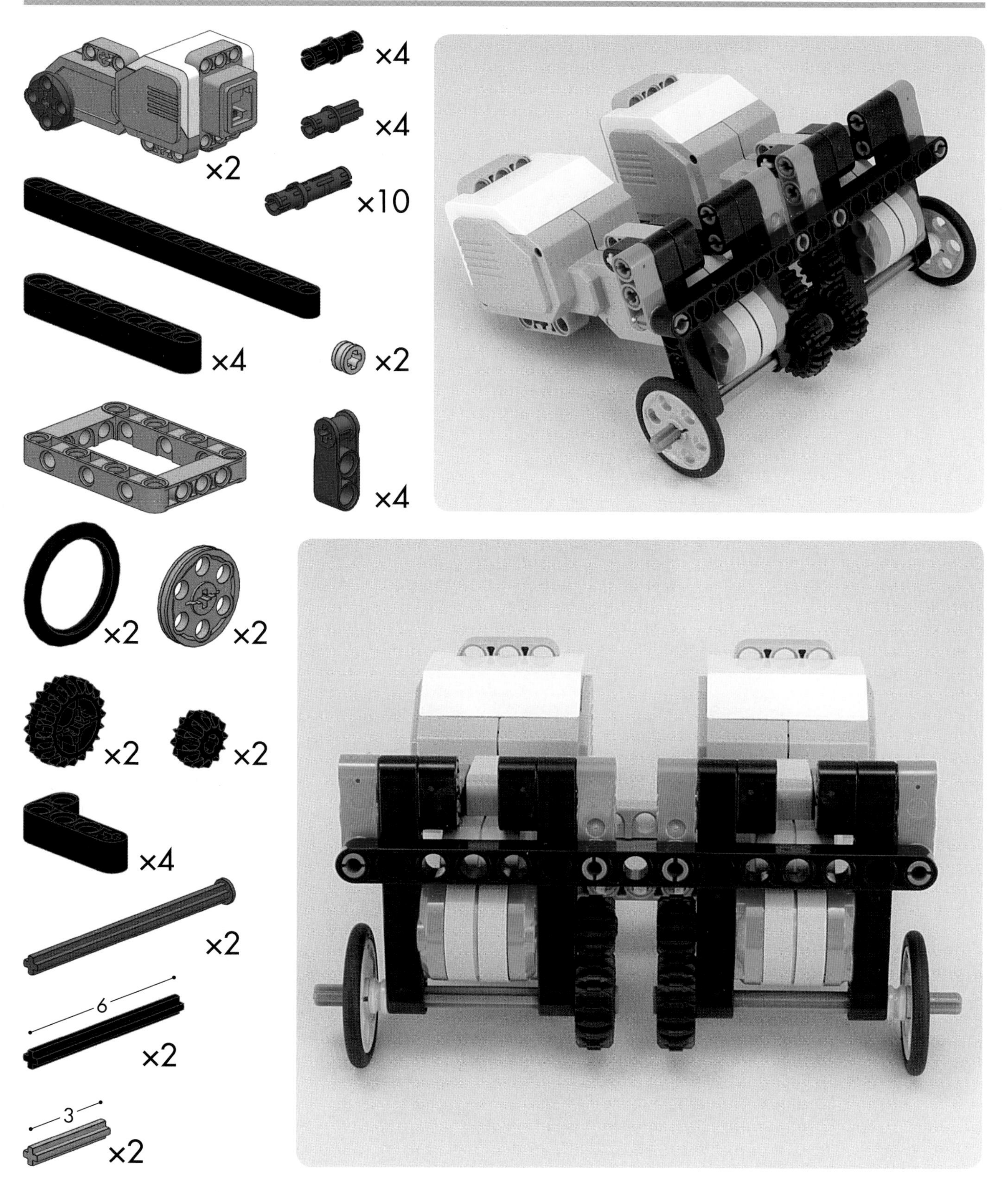
×2
×4
×4
×10
×4
×2
×4
×2
×2
×2
×2
×4
×2
6
×2
3
×2

12:20 = 3:5

#110

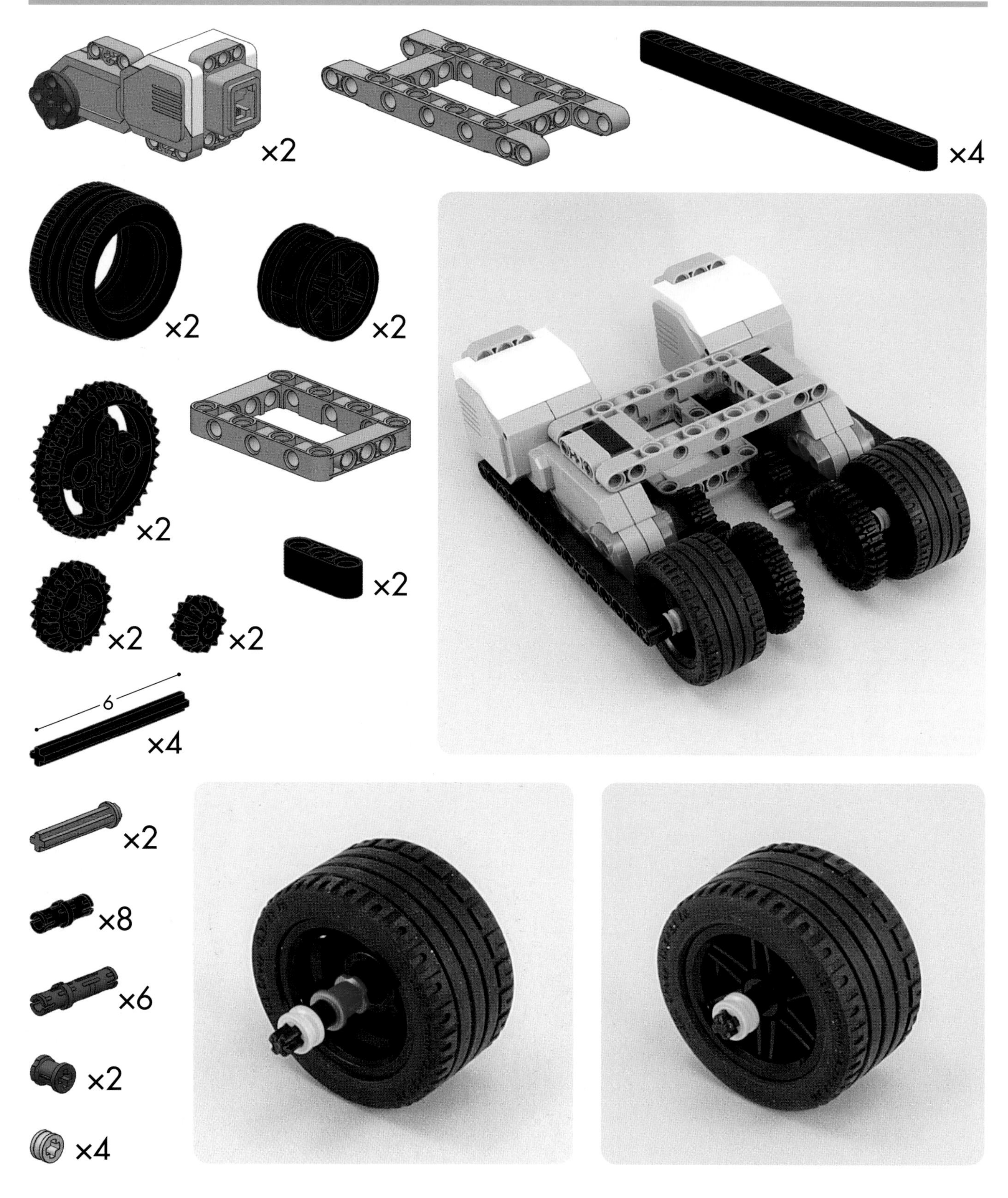
×2
×4
×2
×2
×2
×2
×2
×2
6
×4
×2
×8
×6
×2
×4

20:36 = 5:9

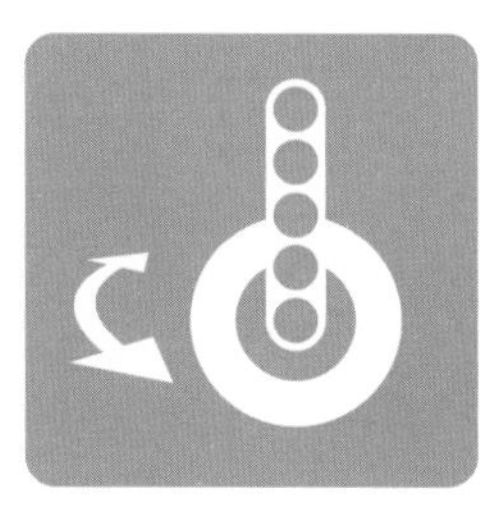

Caster wheels

#111

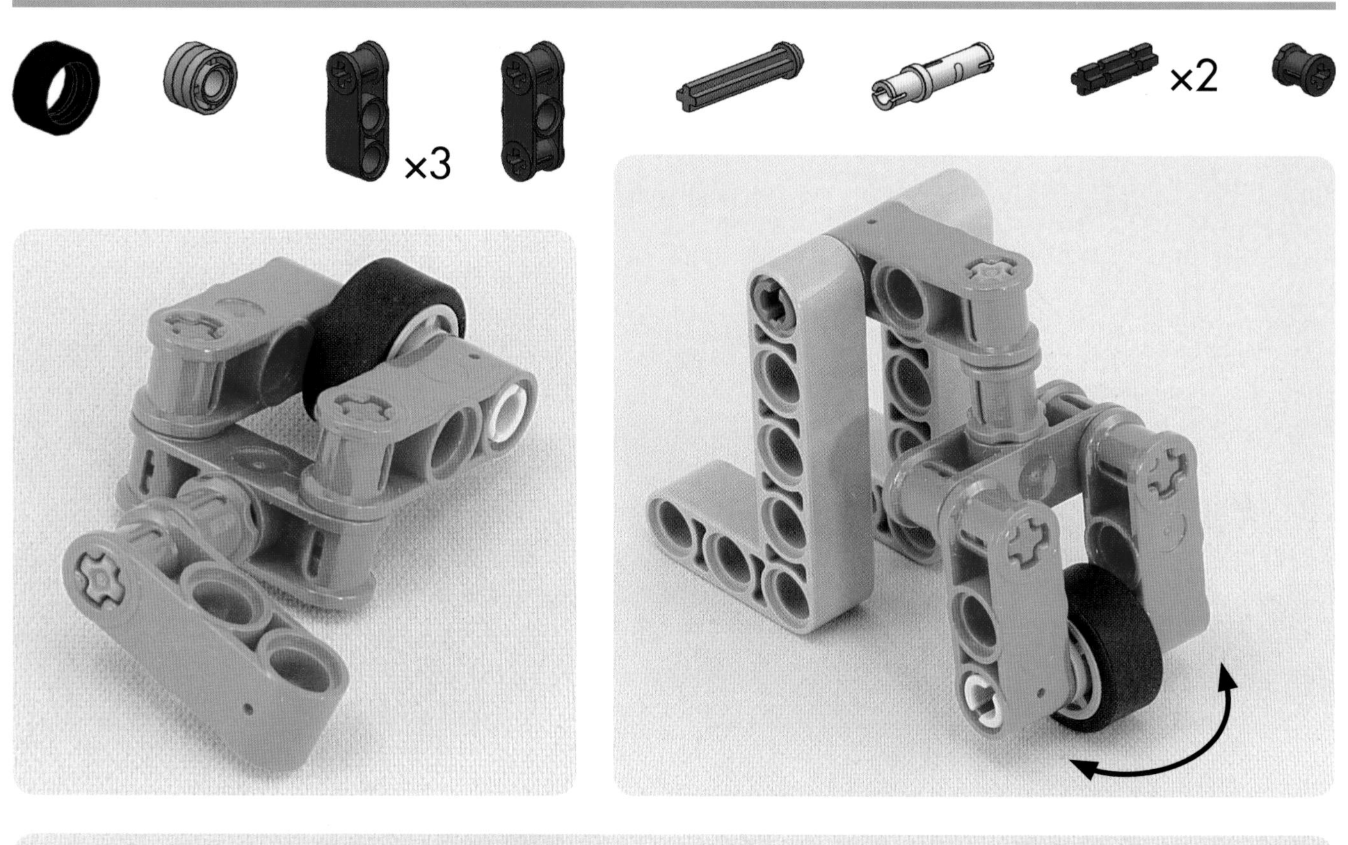
×3
×2

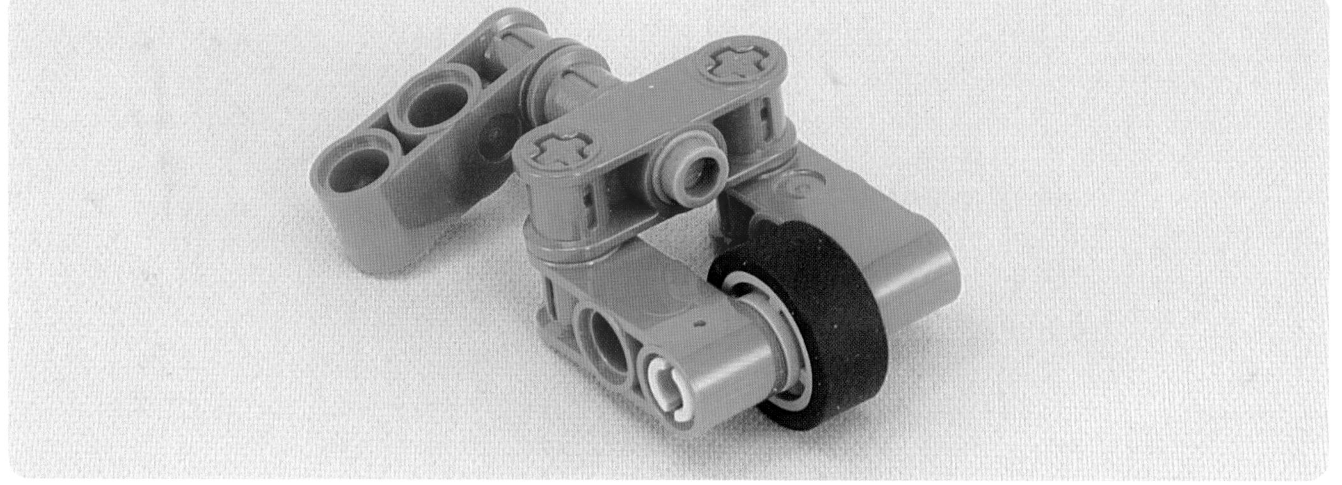

#112

×2
×2

#113

×2
×2
×2
3
×2

#114

×2
×2
3
×3
×2
×3

Crawlers

#115

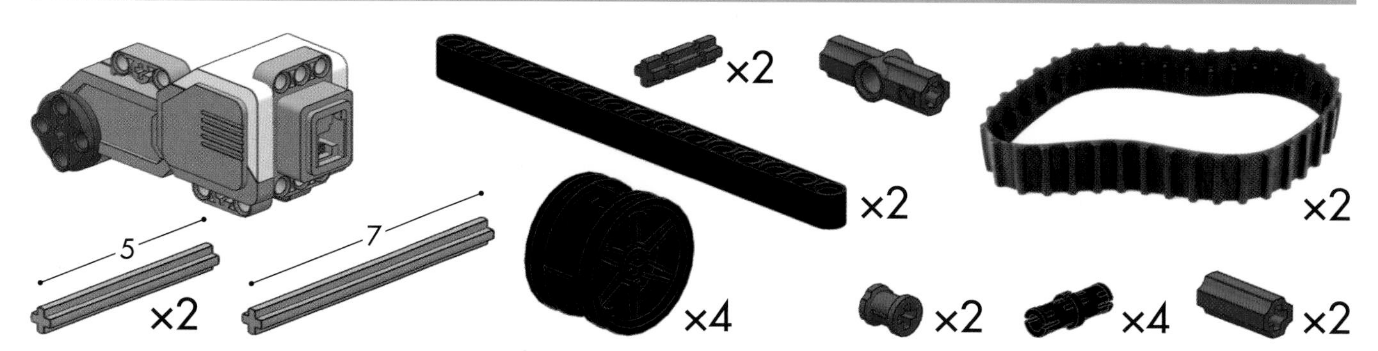
×2
×2
×2
5
7
×2
×4
×2
×4
×2
×2

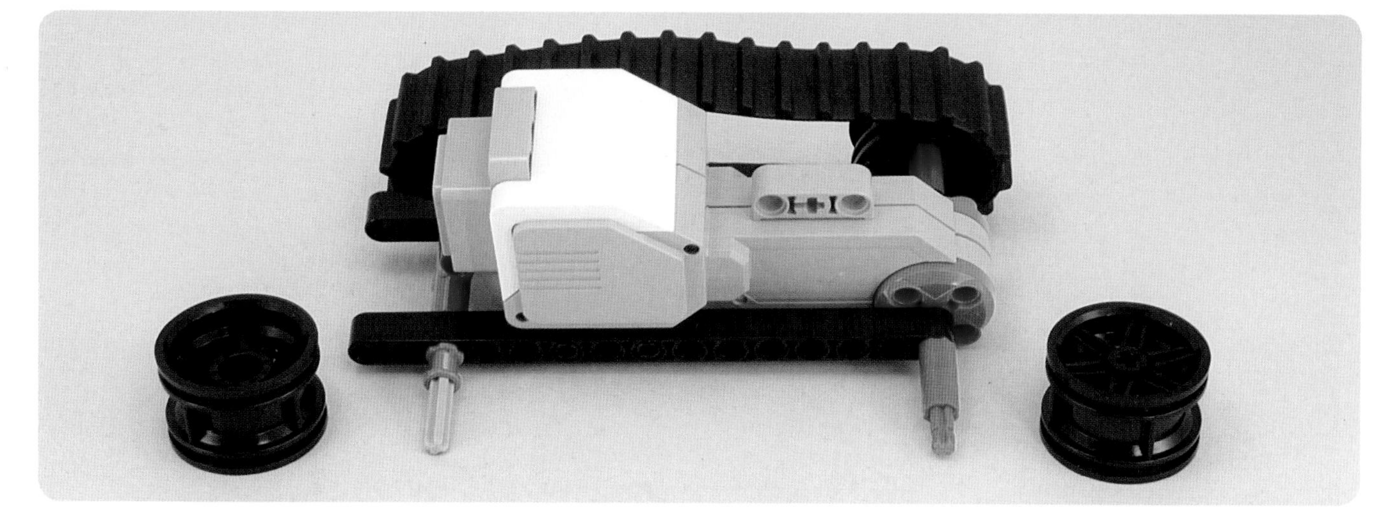

#116

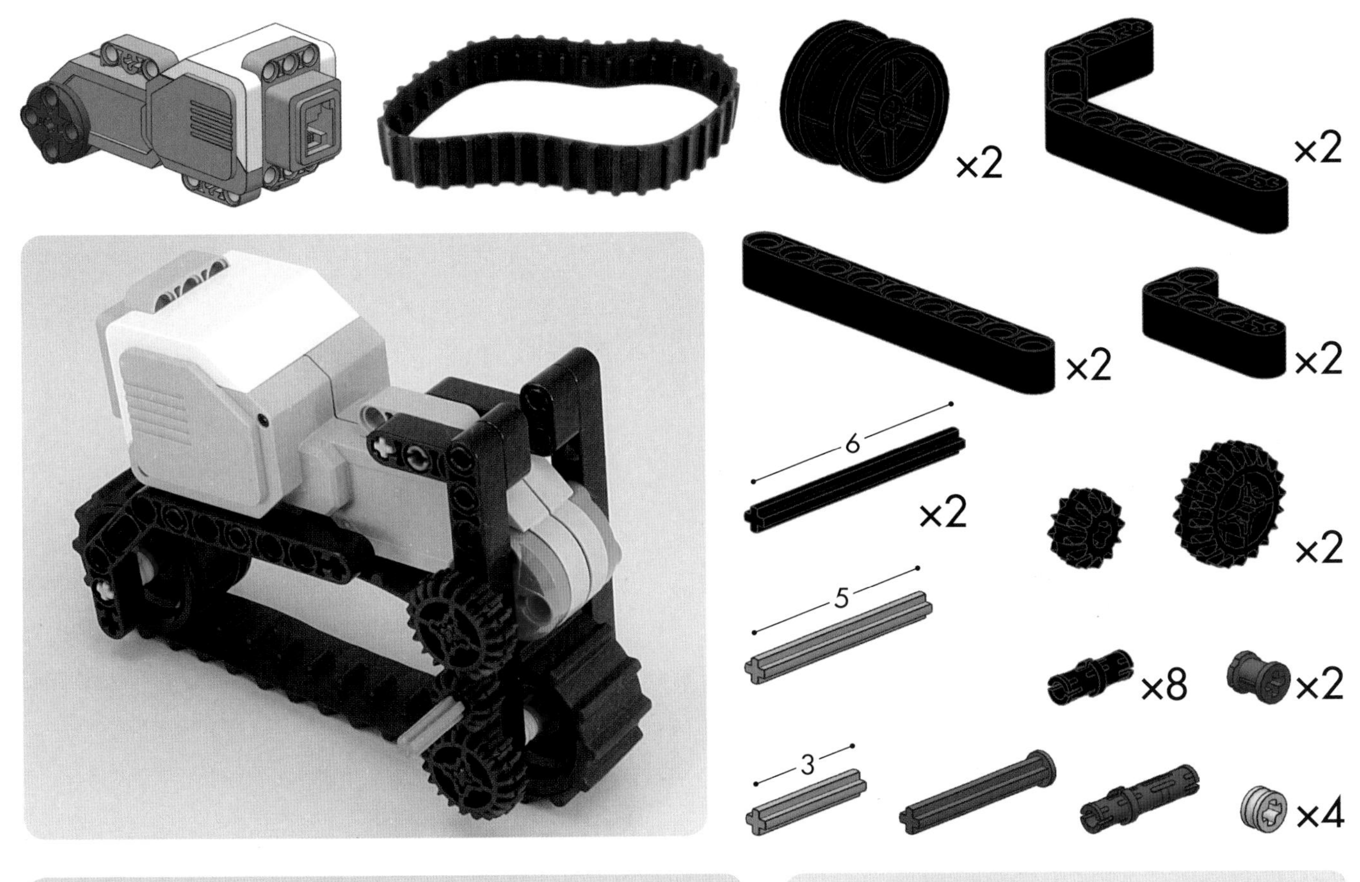

×2
×2
×2
×2
6
×2
×2
5
×8
×2
3
×4

20:12:20 = 5:3:5

#117

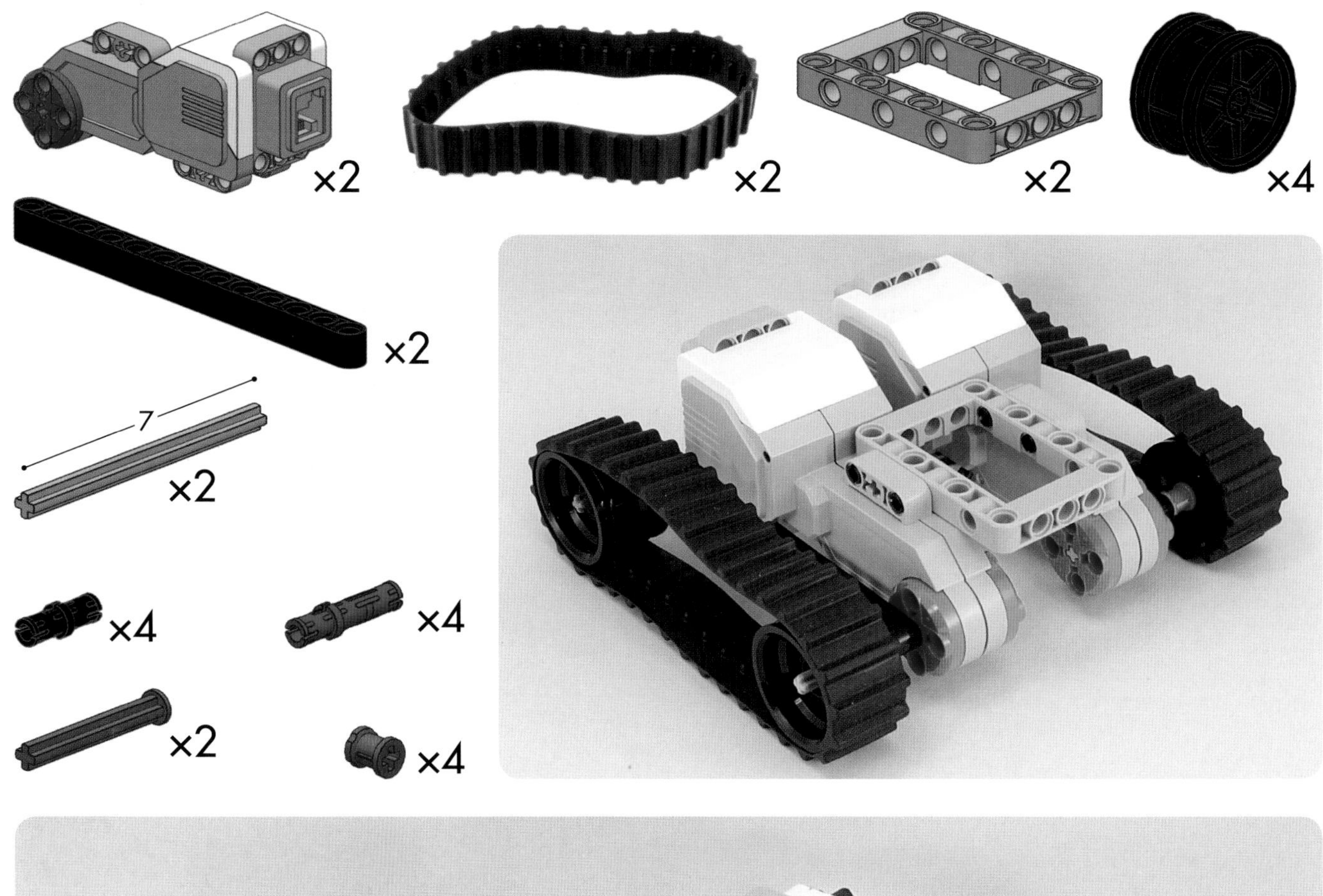
×2
×2
×2
×4
×2
7
×2
×4
×4
×2
×4

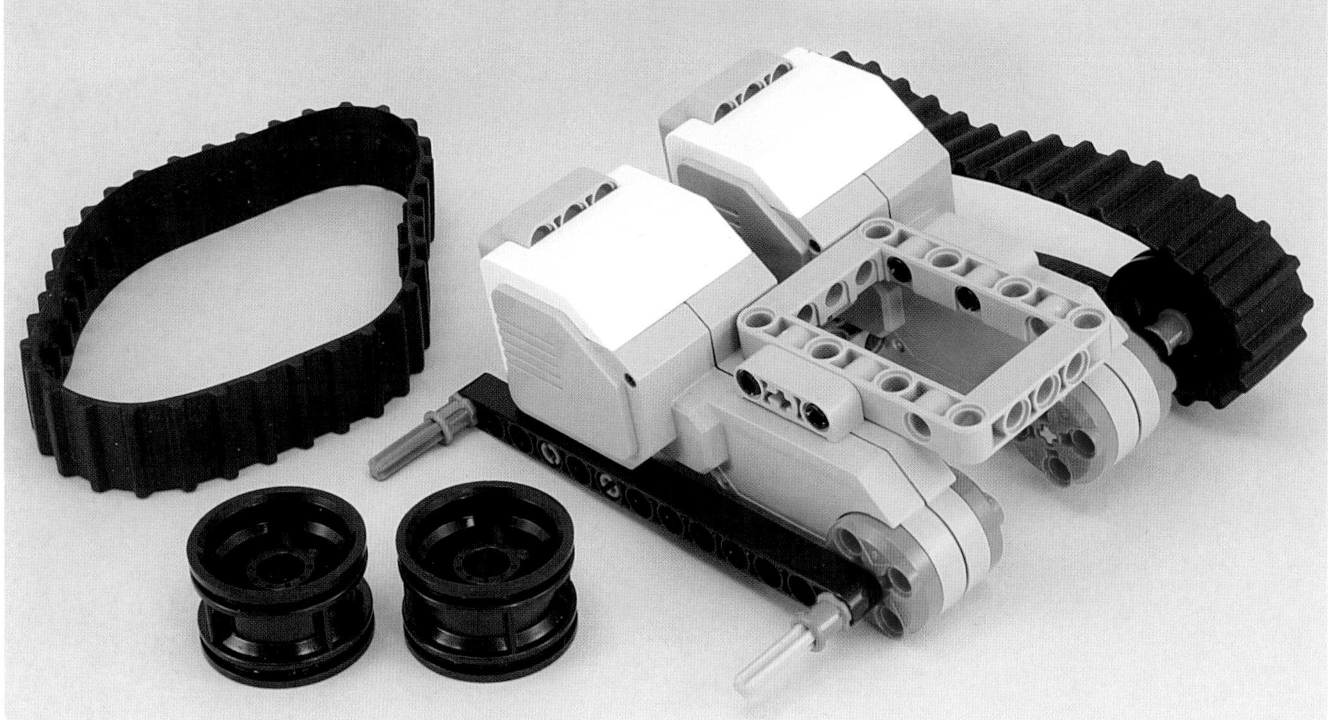

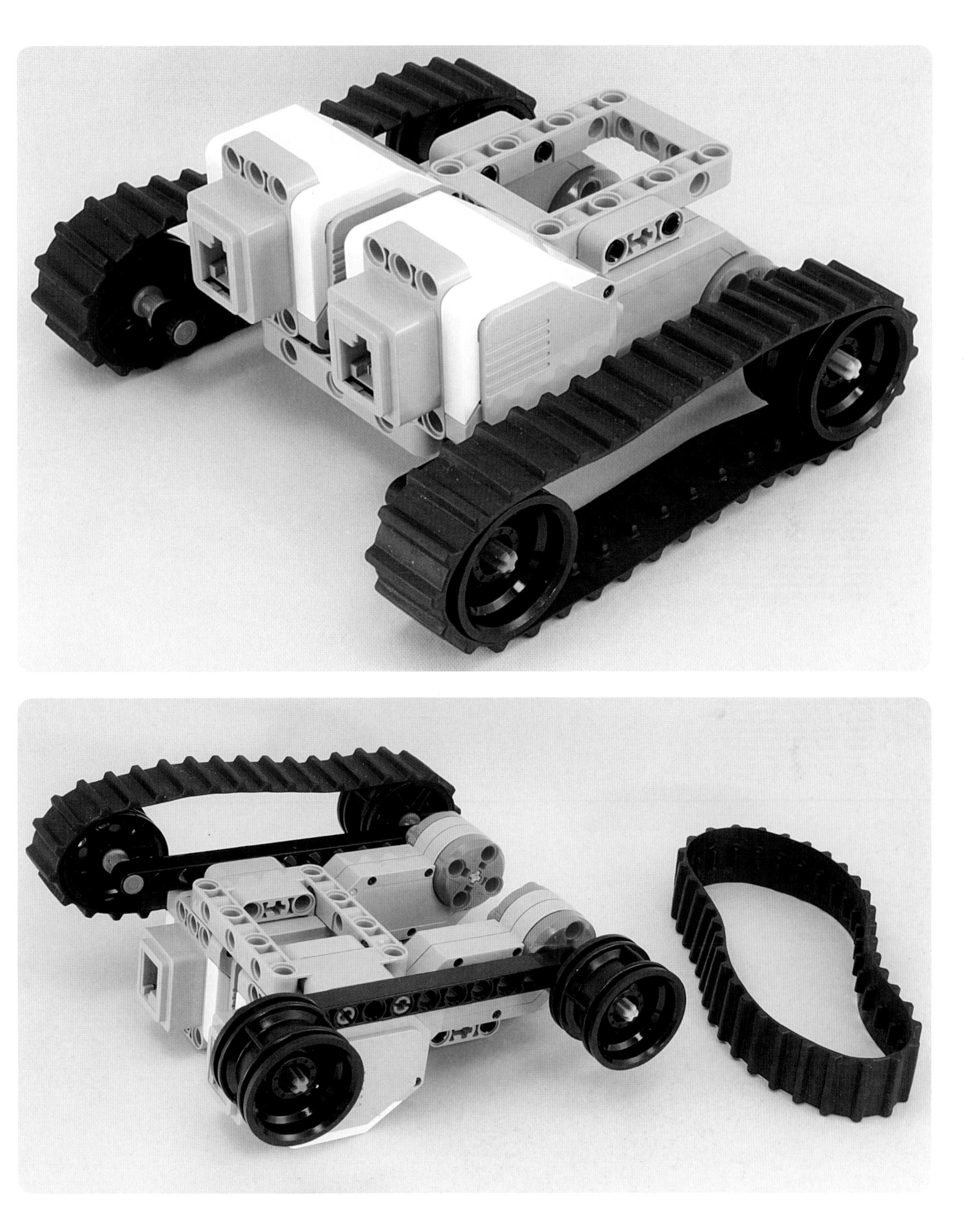

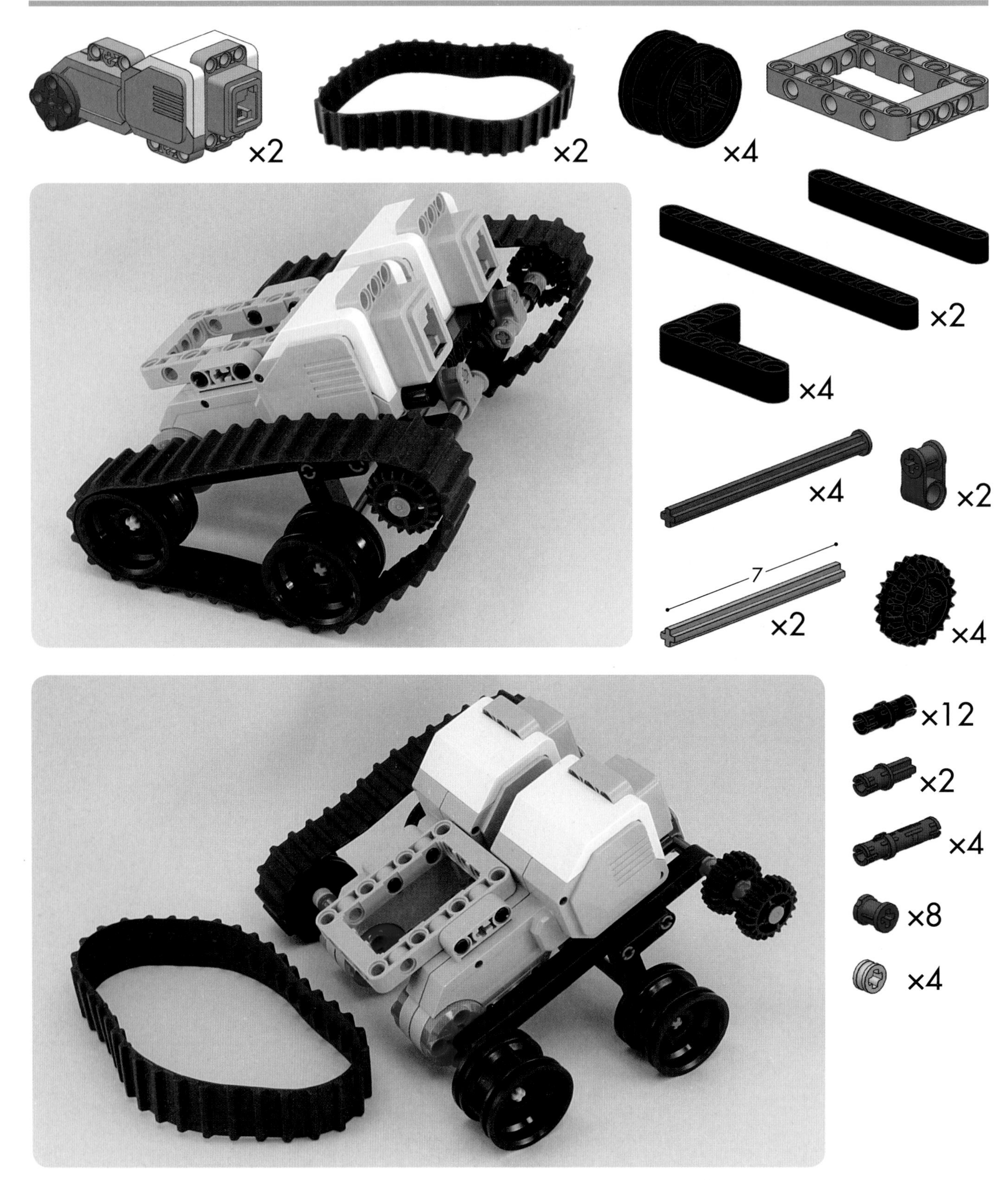
×2
×2
×4
×2
×4
×4
×2
7
×2
×4
×12
×2
×4
×8
×4

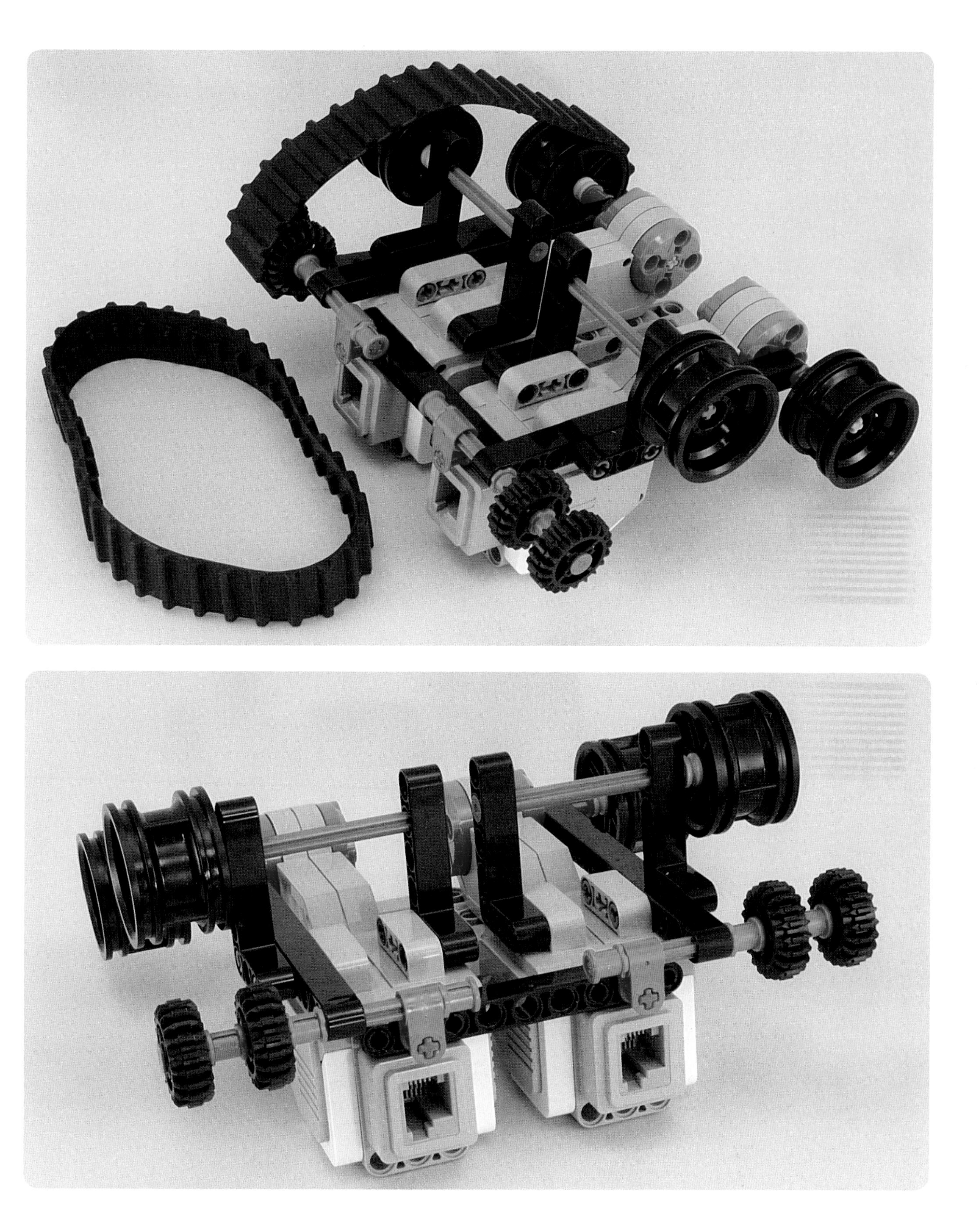

Suspended wheels

#119

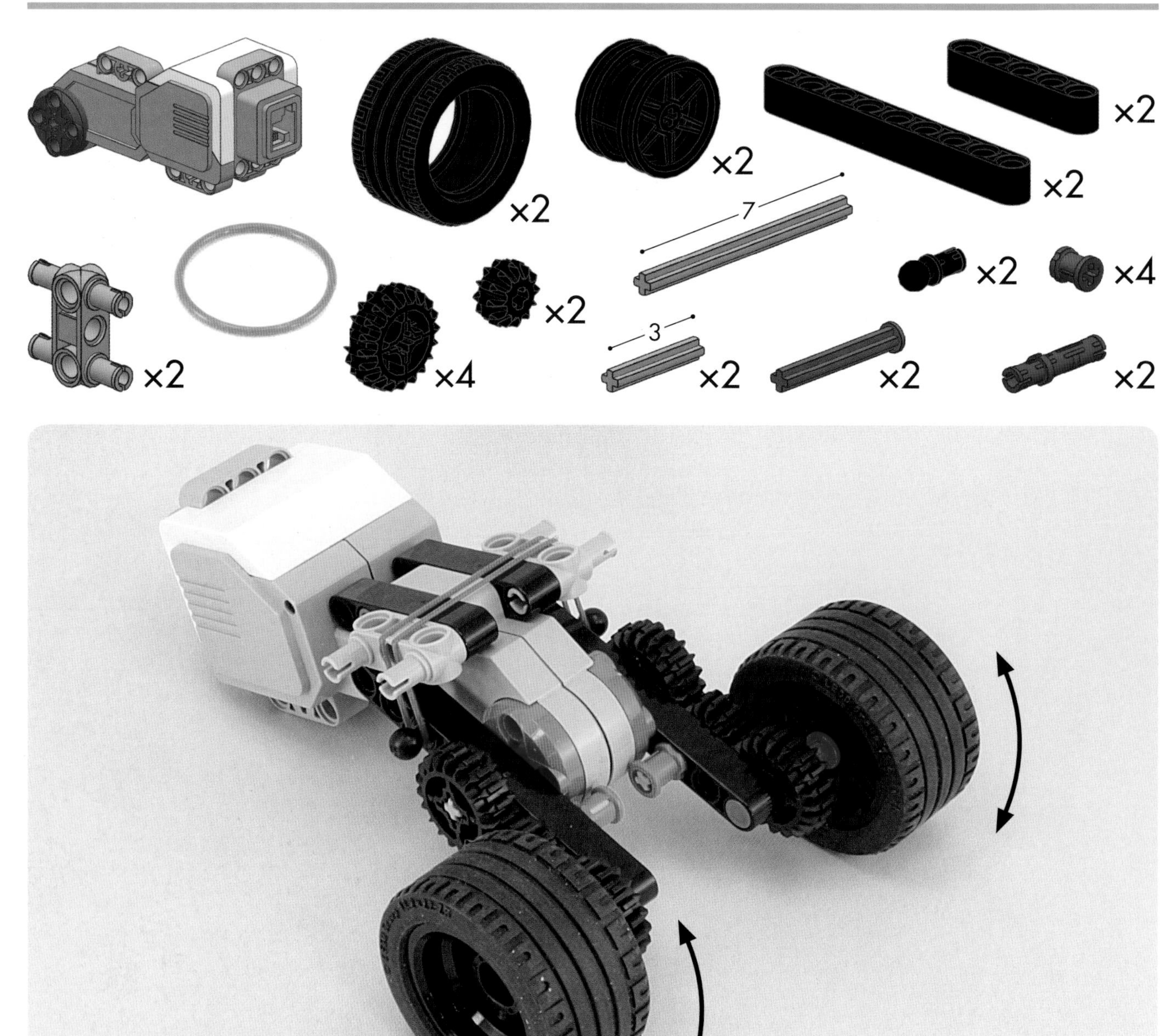

×2
×2
×2
×2
7
×2
×4
×2
×2
3
×2
×4
×2
×2
×2

#120

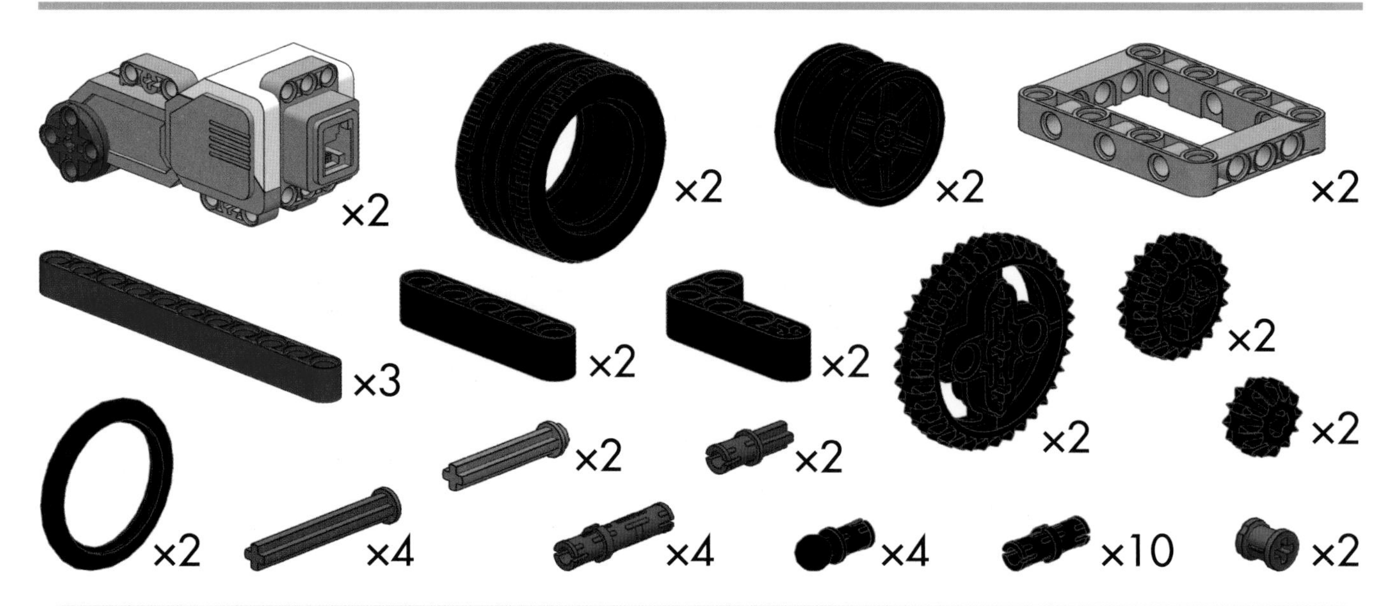
×2
×2
×2
×2
×3
×2
×2
×2
×2
×2
×2
×2
×2
×4
×4
×4
×10
×2

Steering

#121

×2
×2
×2
×5
×2
×2
×3
3
×5
×2
5
×14
×5
×2

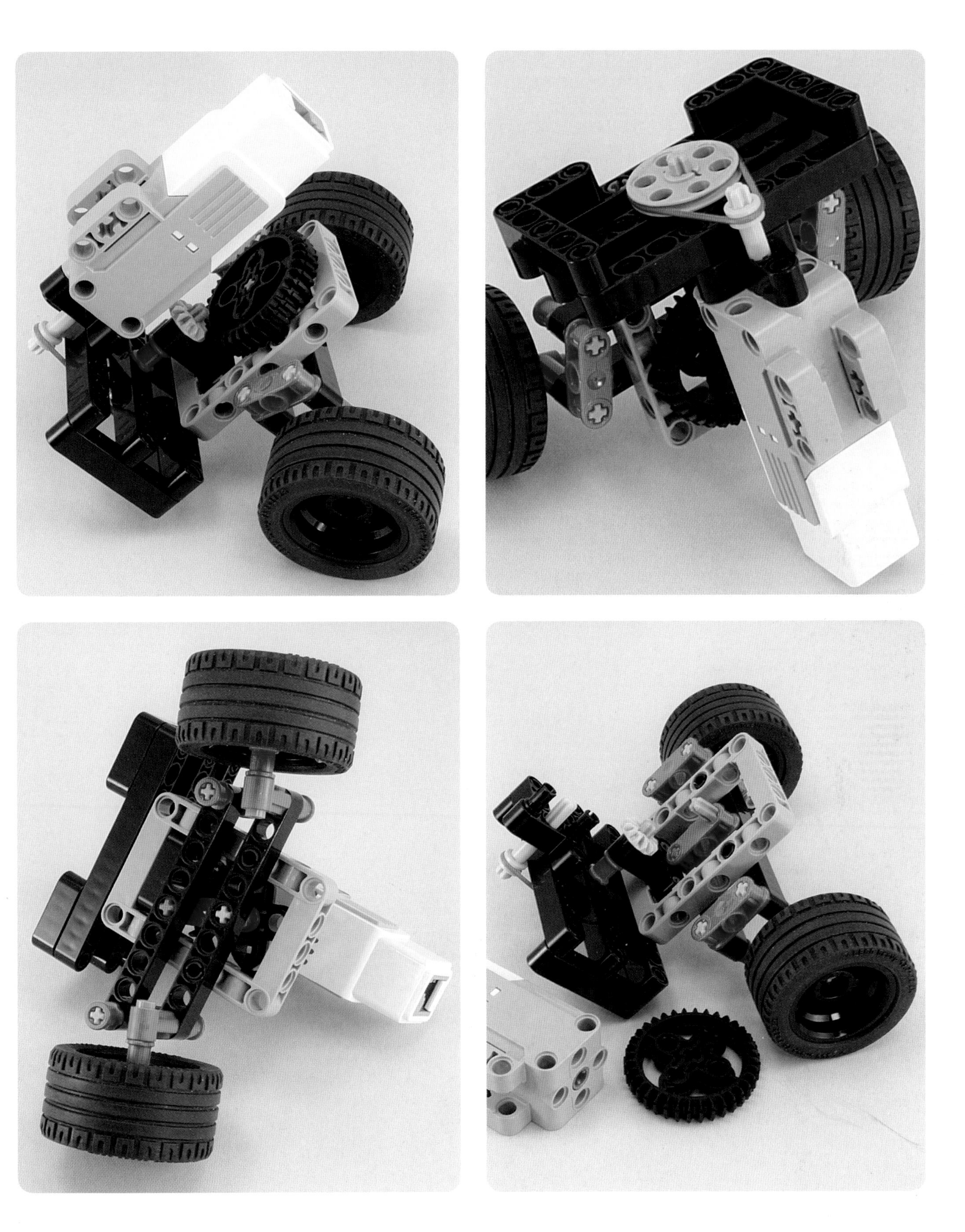

EV3
LEGO
1
2
3
4
6
×2
×2
×2
×2
×10
×2
×2
×2
×2

EV3
USB
SD

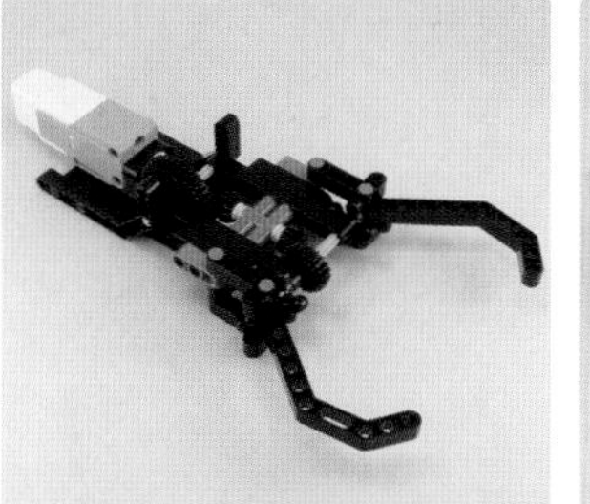

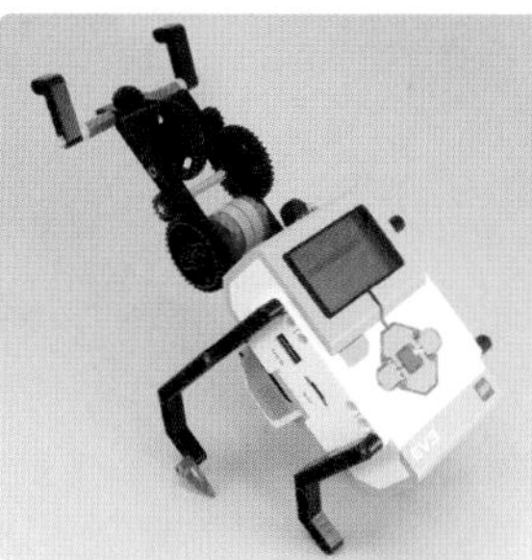

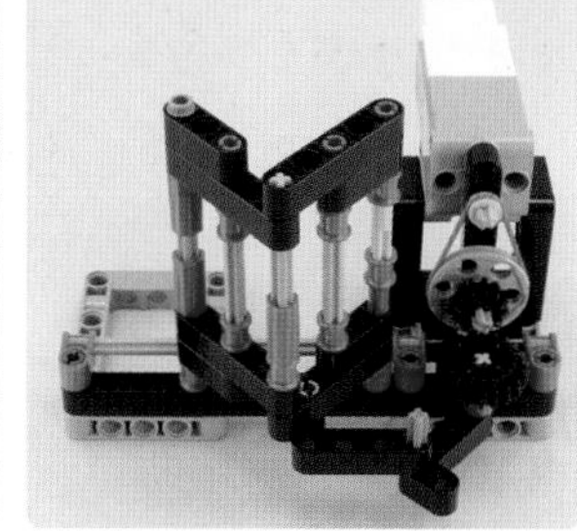
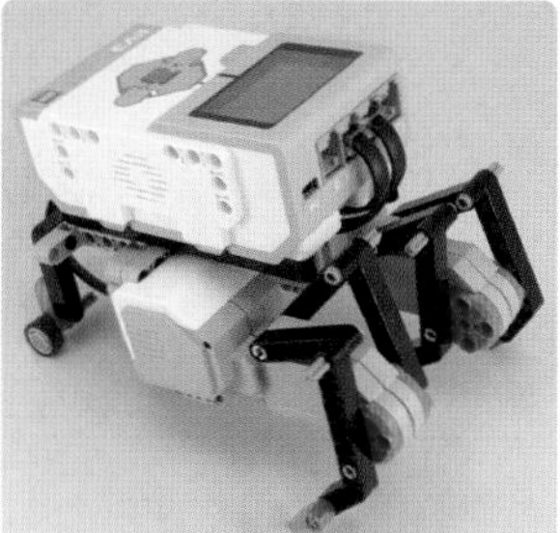
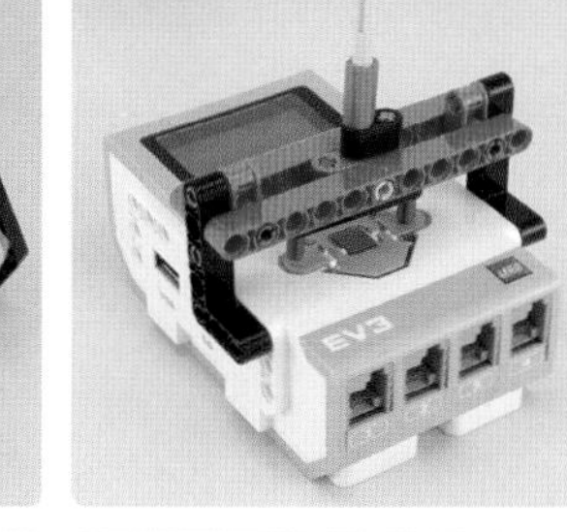

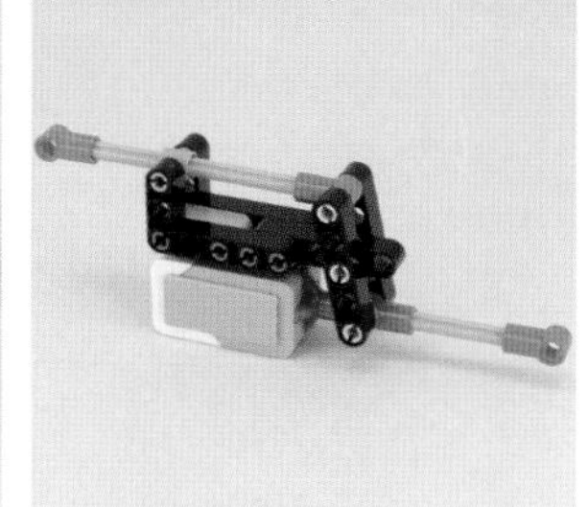

PART 3

Moving Without Tires

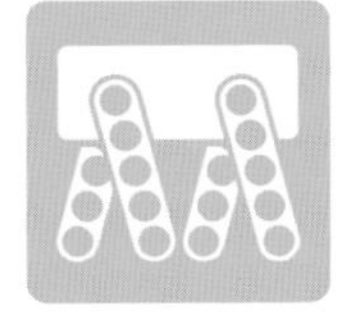
110

122

126

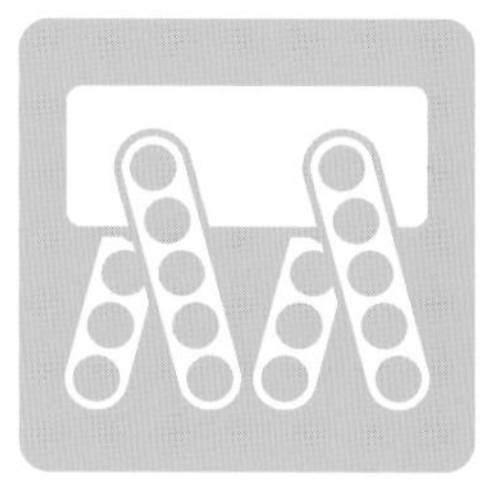

Walking machines

#122

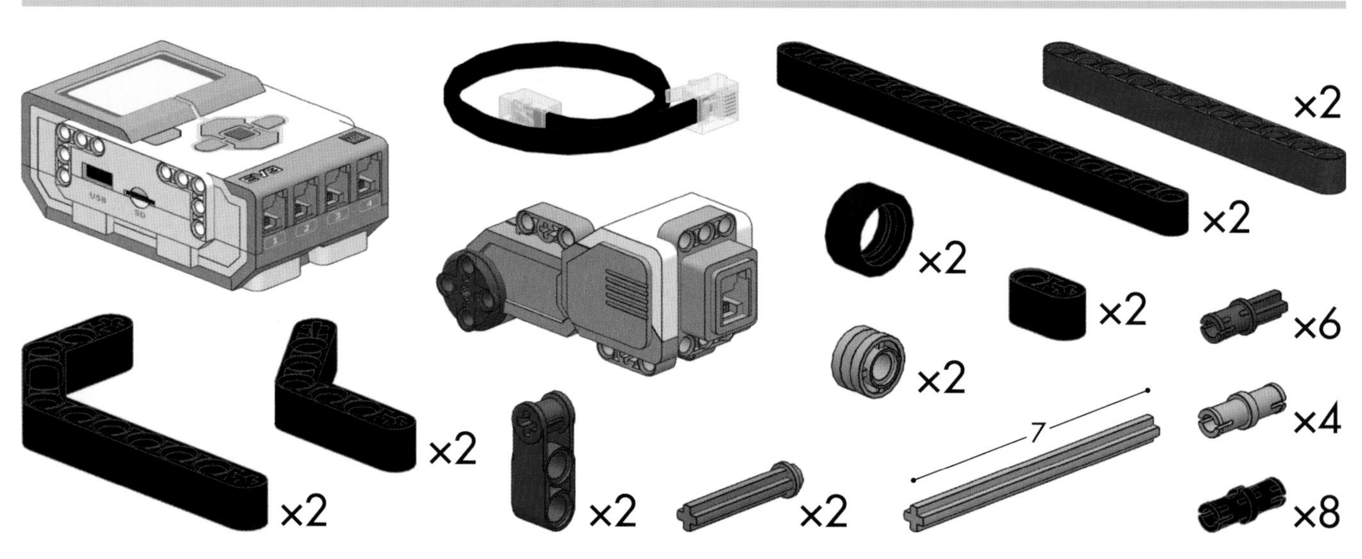
×2
×2
×2
×2
×6
×2
×4
×2
×2
×2
×8
7

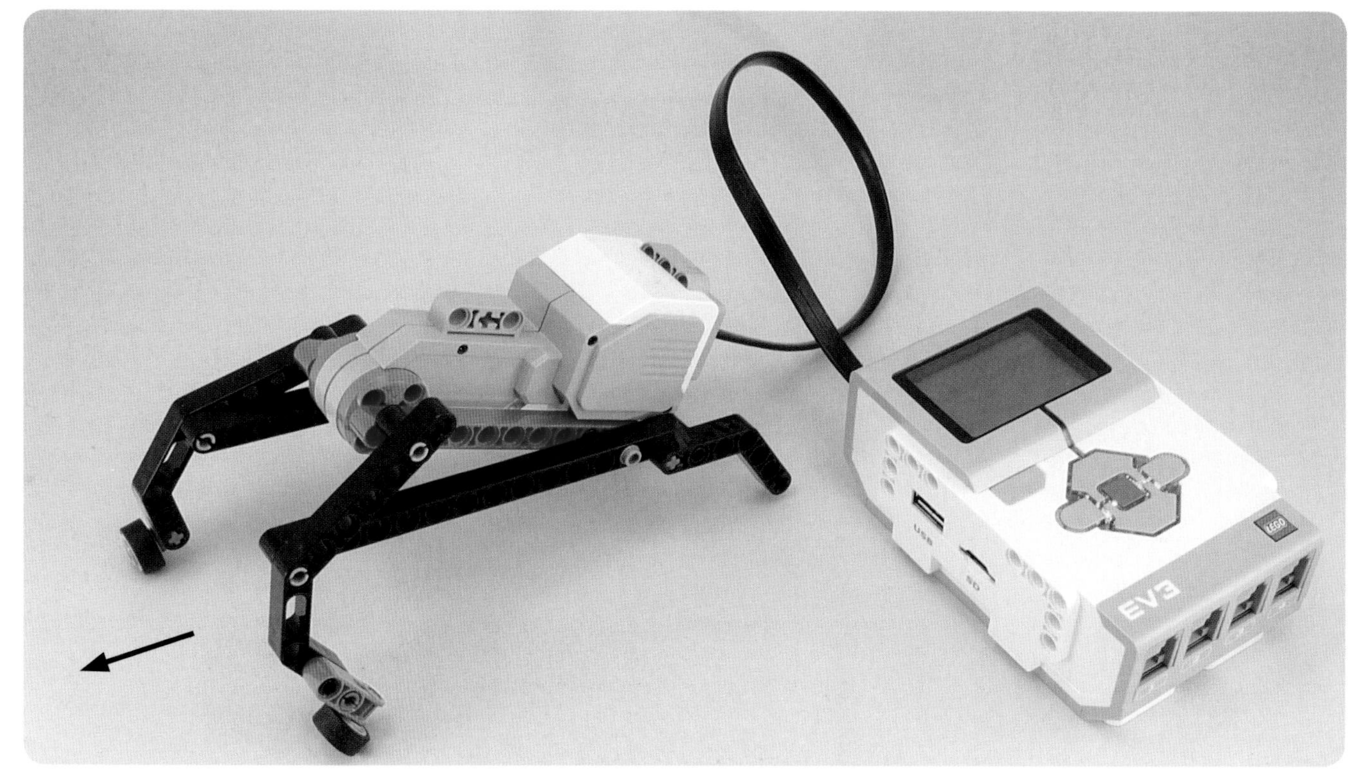

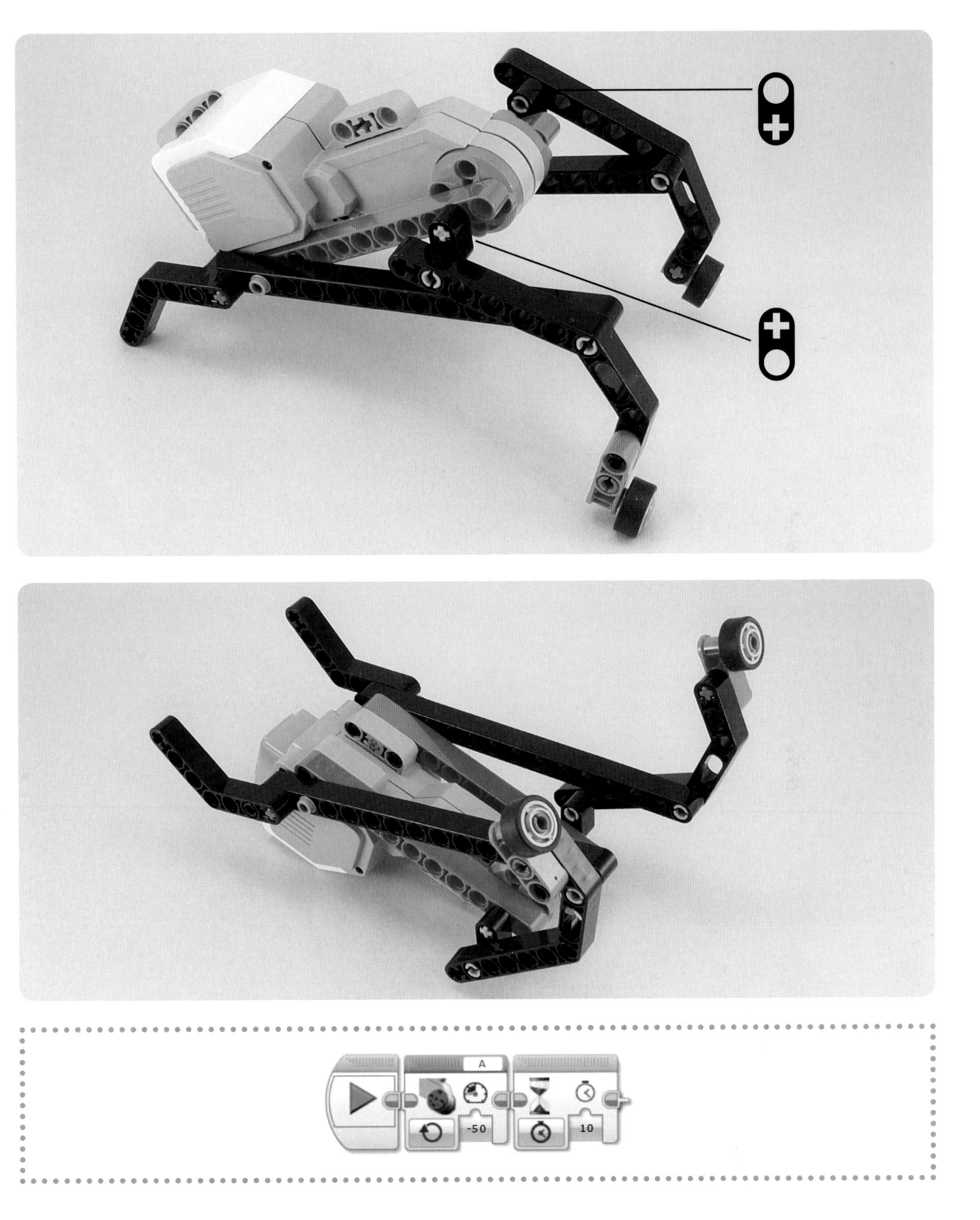
A
-50
10

#123

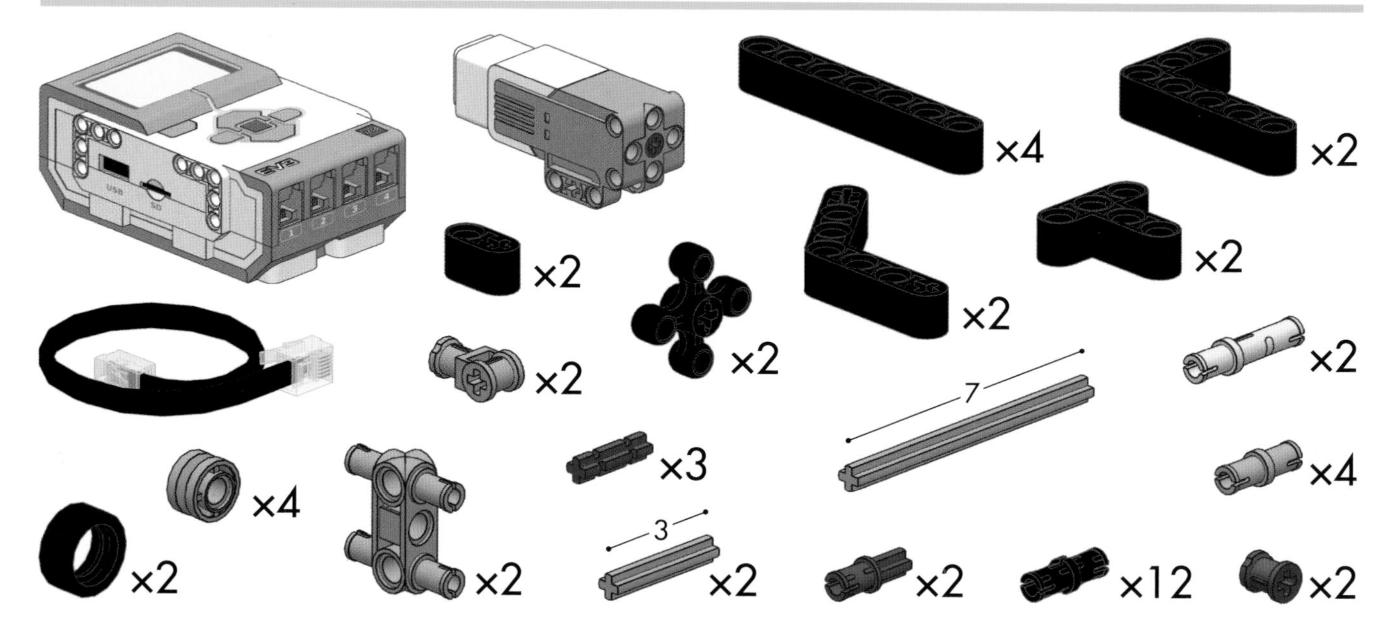
×4
×2
×2
×2
×2
×2
×2
×2
7
×2
×3
×4
×4
×2
×2
3
×2
×2
×12
×2

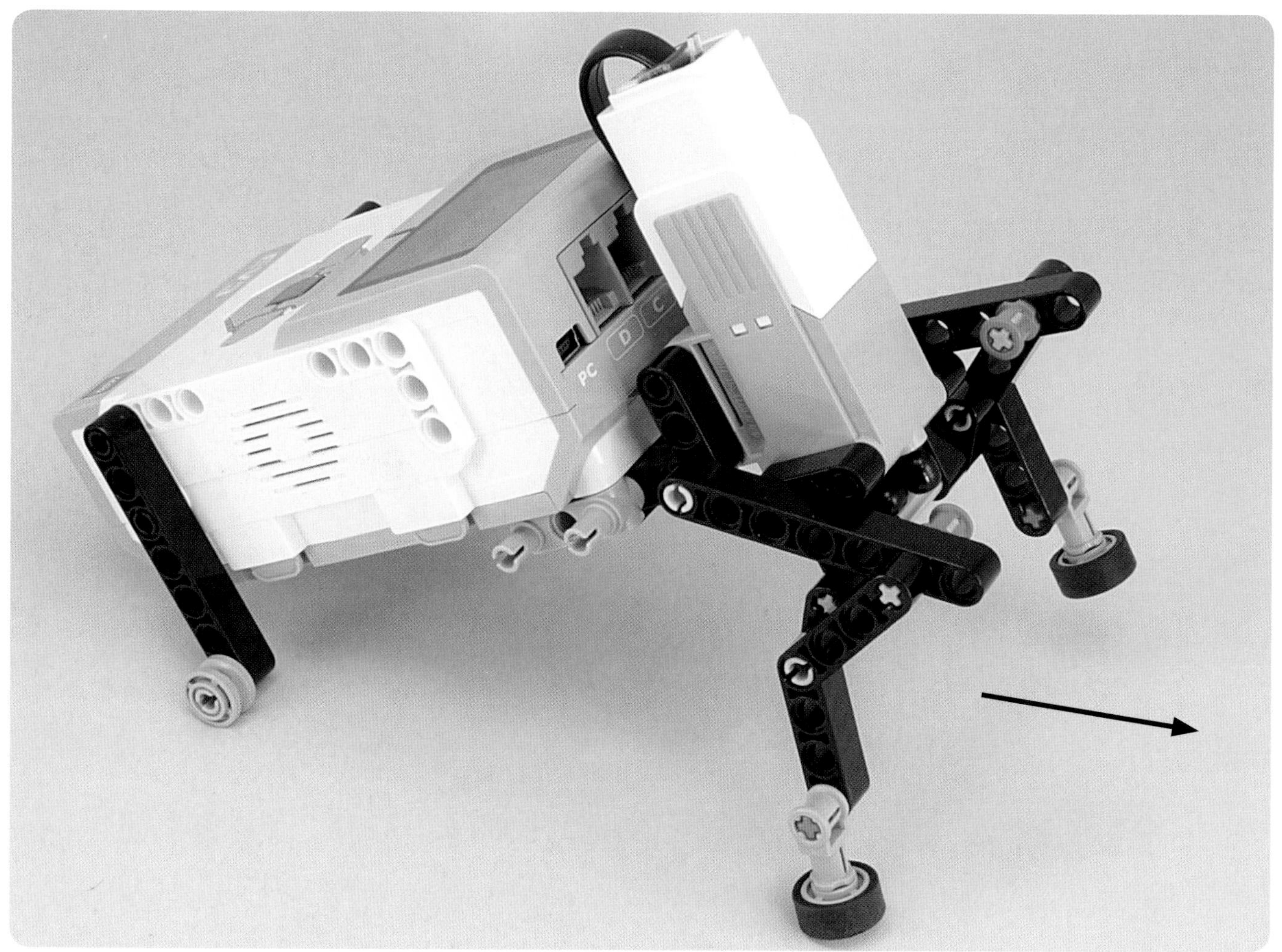
PC
D
C

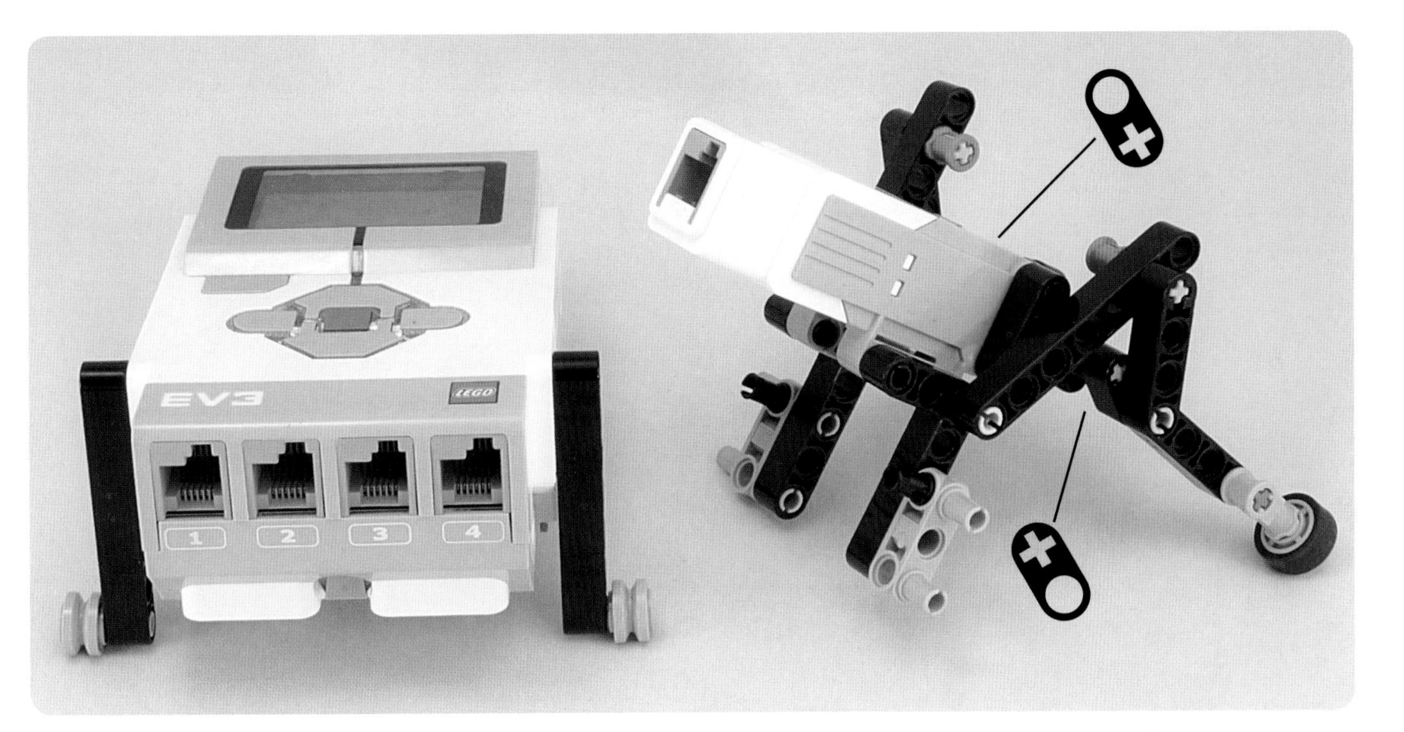
EV3
LEGO
1
2
3
4

#124

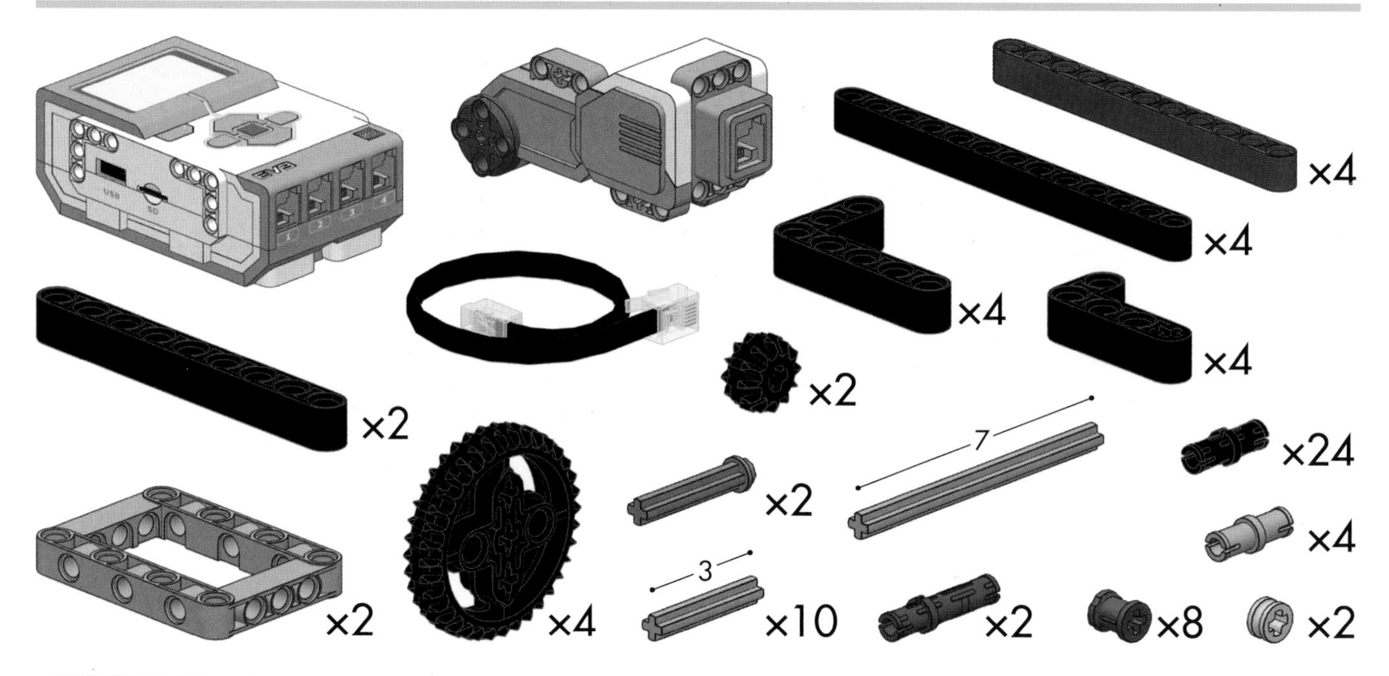
×4
×4
×4
×4
×2
×2
7
×24
×2
×4
3
×2
×4
×10
×2
×8
×2

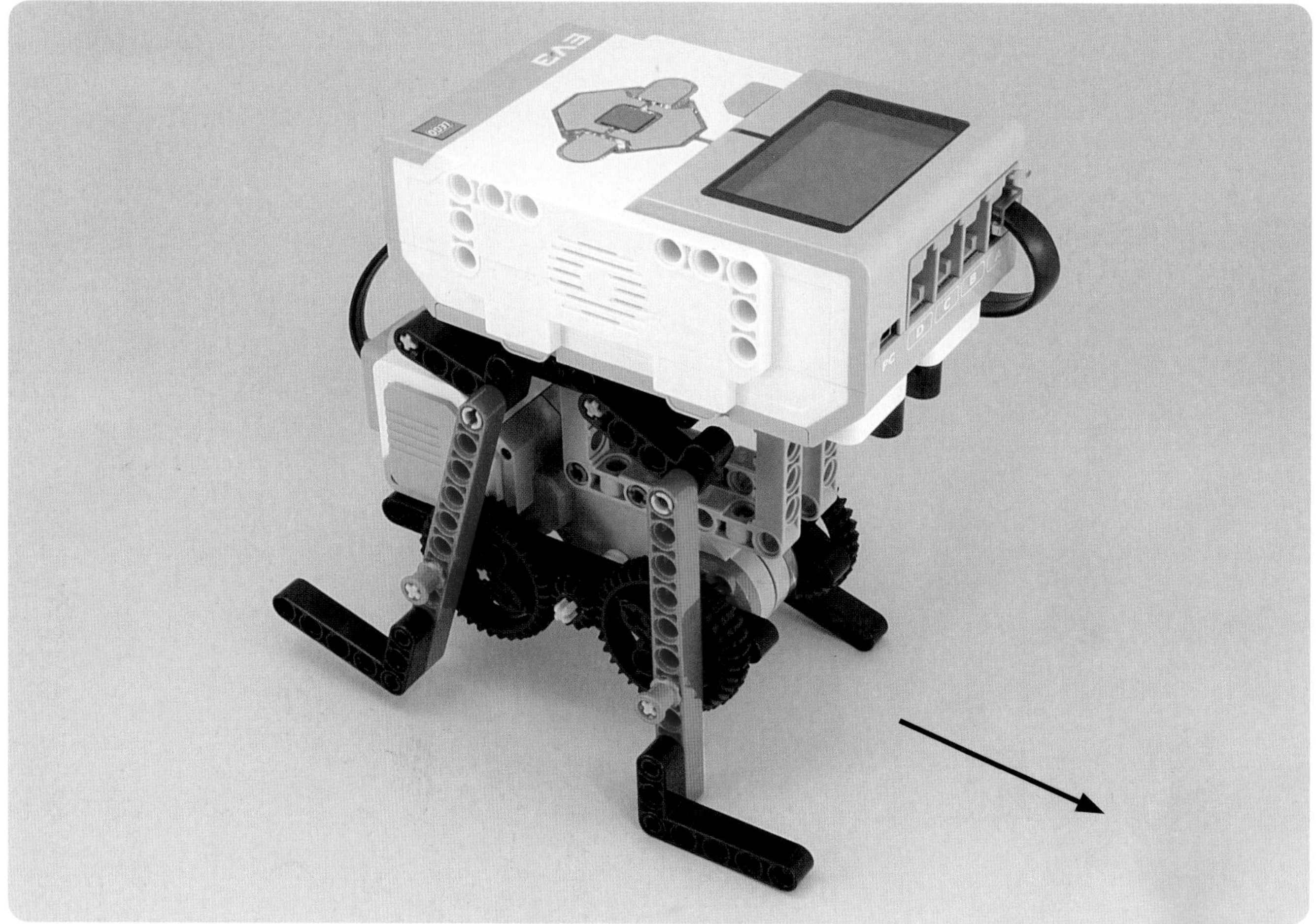

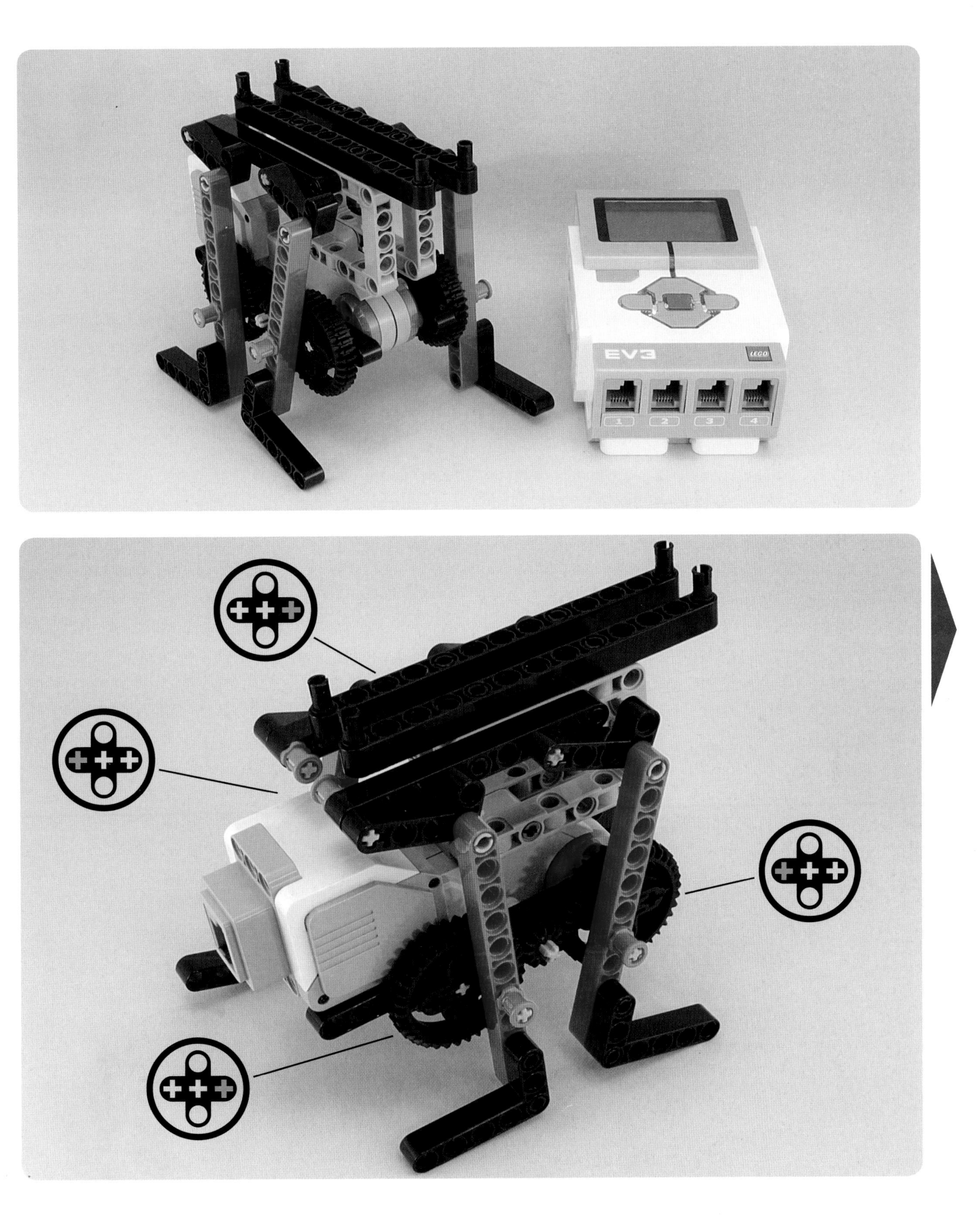
EV3
LEGO
1
2
3
4

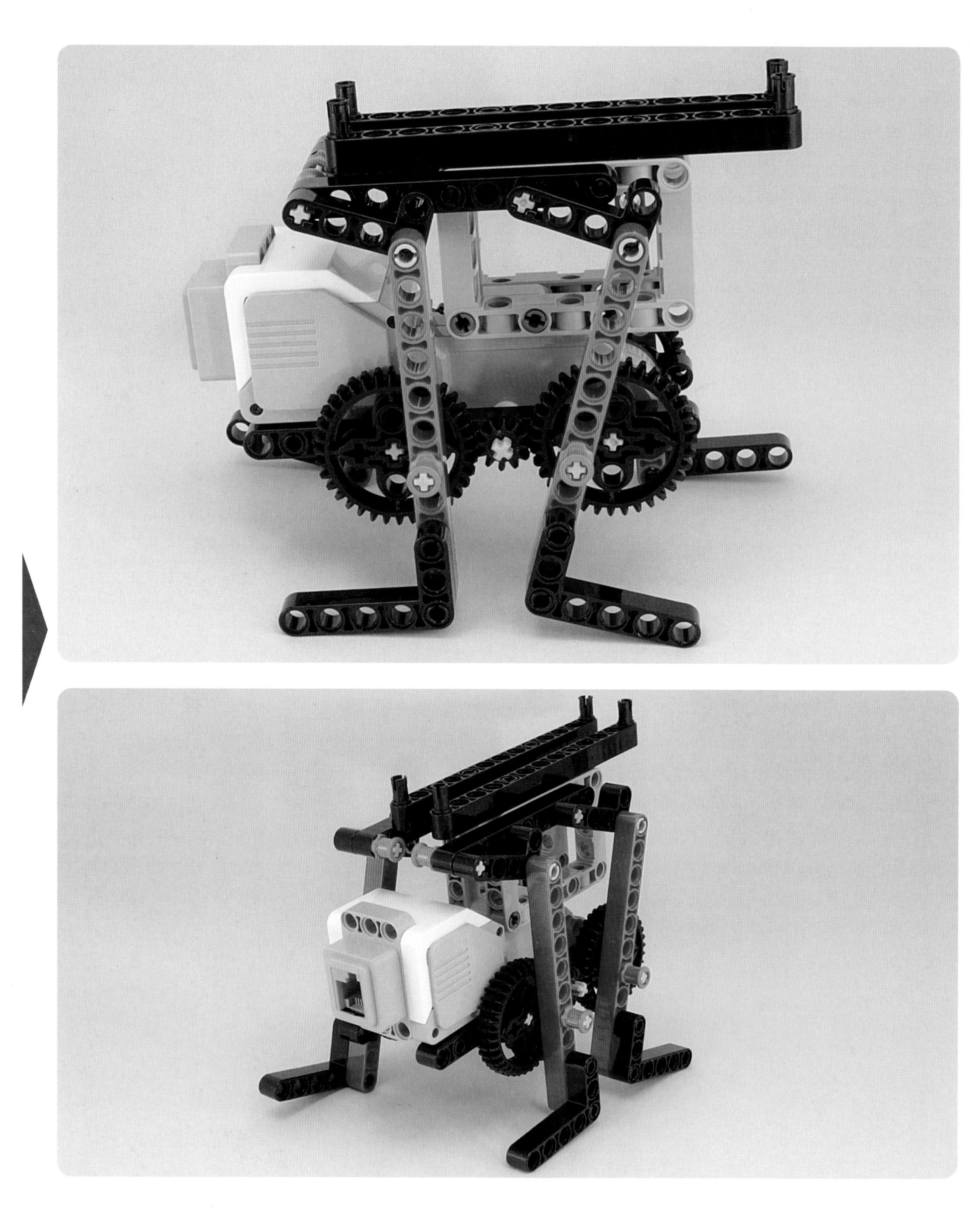

#125

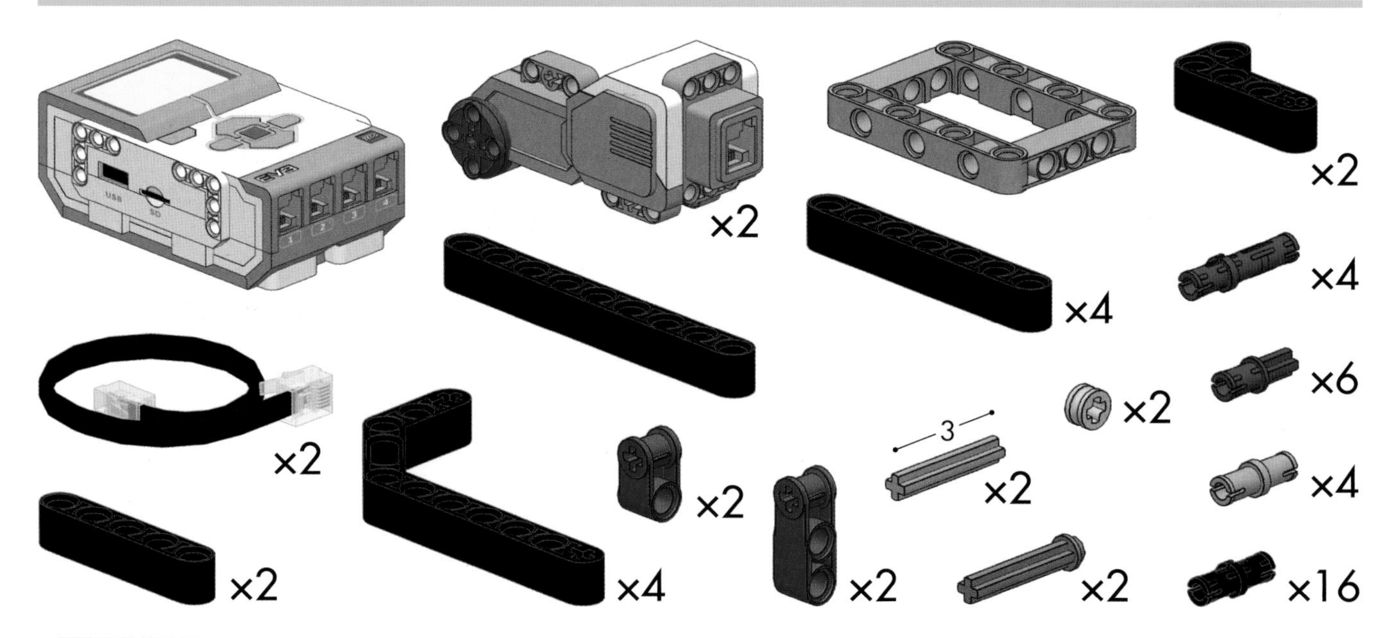
×2
×2
×4
×4
×6
×2
×2
×2
×4
×2
×2
×4
×2
×2
×2
×16
3
EV3
USB
SD

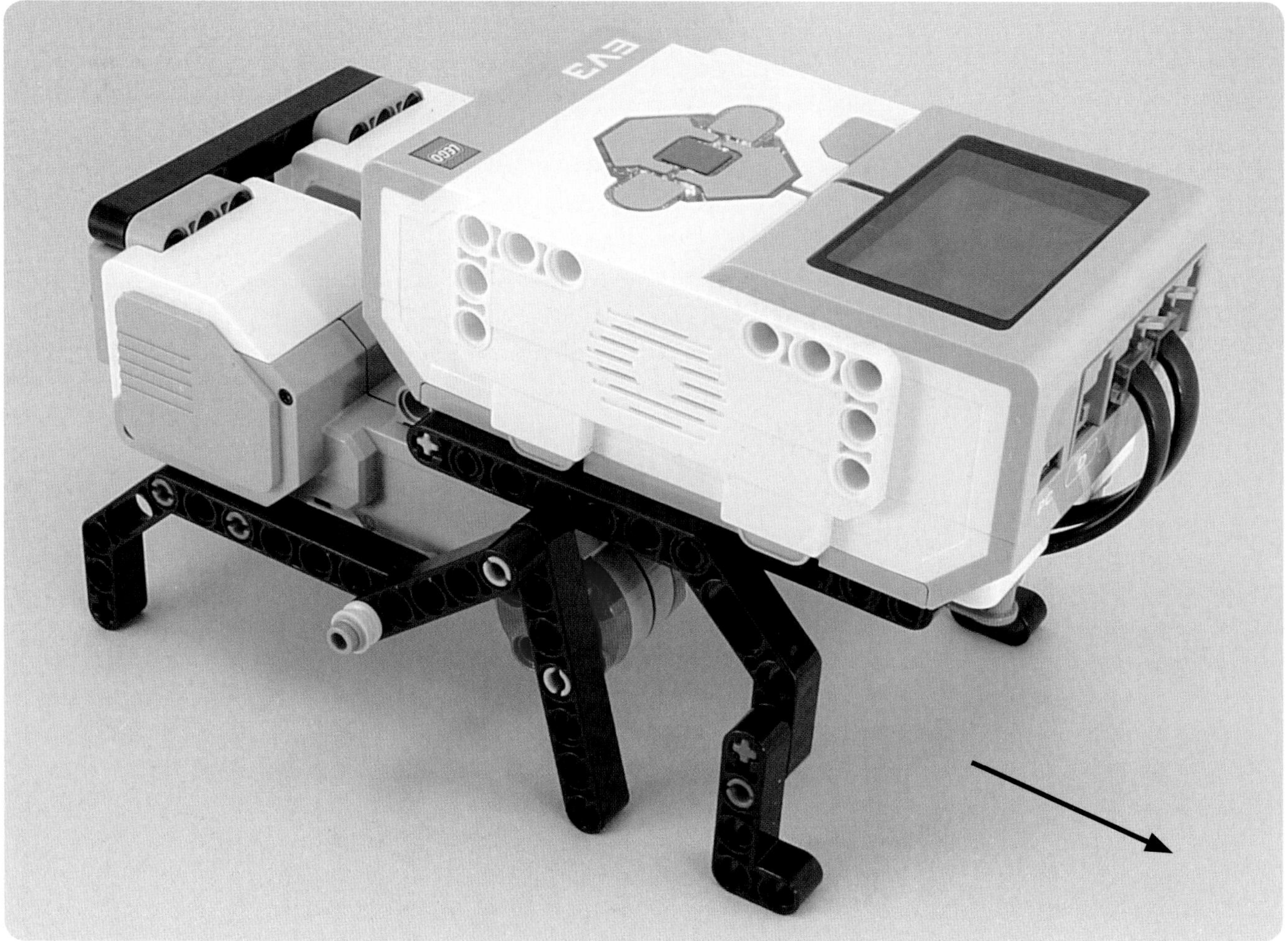
EV3

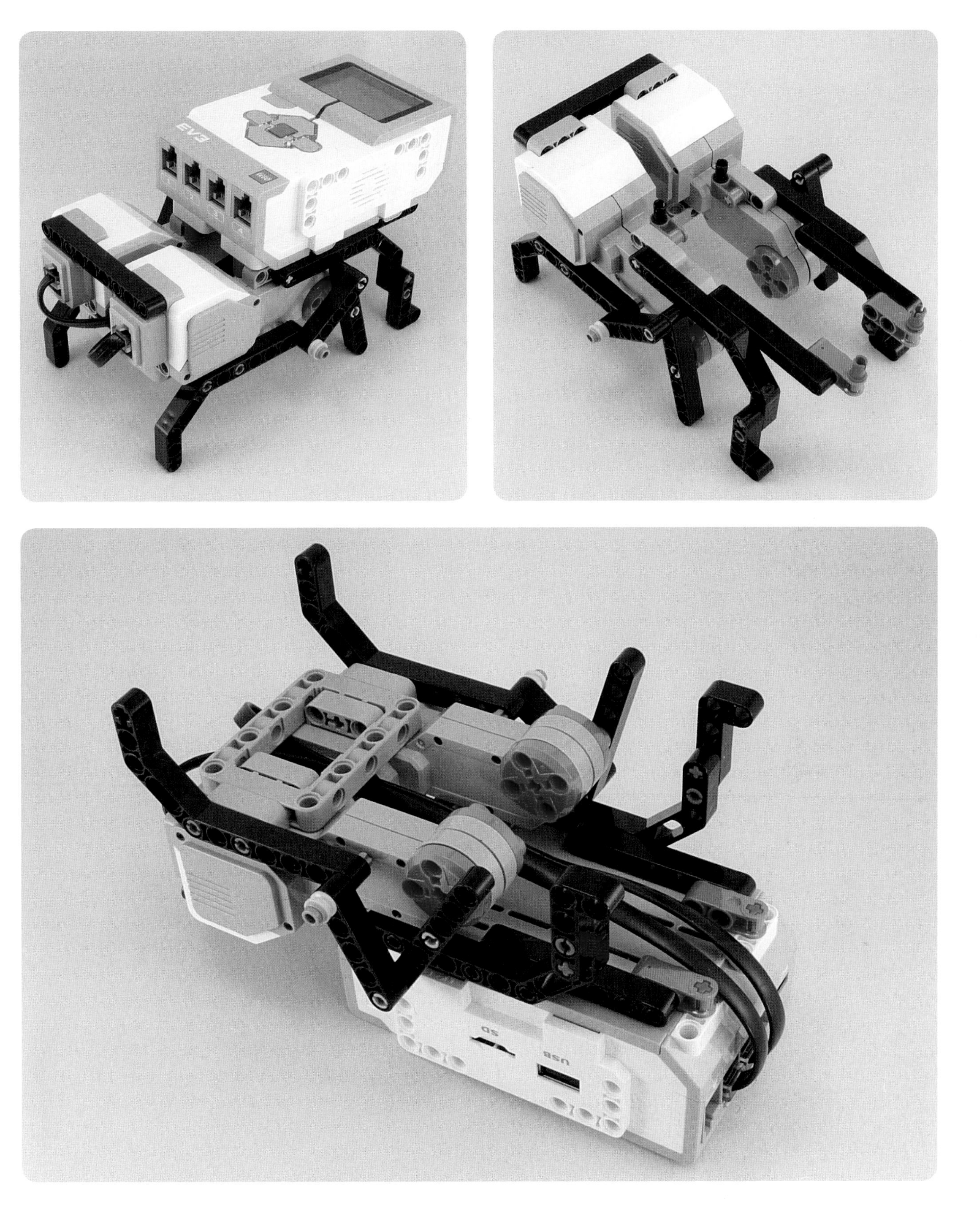
EV3
SD
USB

#126

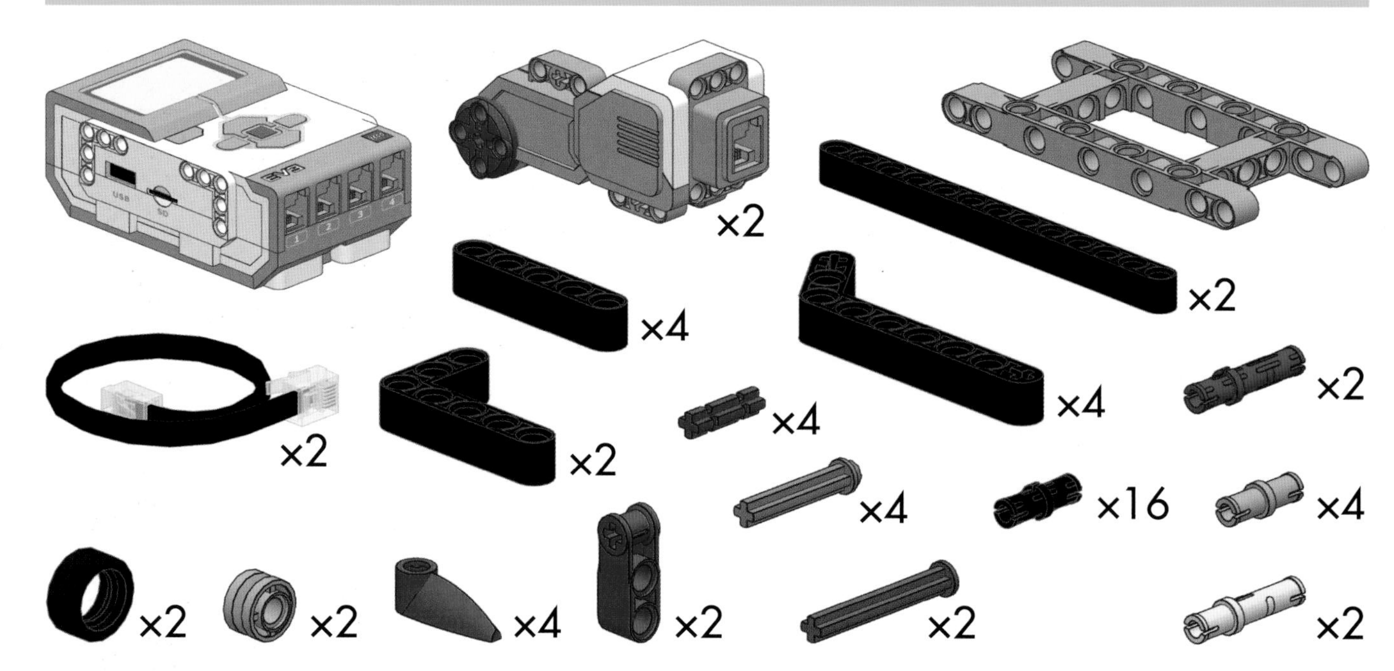
×2
×2
×4
×4
×2
×2
×4
×2
×4
×16
×4
×2
×2
×4
×2
×2
×2

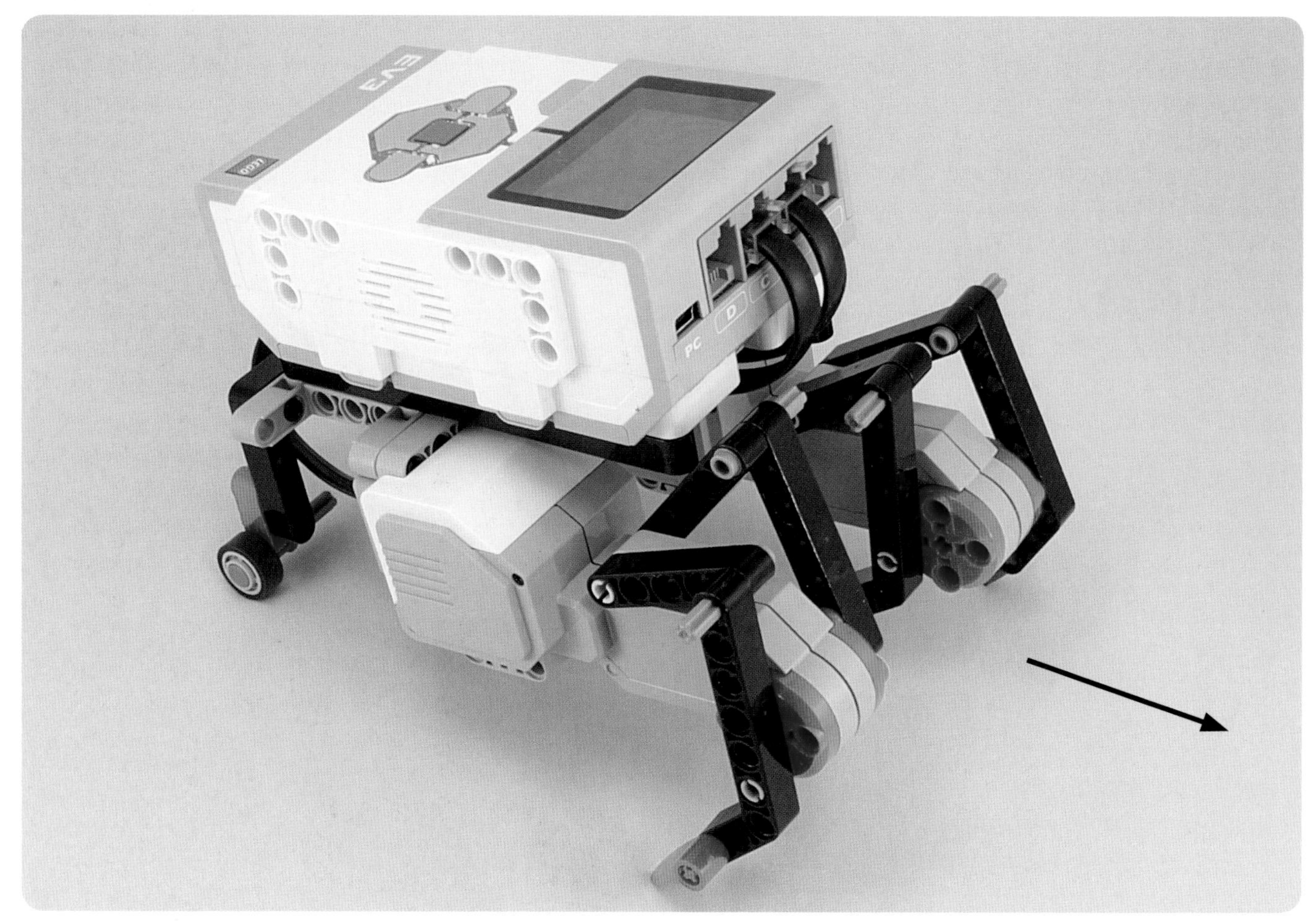

Moving like an inchworm

#127

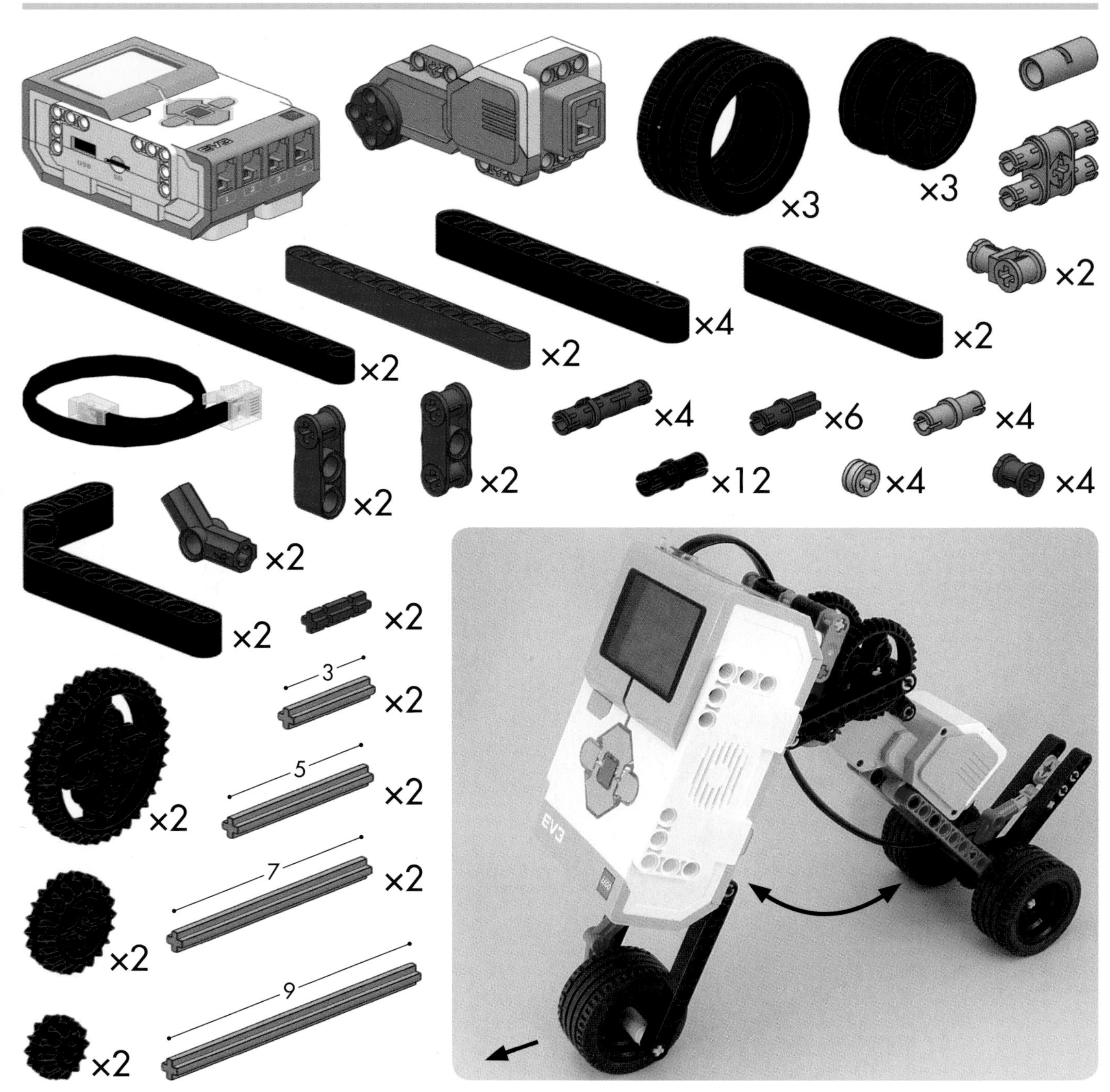

×3
×3
×2
×4
×2
×2
×2
×4
×6
×4
×2
×2
×12
×4
×4
×2
×2
×2
3
×2
5
×2
×2
7
×2
×2
9
×2

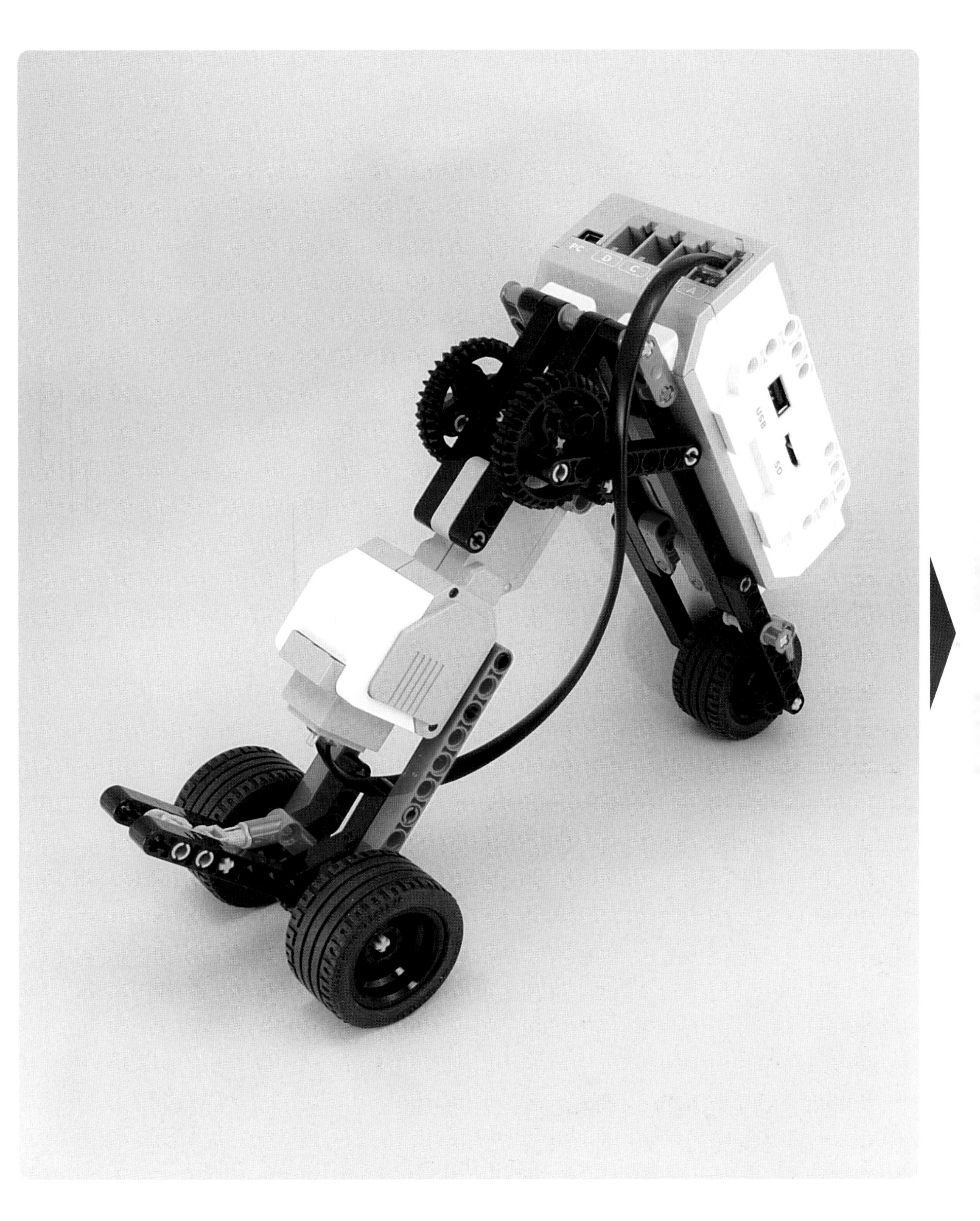
PC
D
C
A
USB
SD

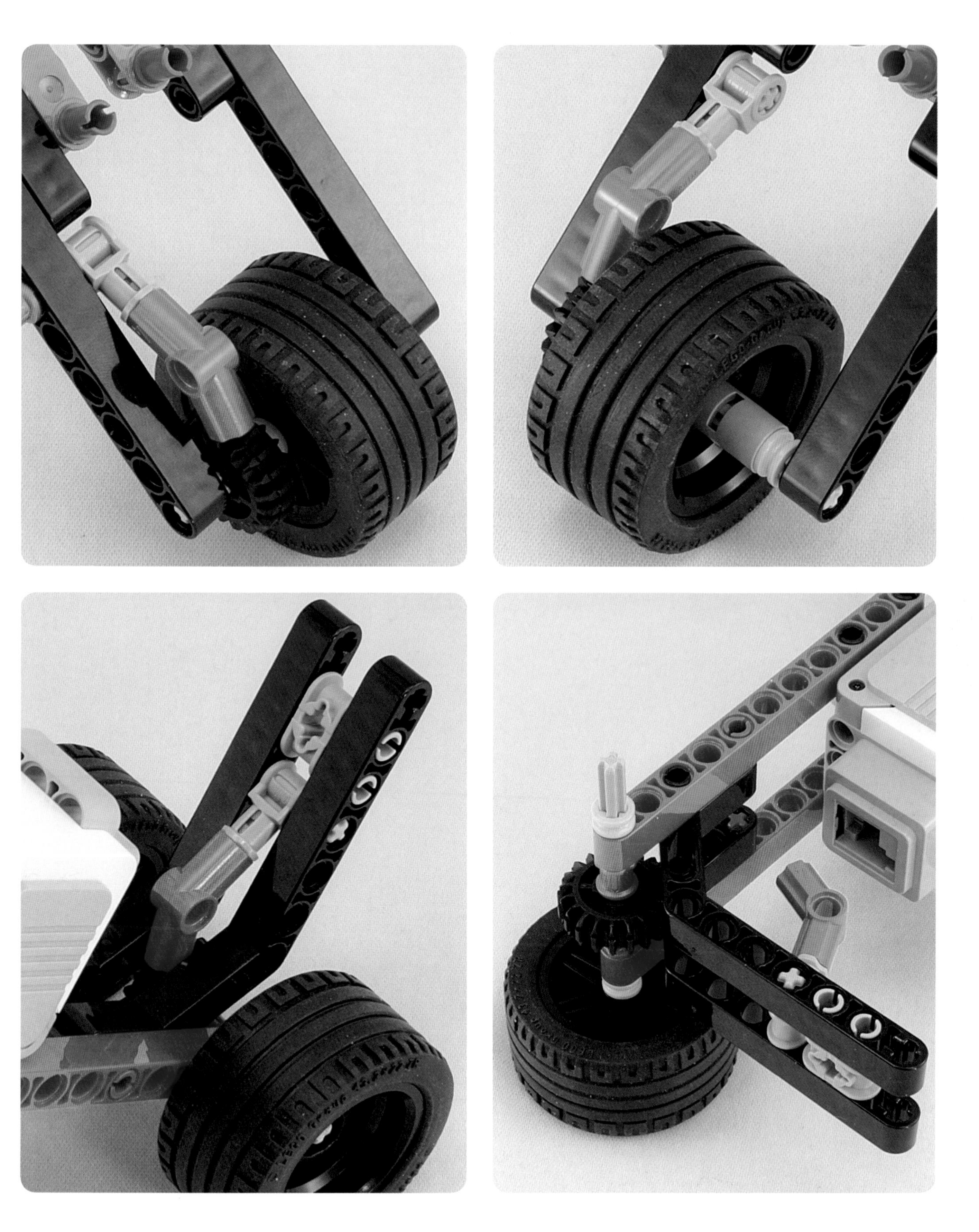

Moving through vibration

#128

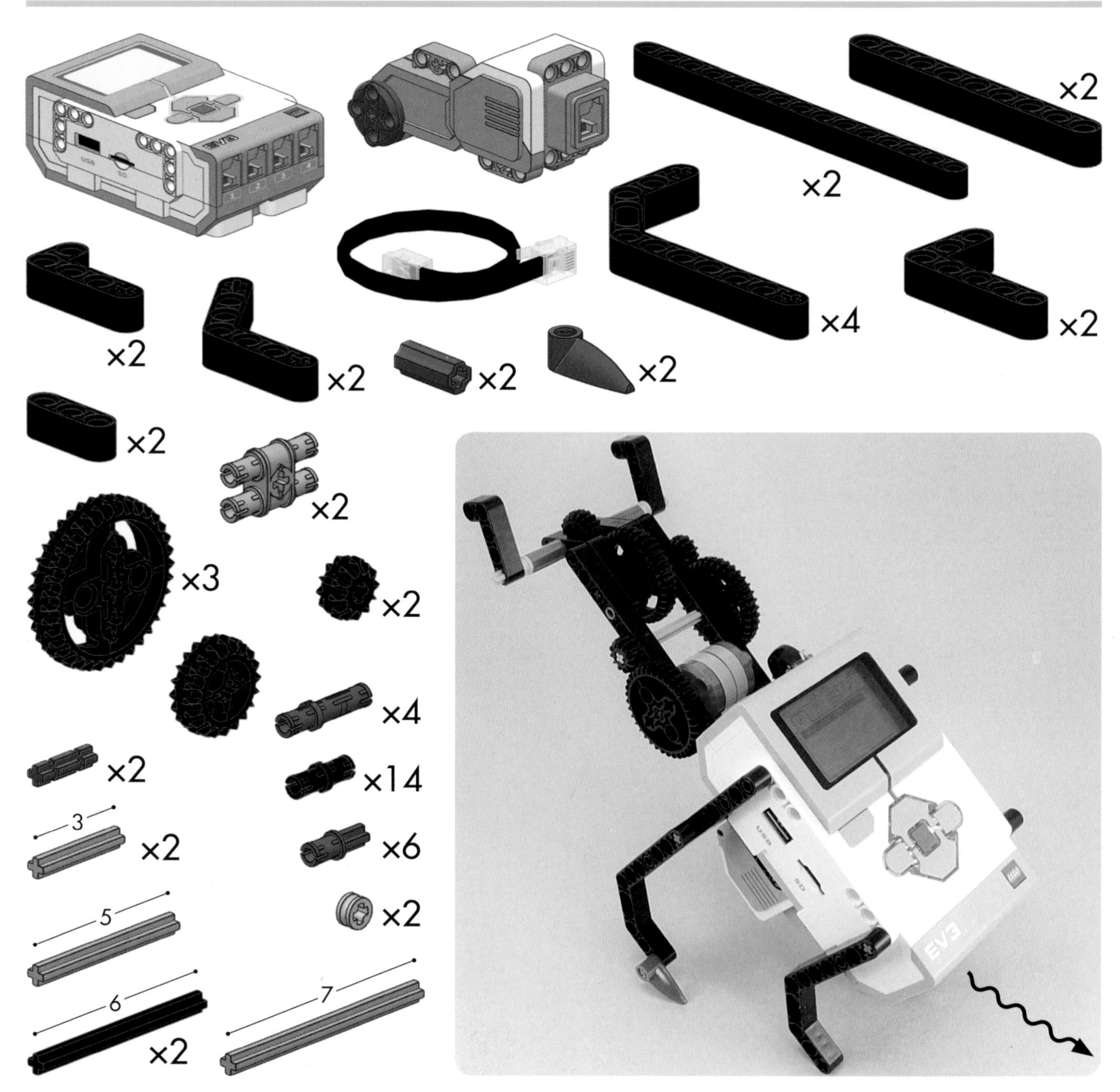

×2
×2
×4
×2
×2
×2
×2
×2
×2
×2
×3
×2
×4
×2
×14
3
×2
×6
5
×2
6
7
×2

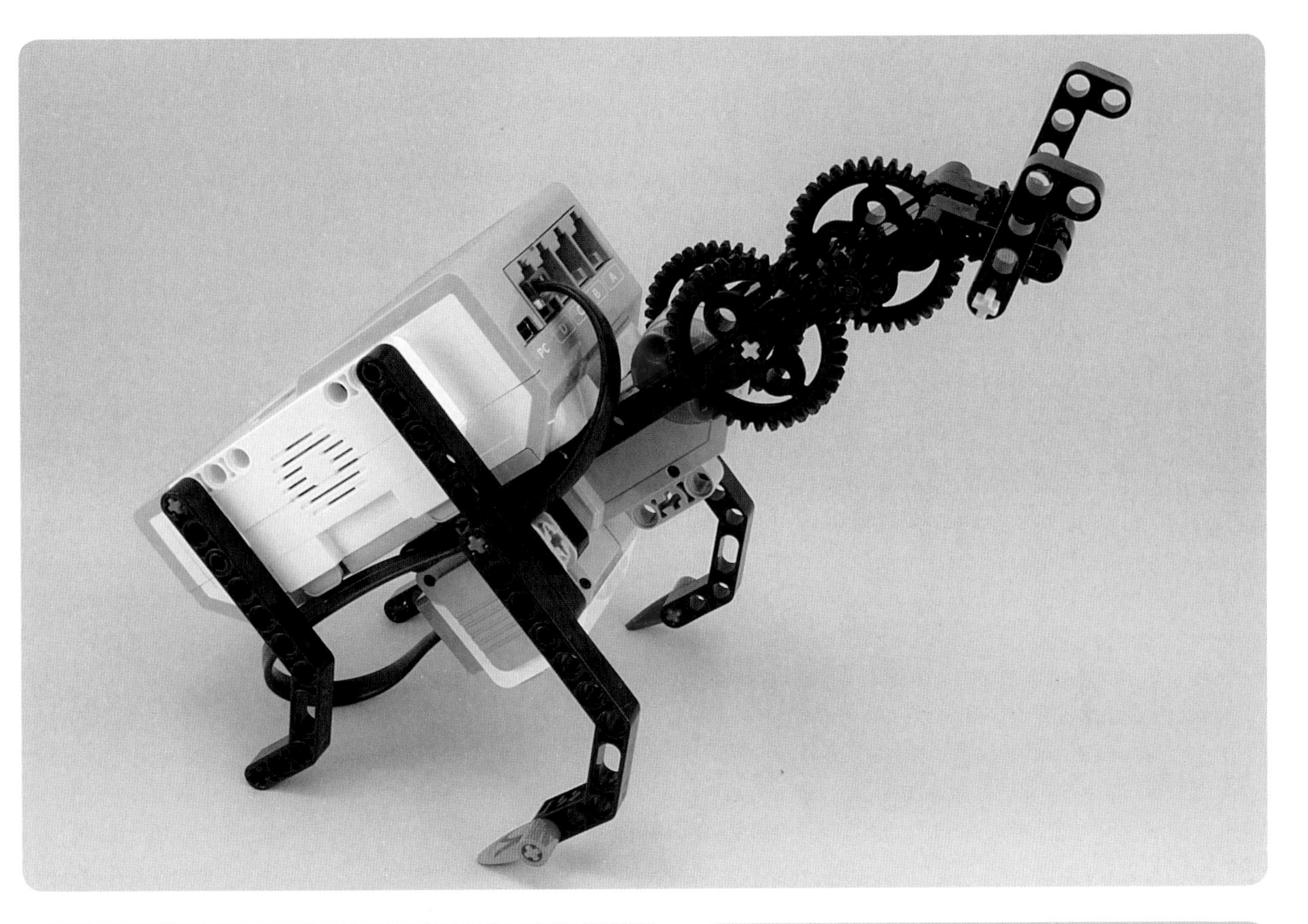

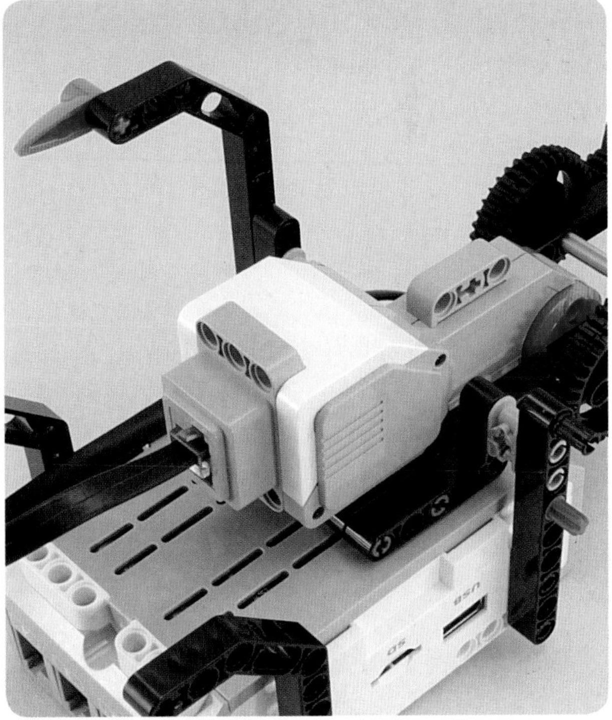

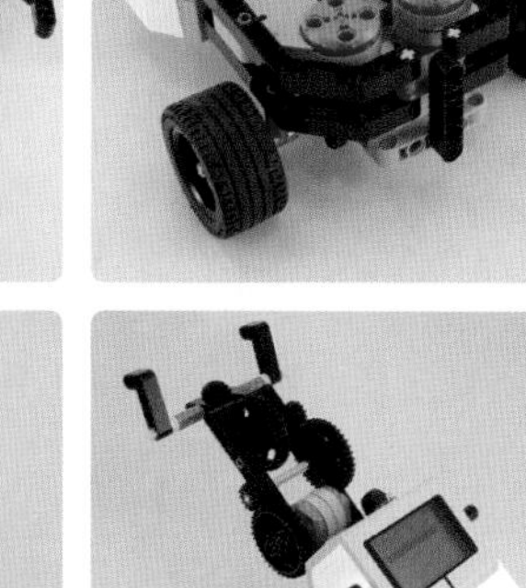

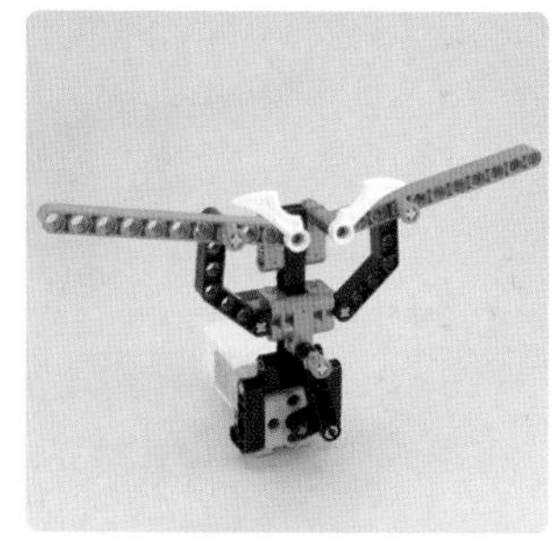

PART 4

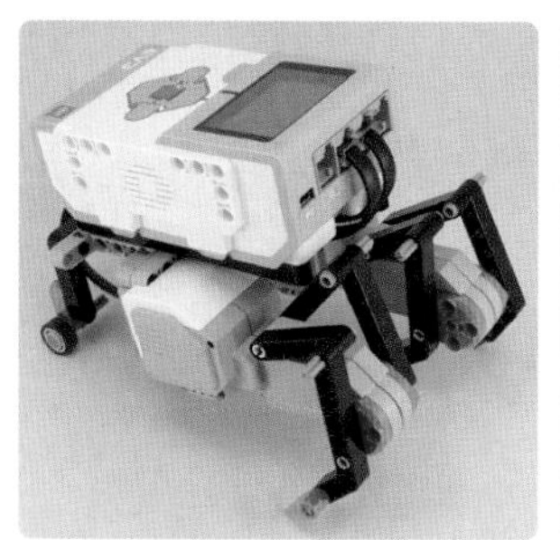

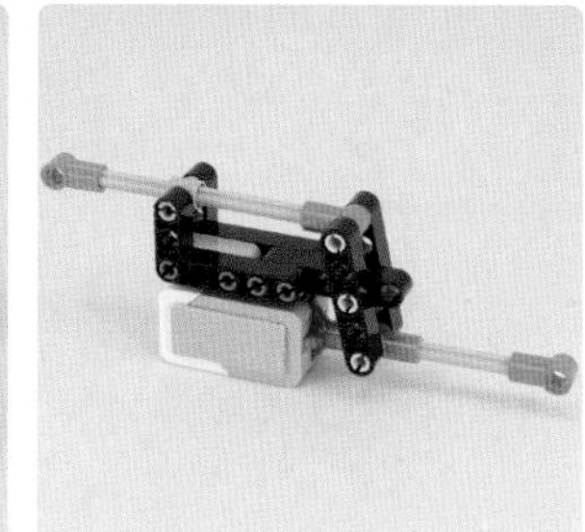

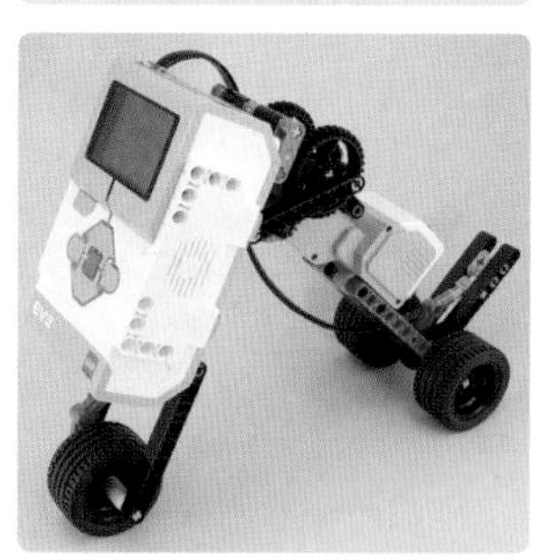

Arms, Wings, and Other Movements

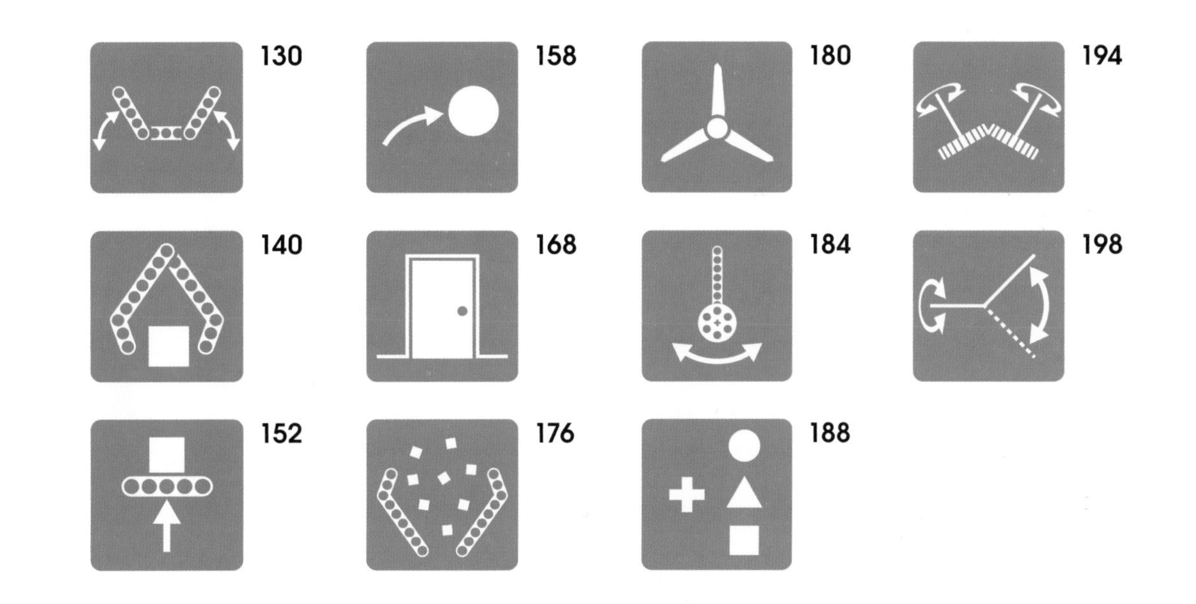

130
158
180
194
140
168
184
198
152
176
188

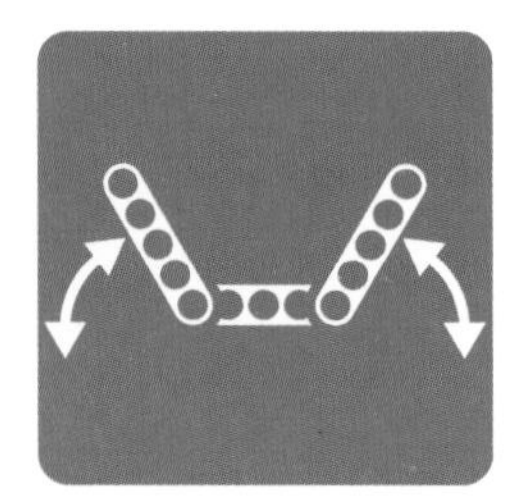

Flapping wings

#129

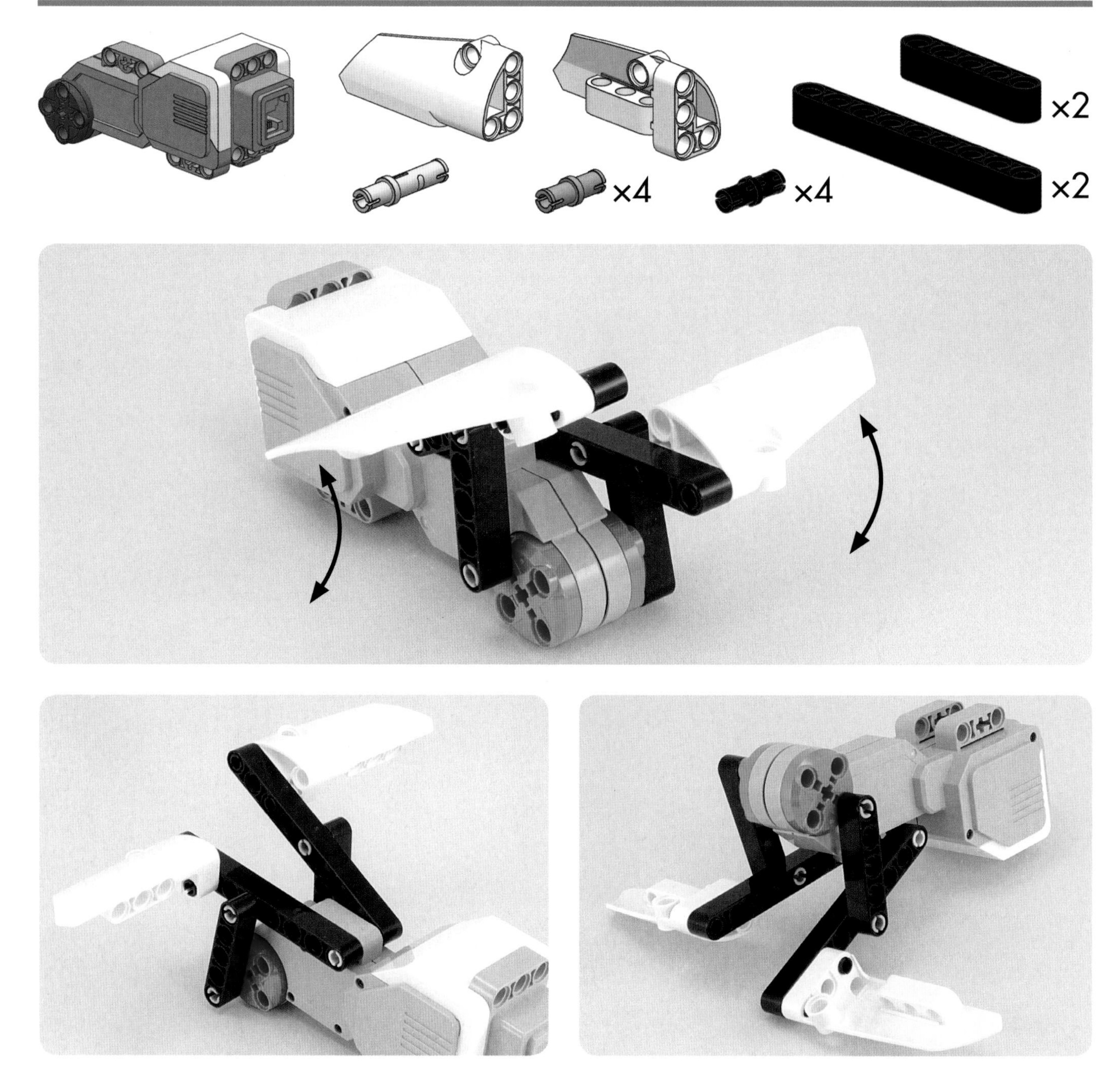
×2
×4
×4
×2

#130

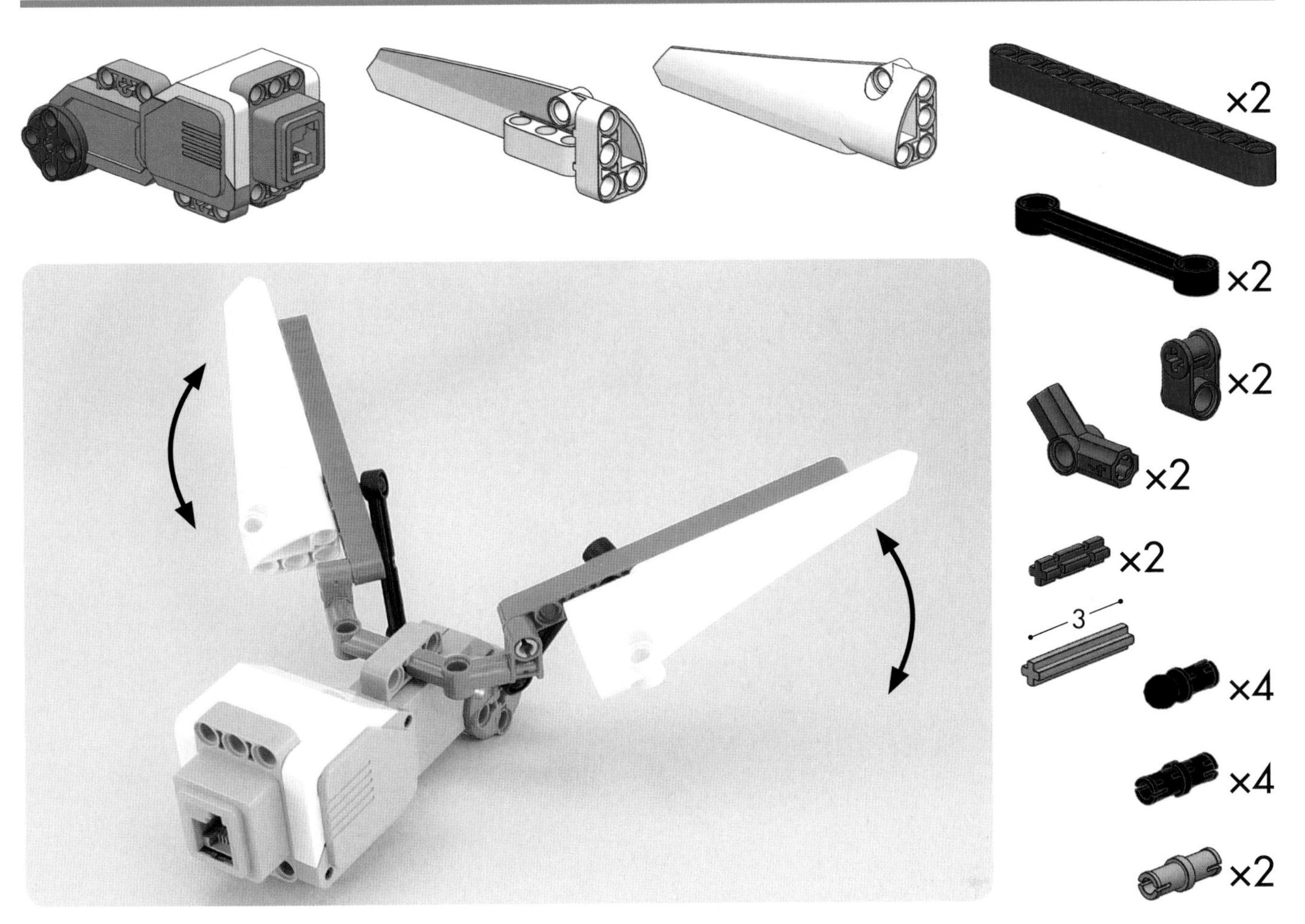
×2
×2
×2
×2
×2
3
×4
×4
×2

#131

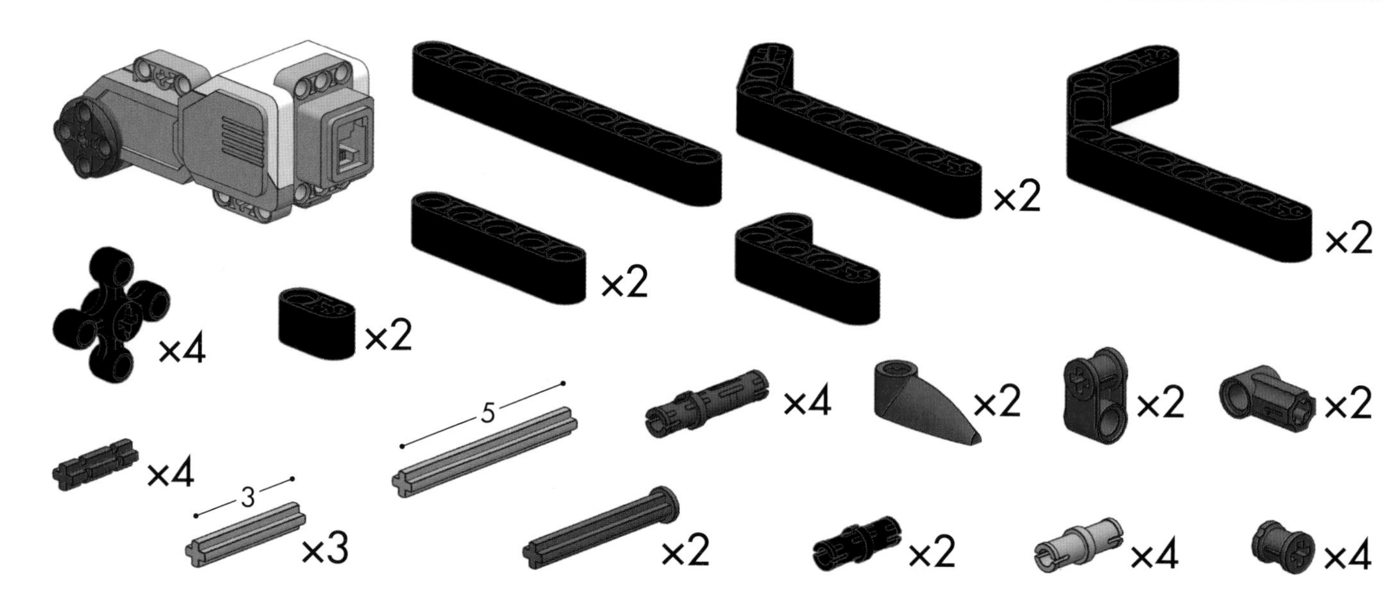
×2
×2
×2
×4
×2
×4
×2
×2
×2
5
×4
3
×3
×2
×2
×4
×4

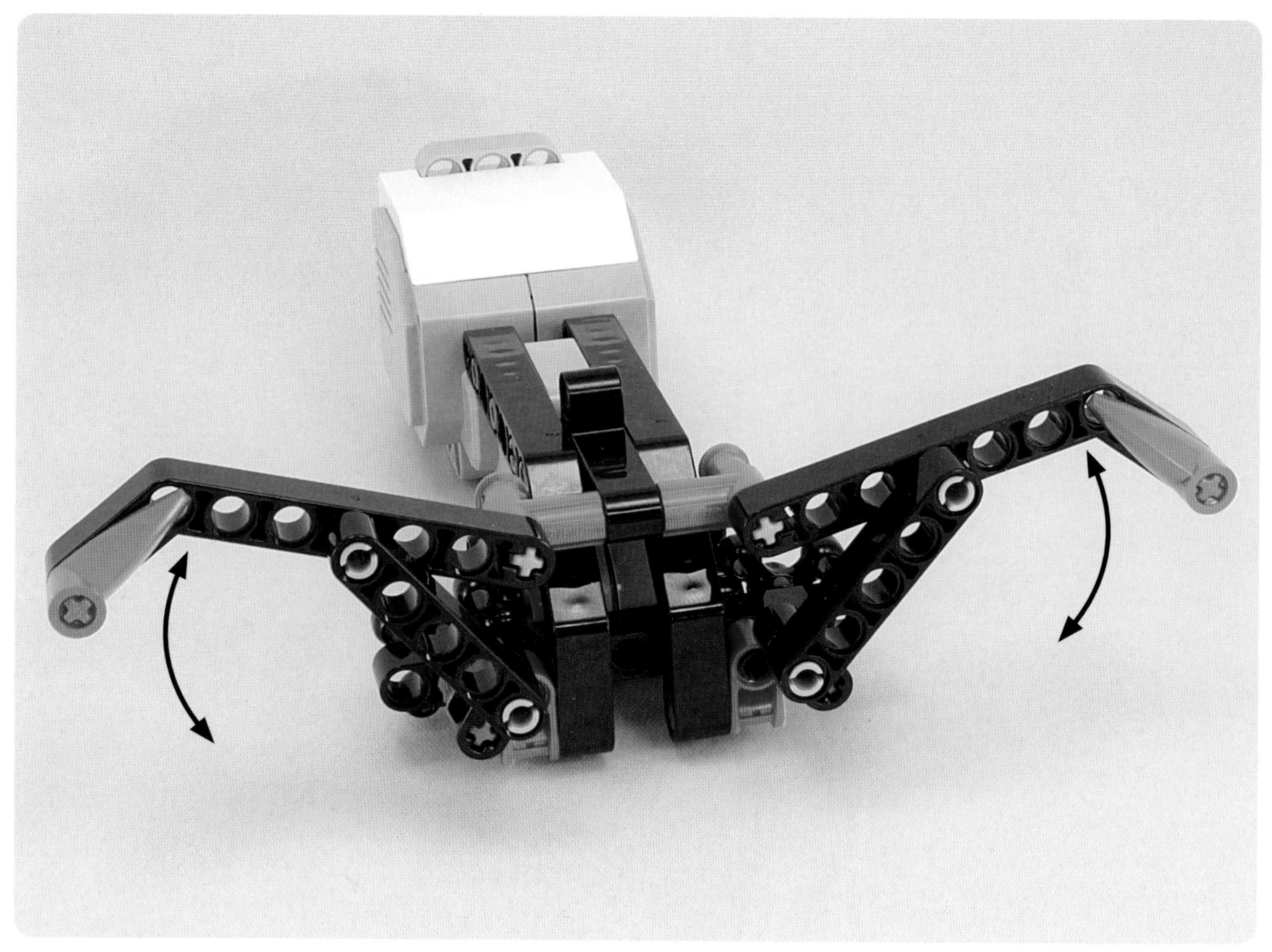

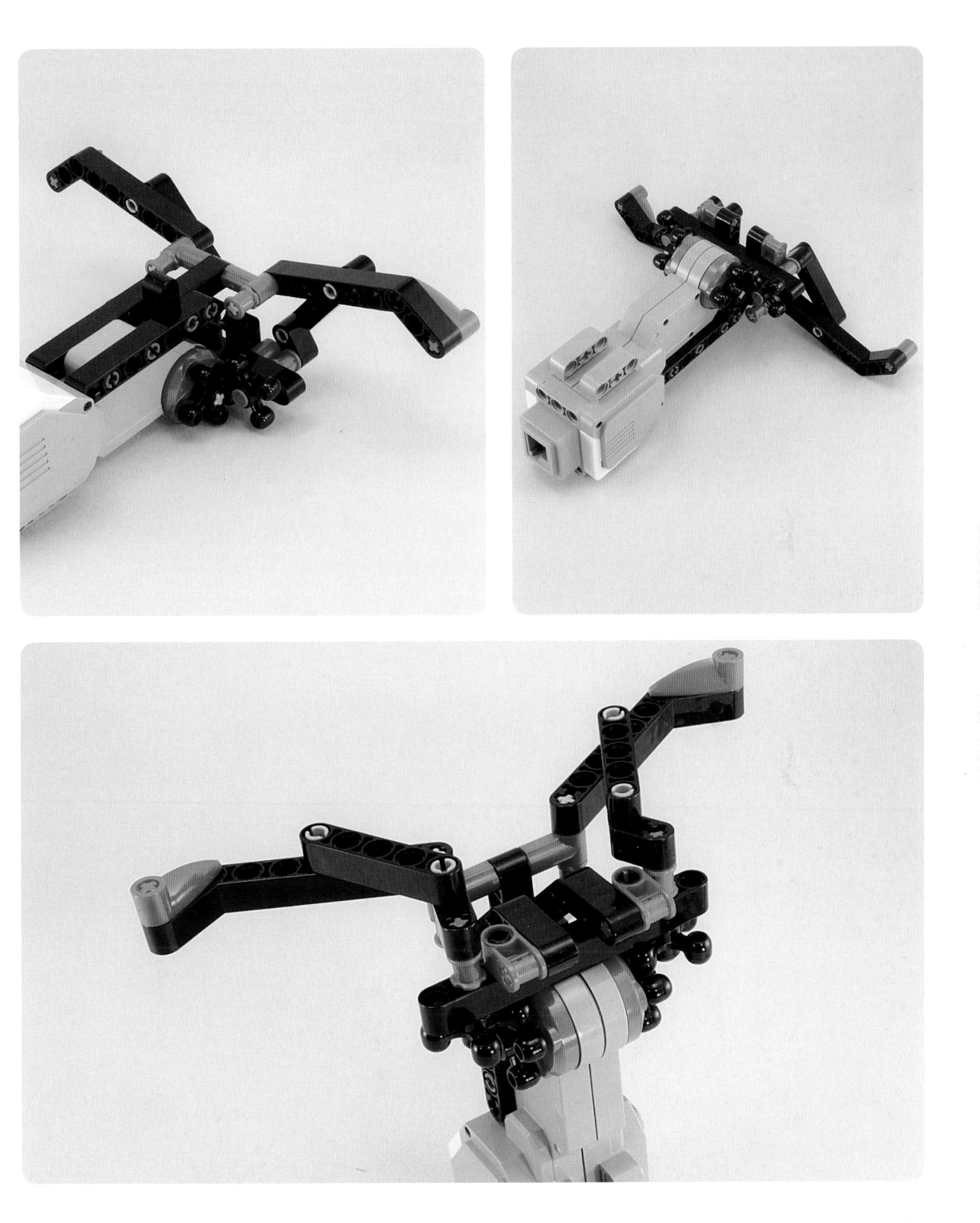

#132

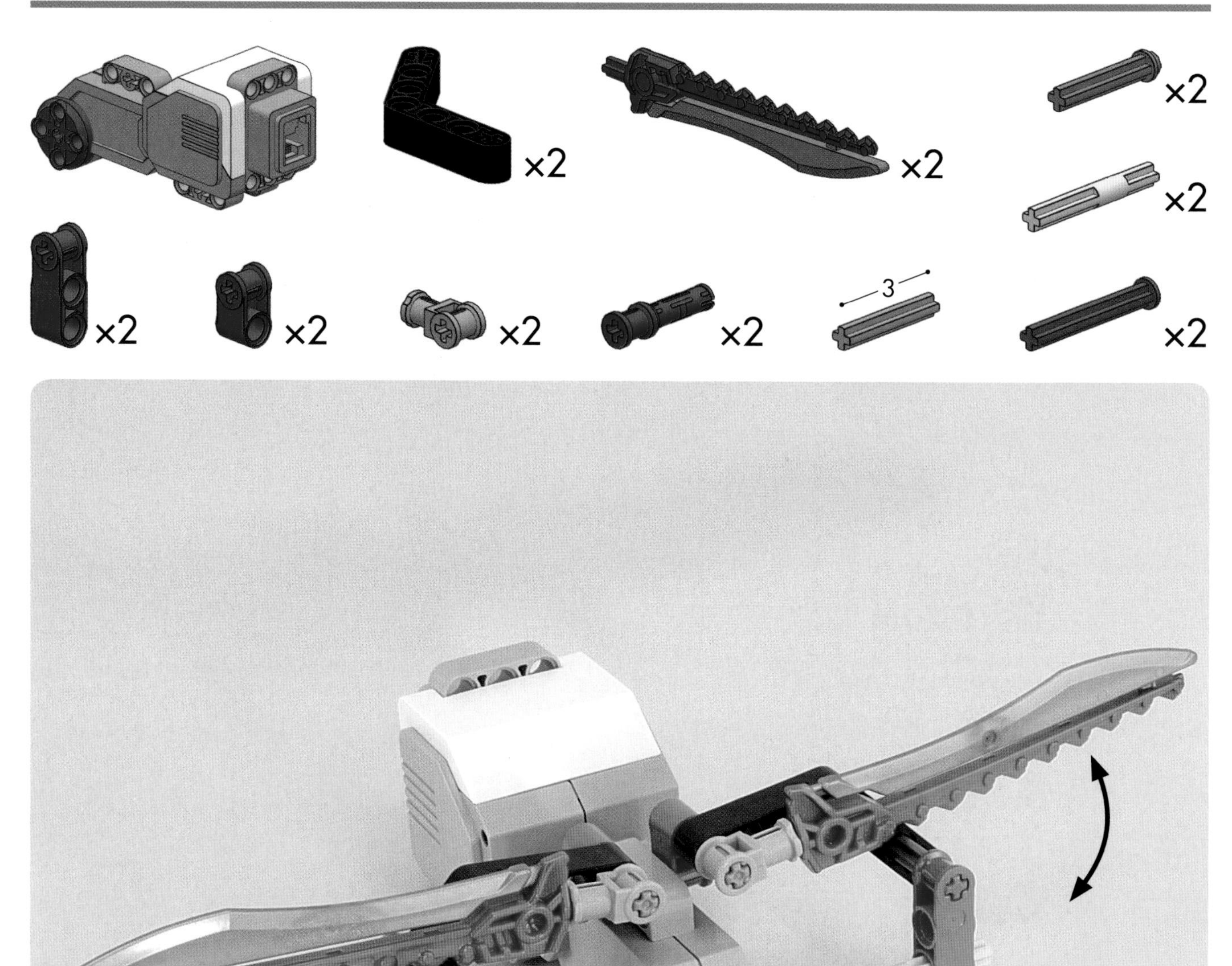
×2
×2
×2
×2
×2
×2
×2
×2
×2
3
×2
×2

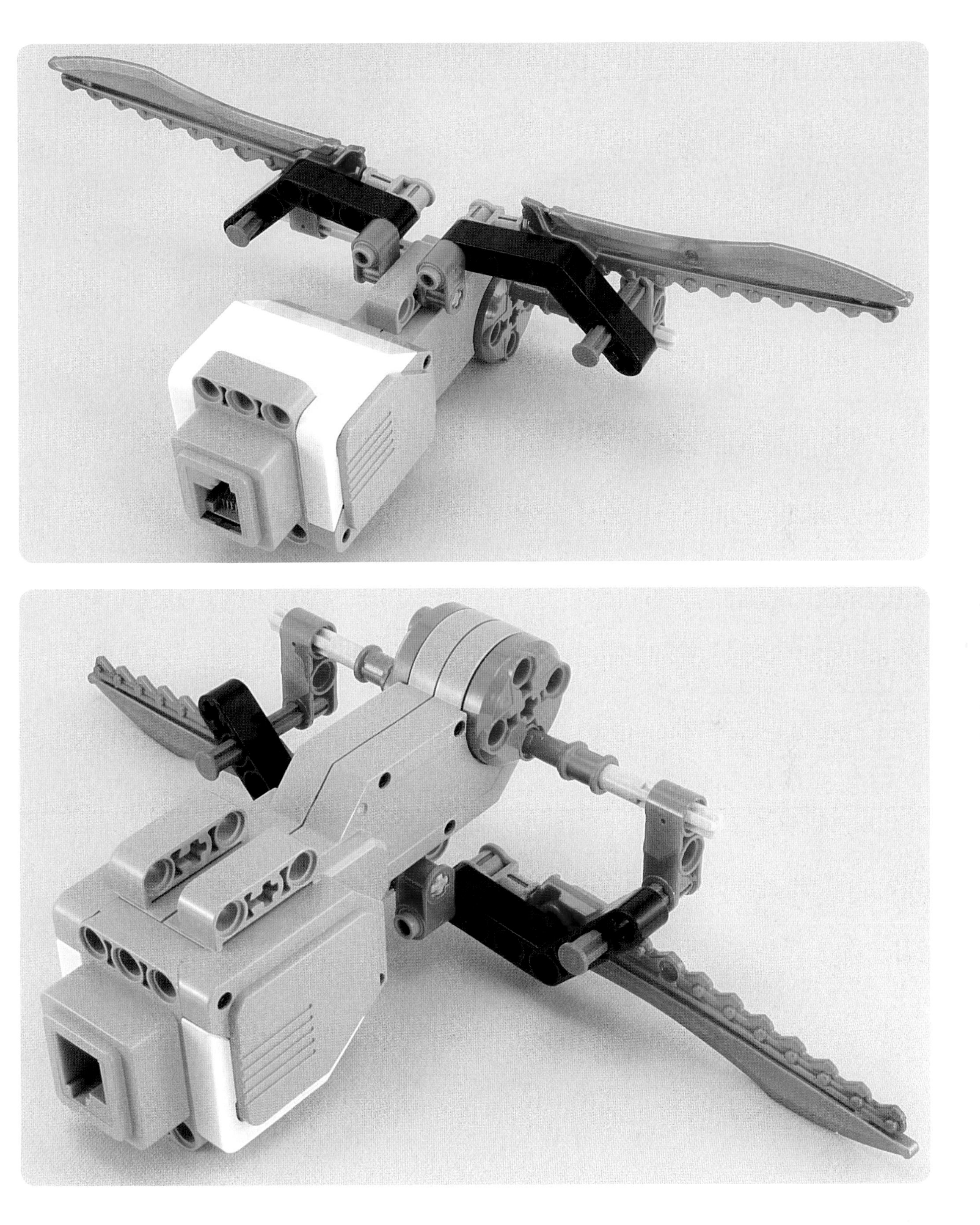

#133

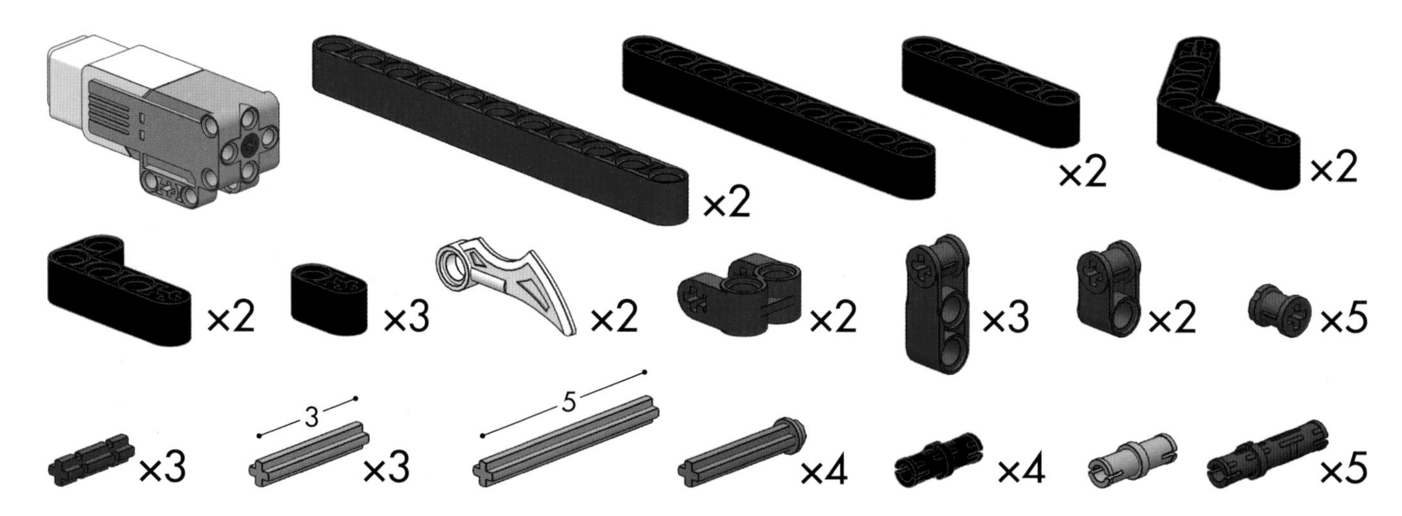
×2
×2
×2
×2
×3
×2
×2
×3
×2
×5
×3
3
×3
5
×4
×4
×5

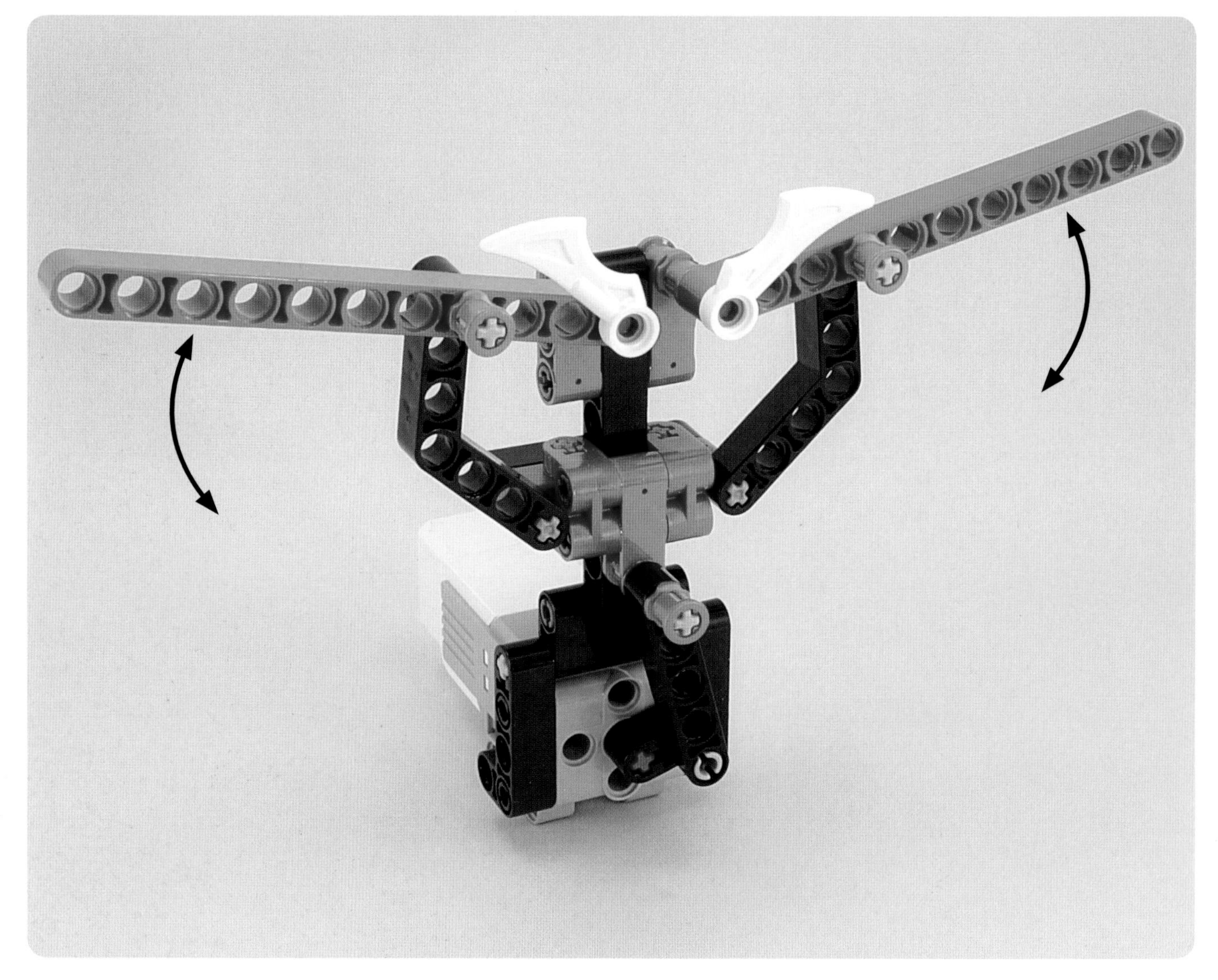

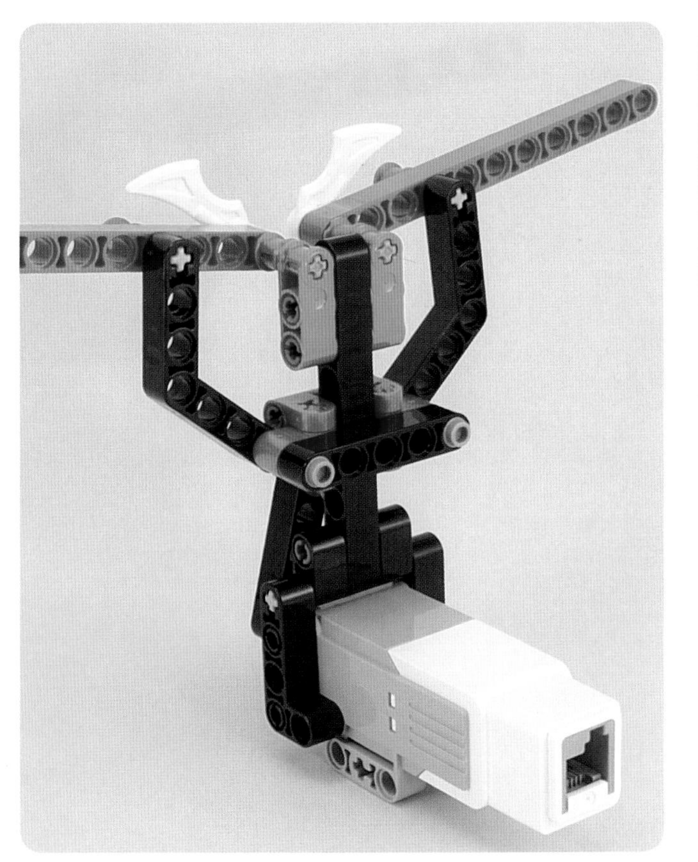

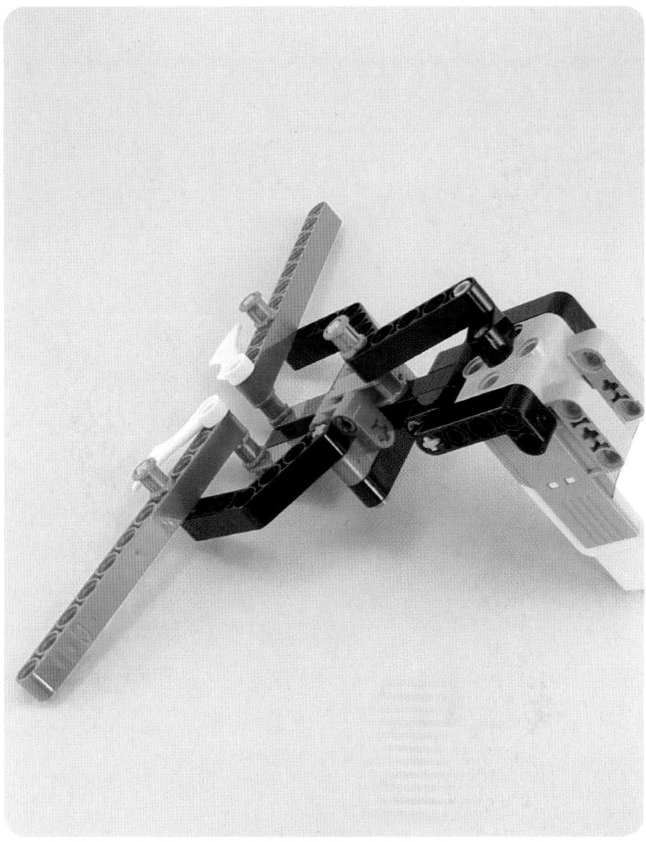

#134

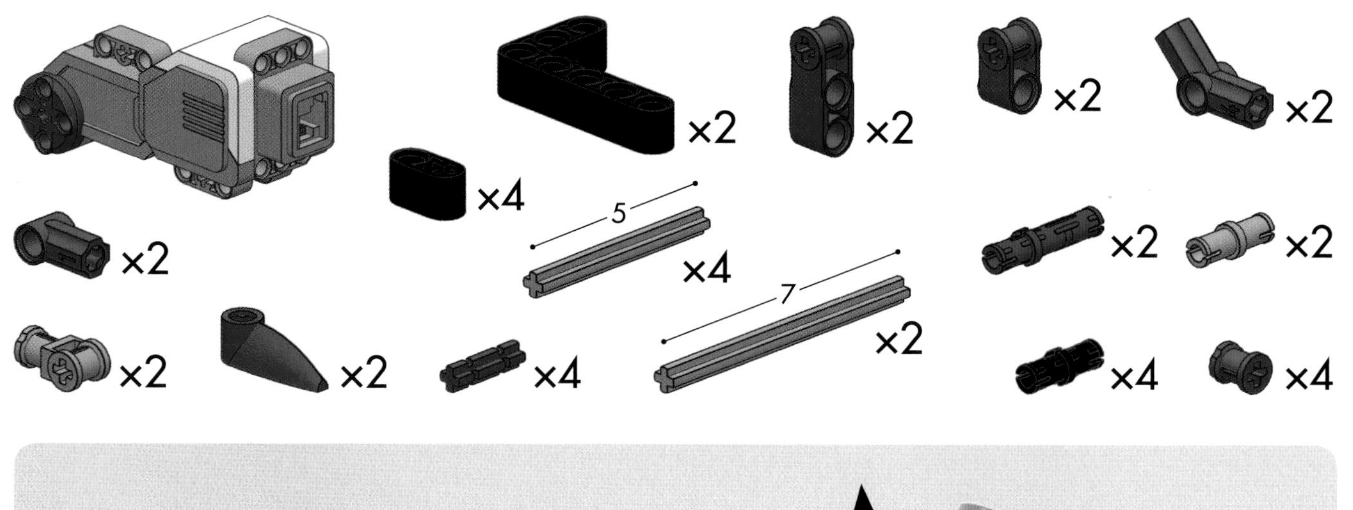
×2
×2
×2
×2
×4
5
×4
×2
×2
×2
7
×2
×2
×2
×4
×4
×4

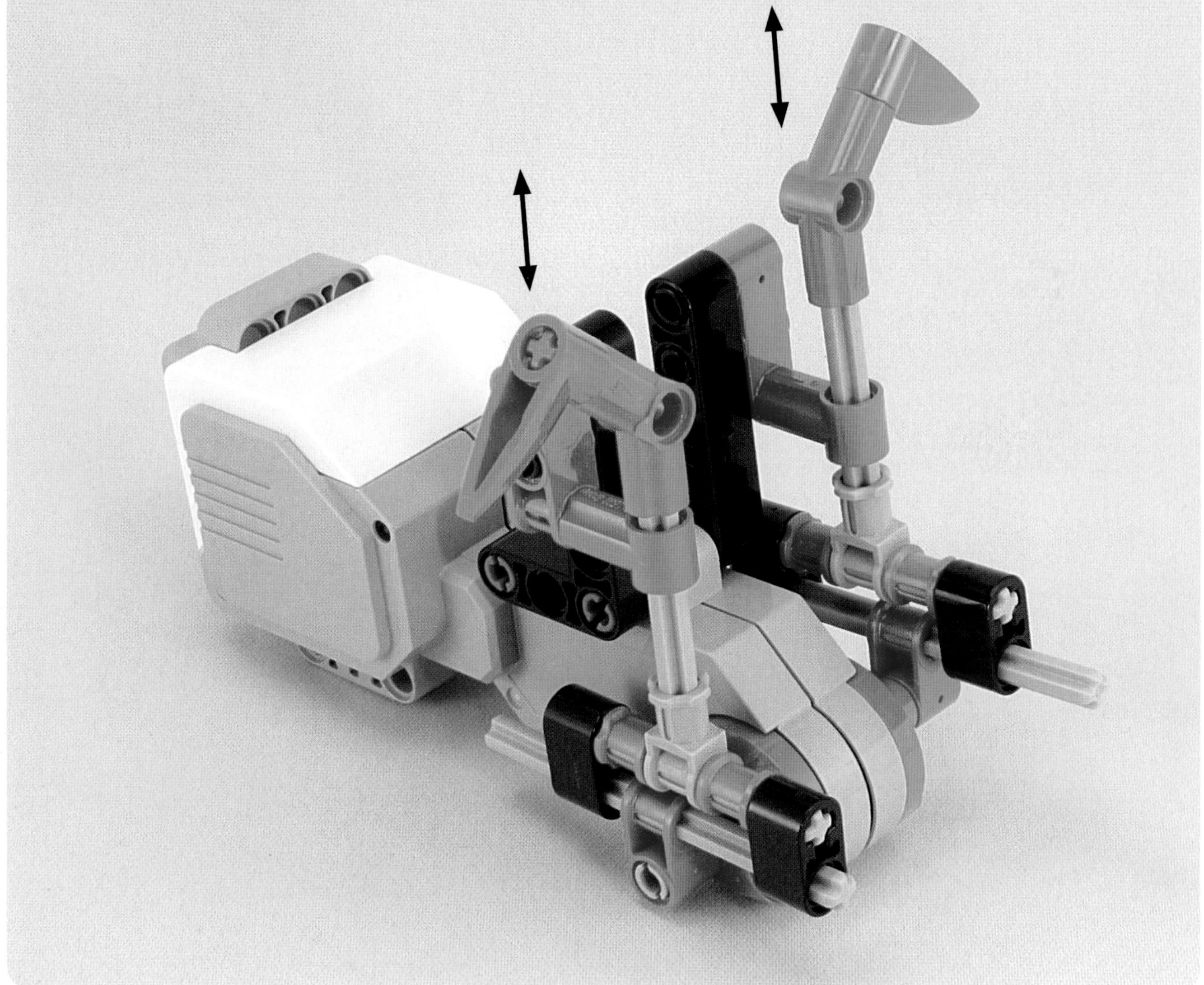

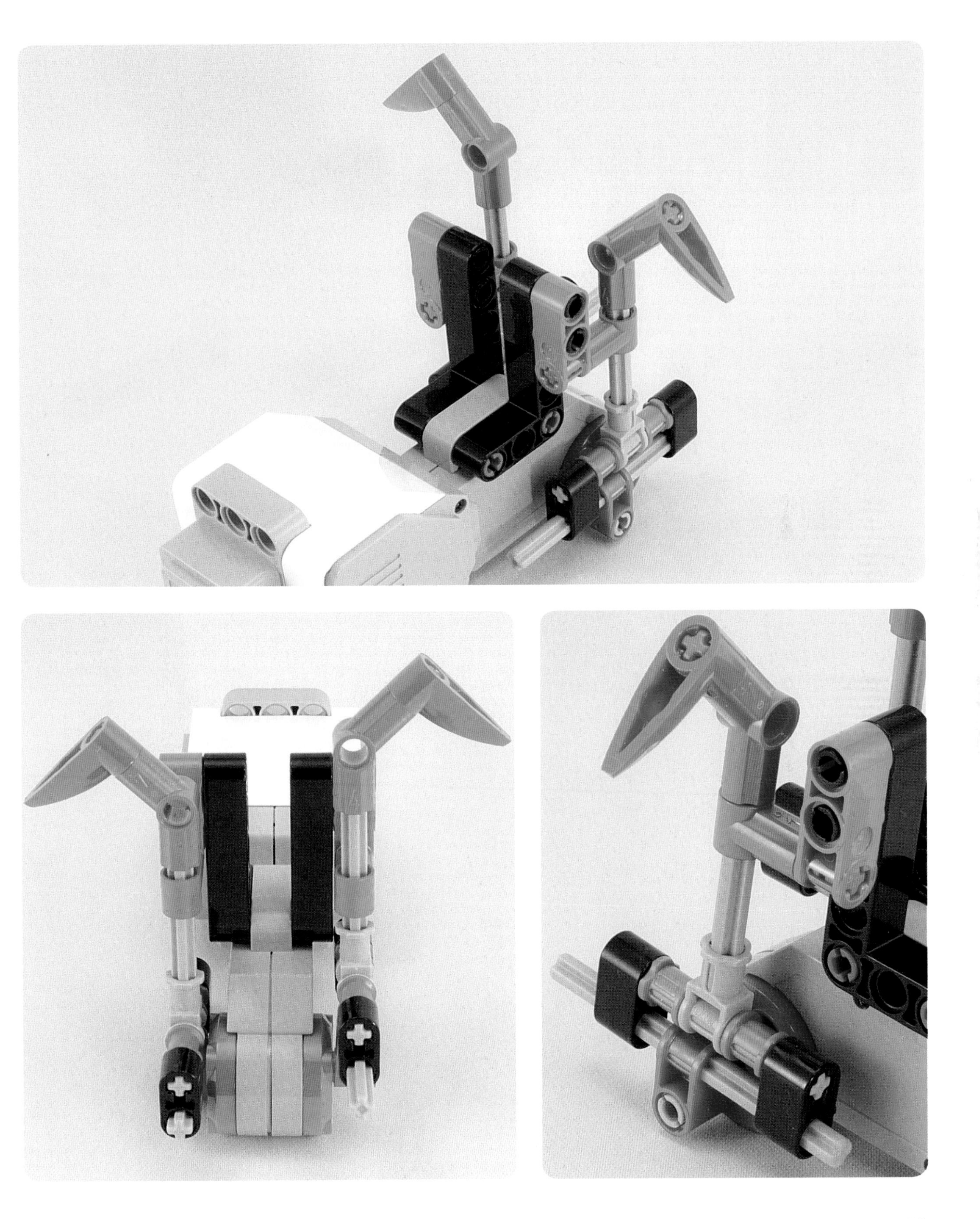

Gripping fingers

#135

×2
×2
×2
5
6
7
×2
×2
×4
×3
×2
×8

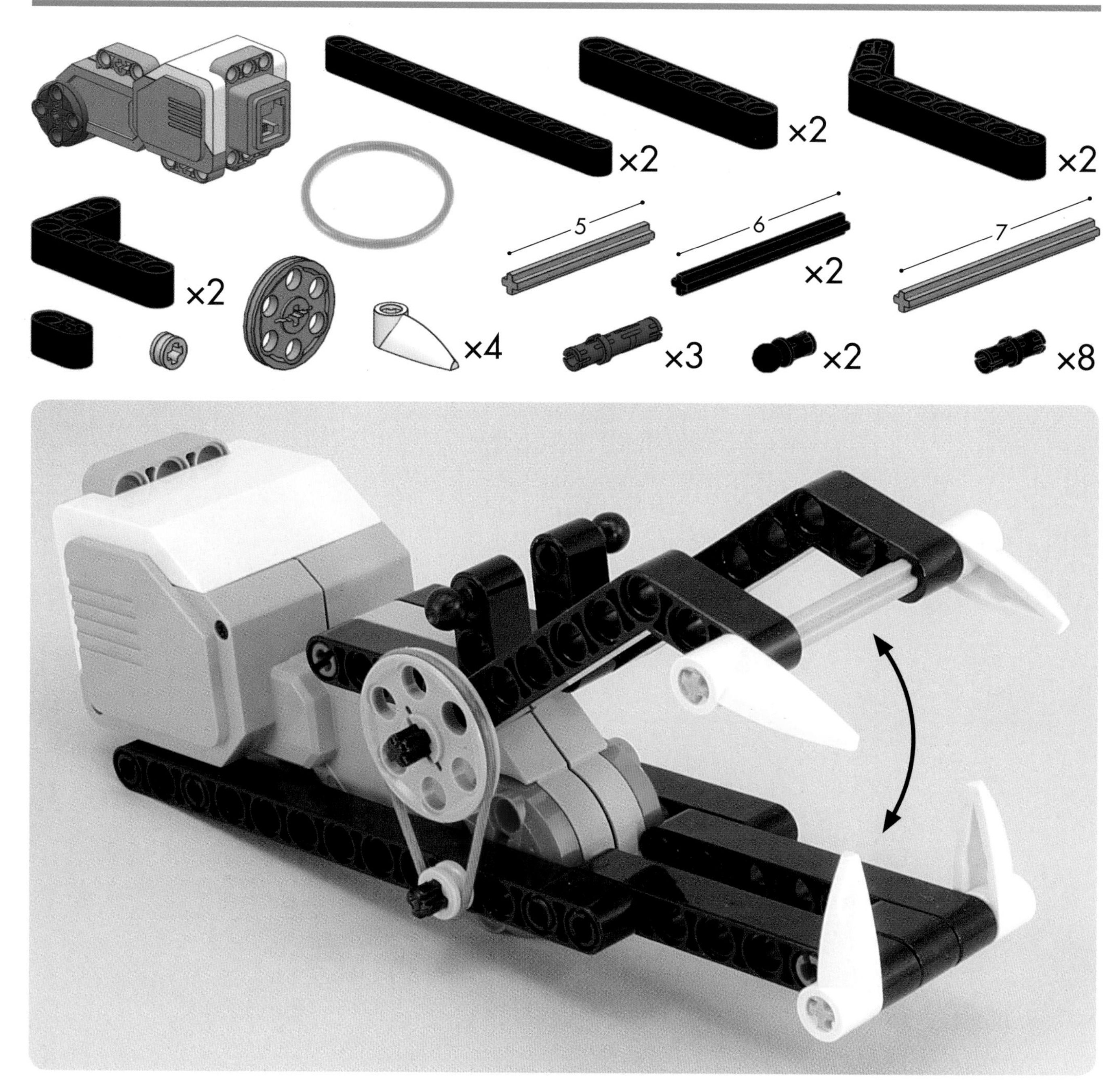

#136

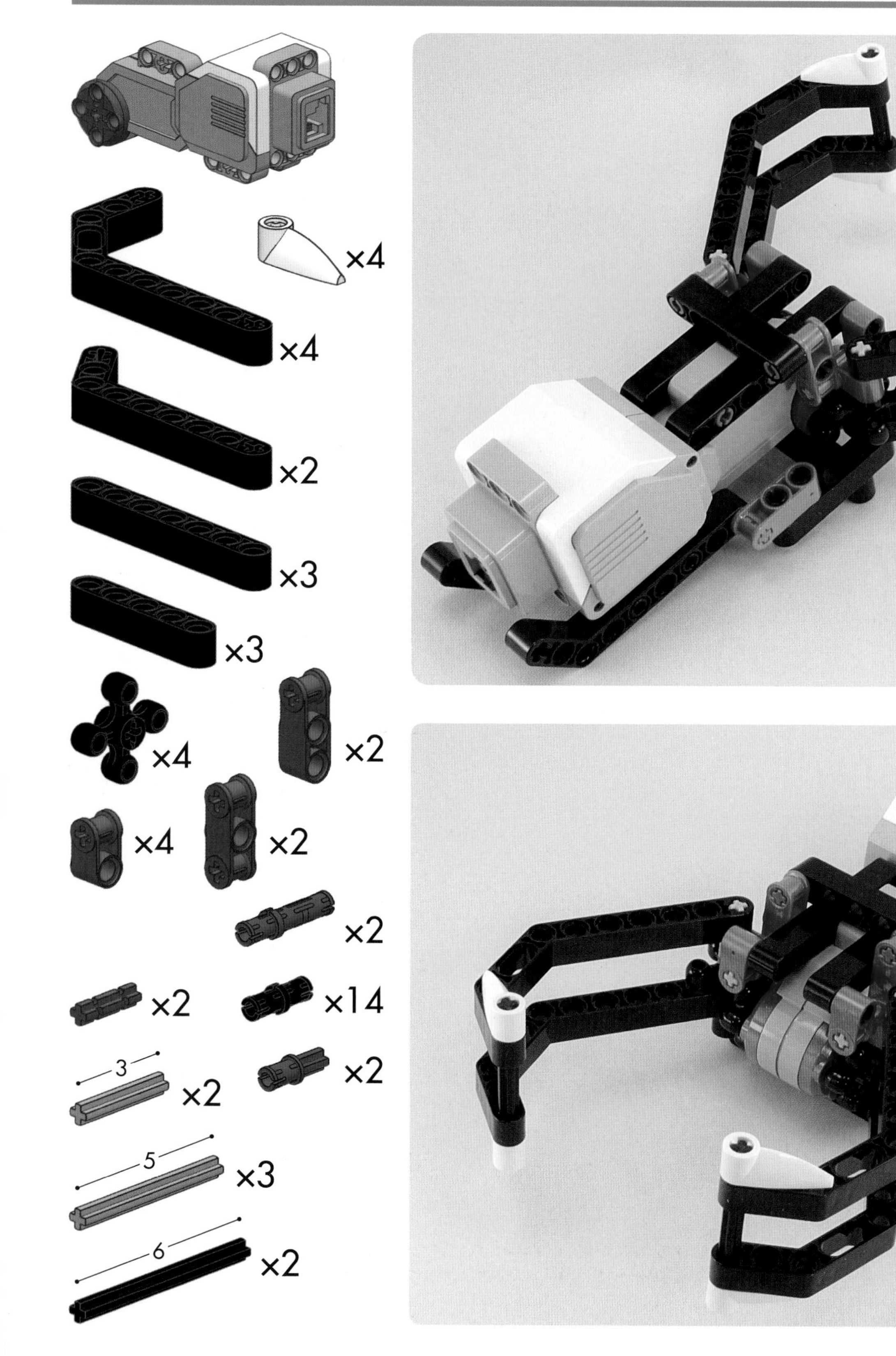
×4
×4
×2
×3
×3
×4
×2
×4
×2
×2
×2
×14
3
×2
×2
5
×3
6
×2

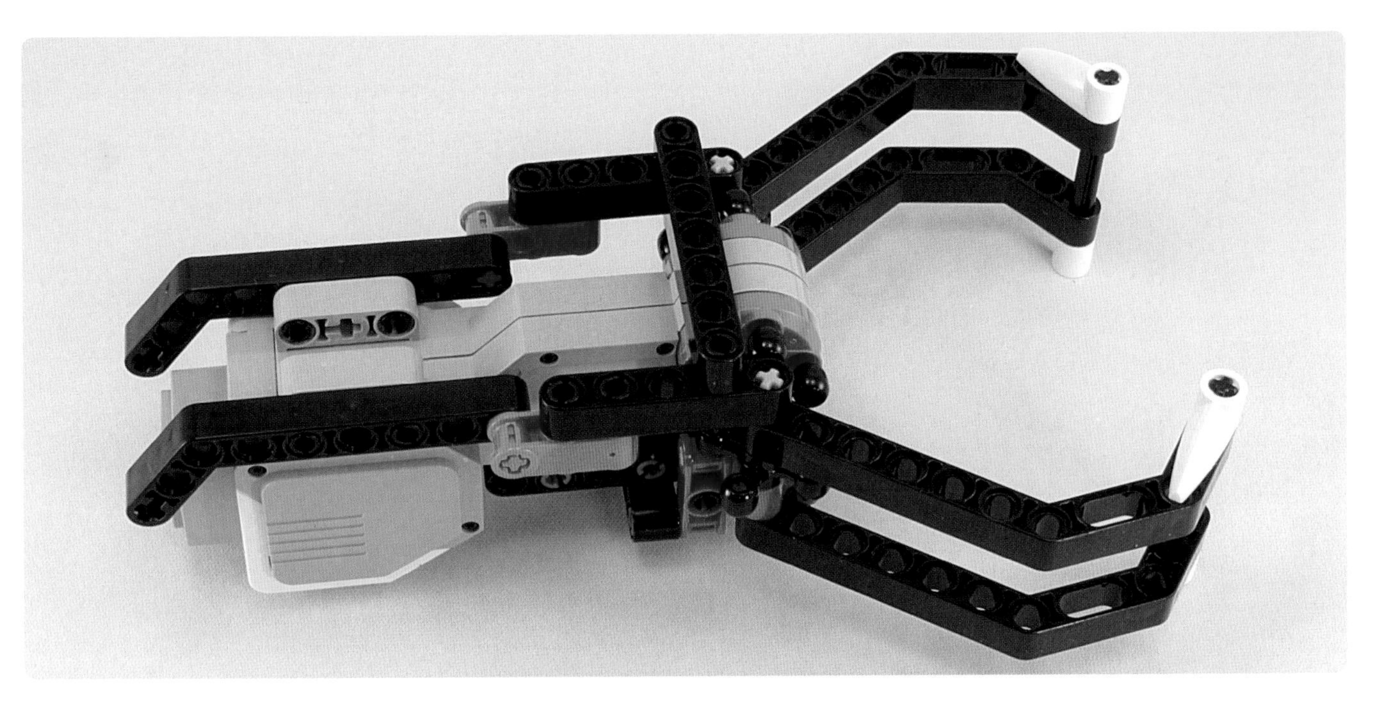

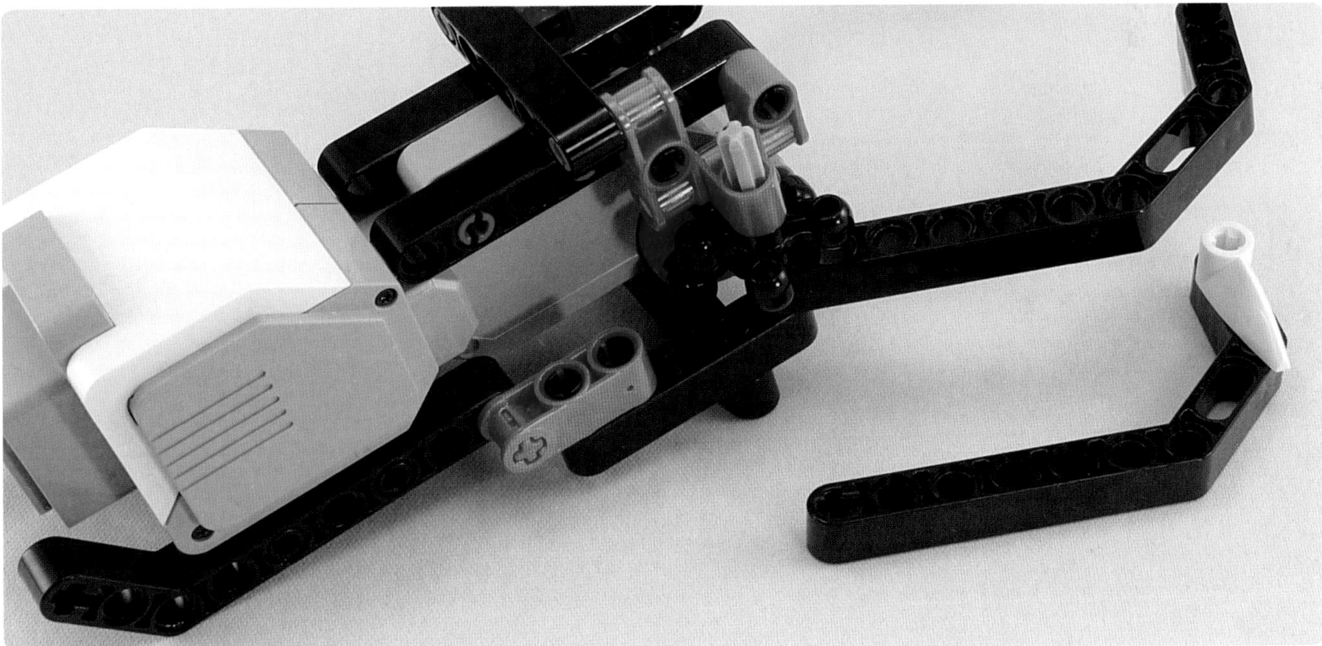

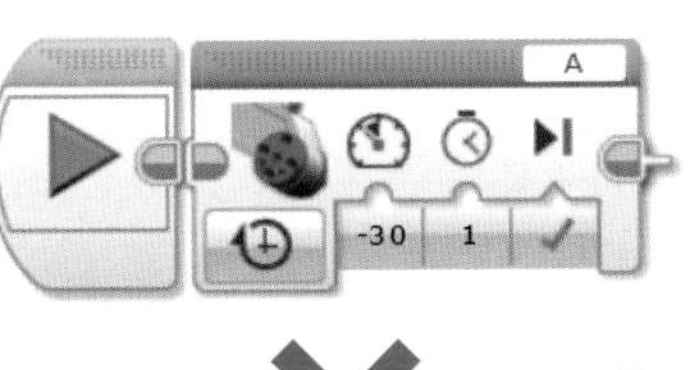
A
-30
1

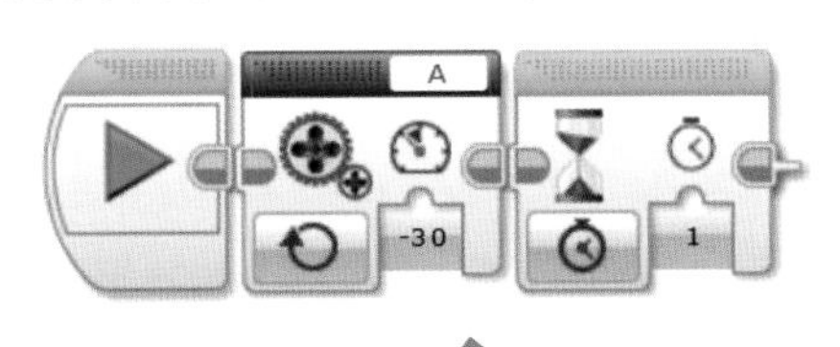
A
-30
1

To avoid overextending the mechanism, use this program instead of the standard one.

#137

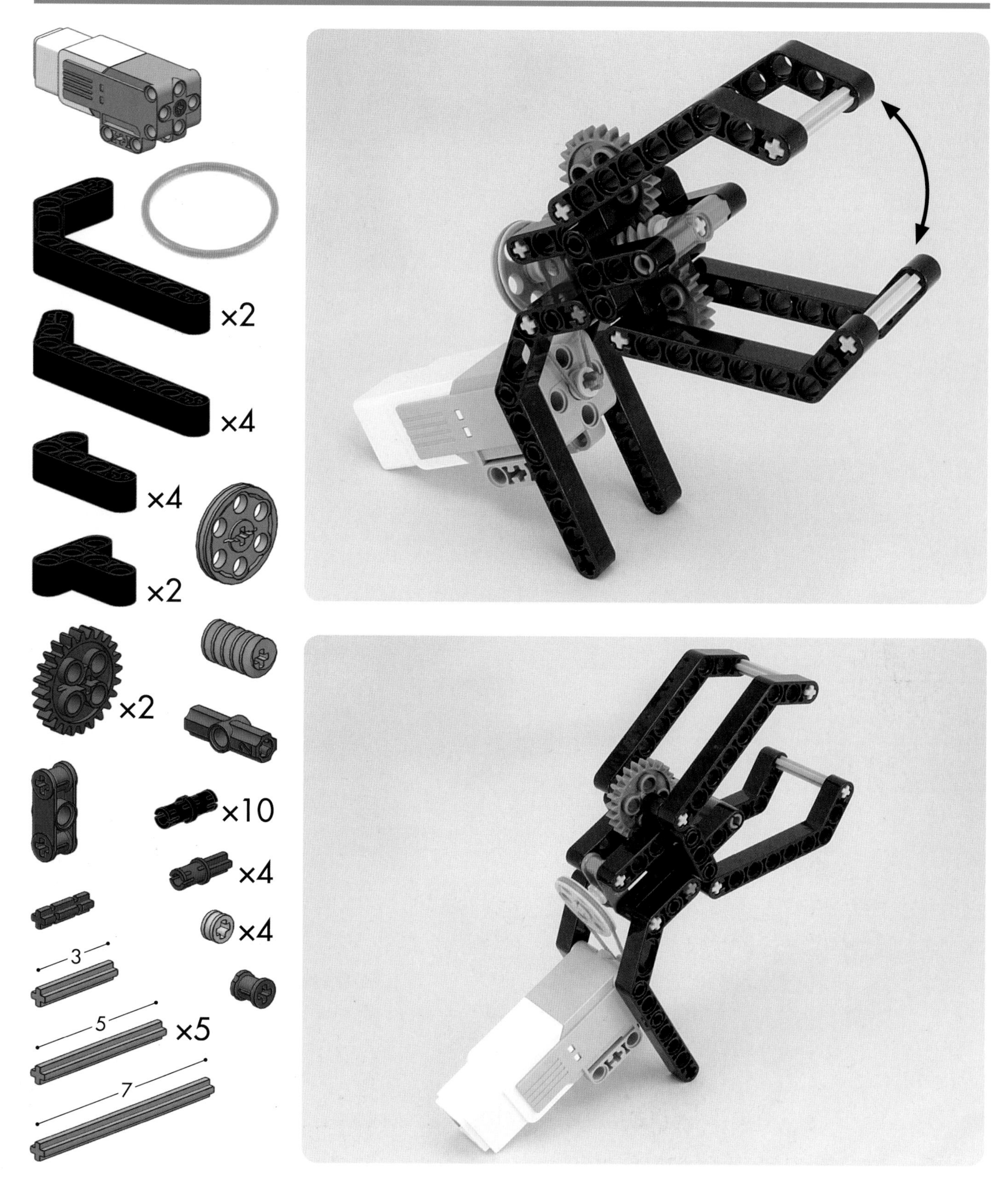
×2
×4
×4
×2
×2
×10
×4
×4
3
5
×5
7

#138

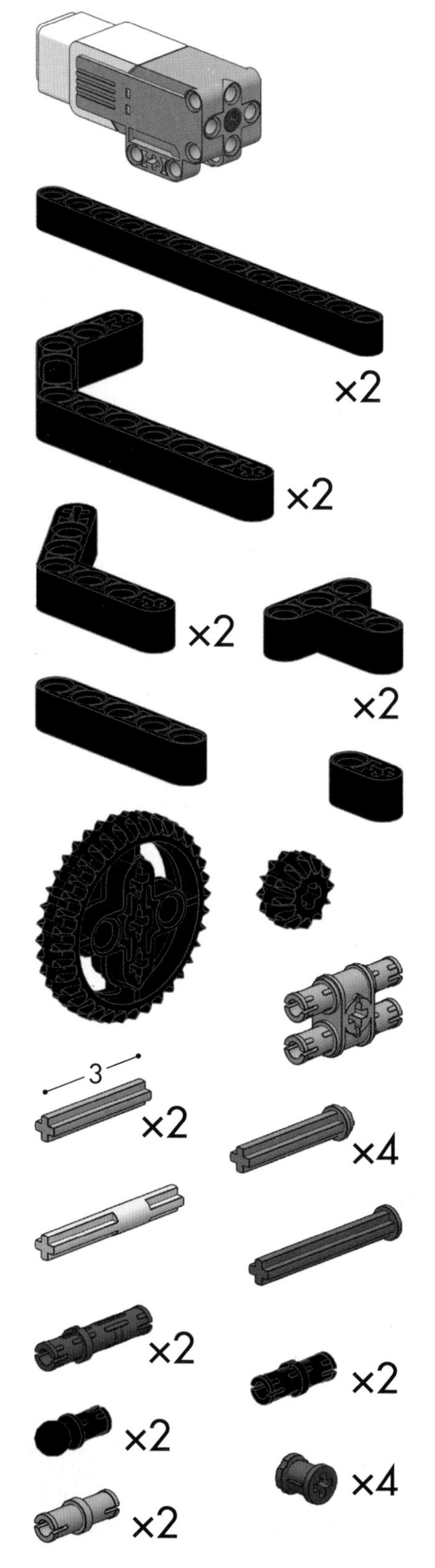
×2
×2
×2
×2
3
×2
×4
×2
×2
×2
×4
×2

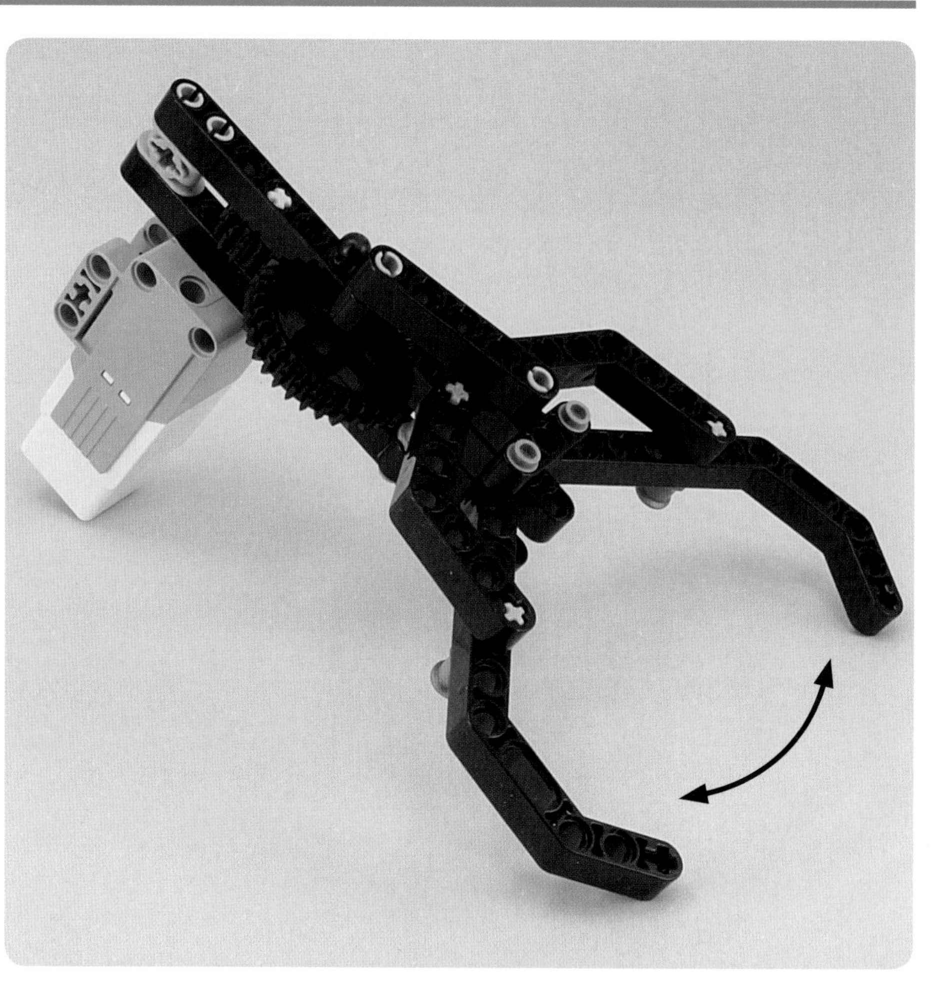

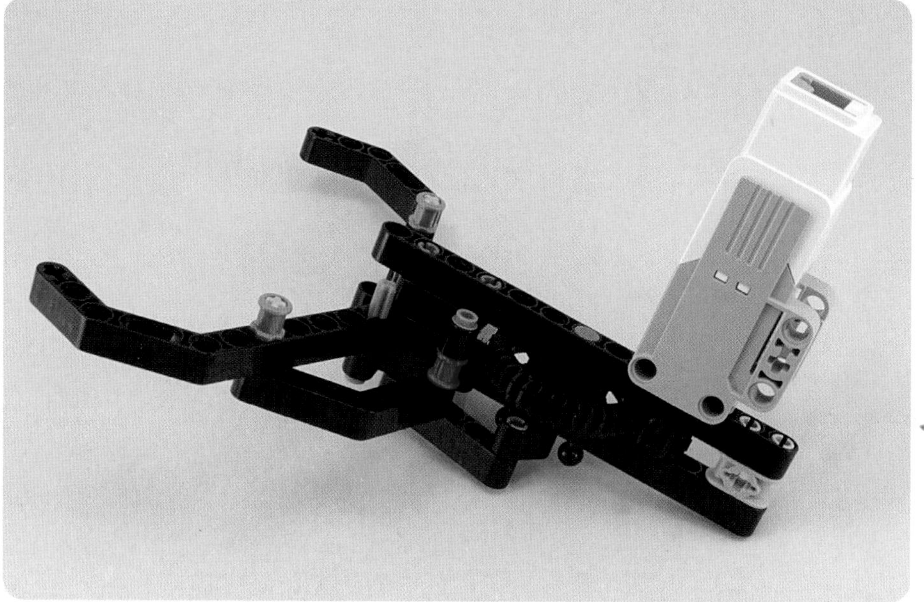

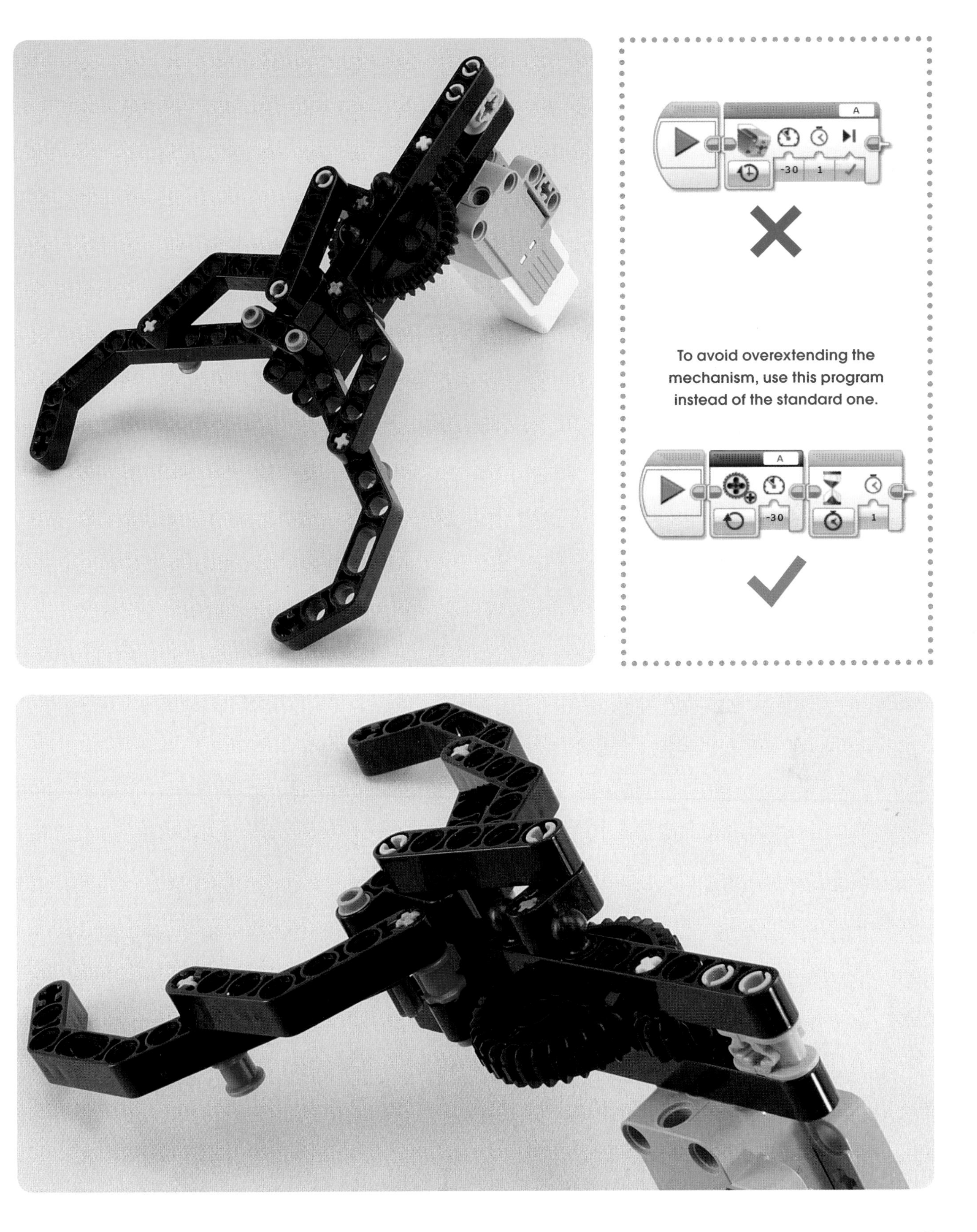
A
-30
1
A
-30
1

To avoid overextending the mechanism, use this program instead of the standard one.

#139

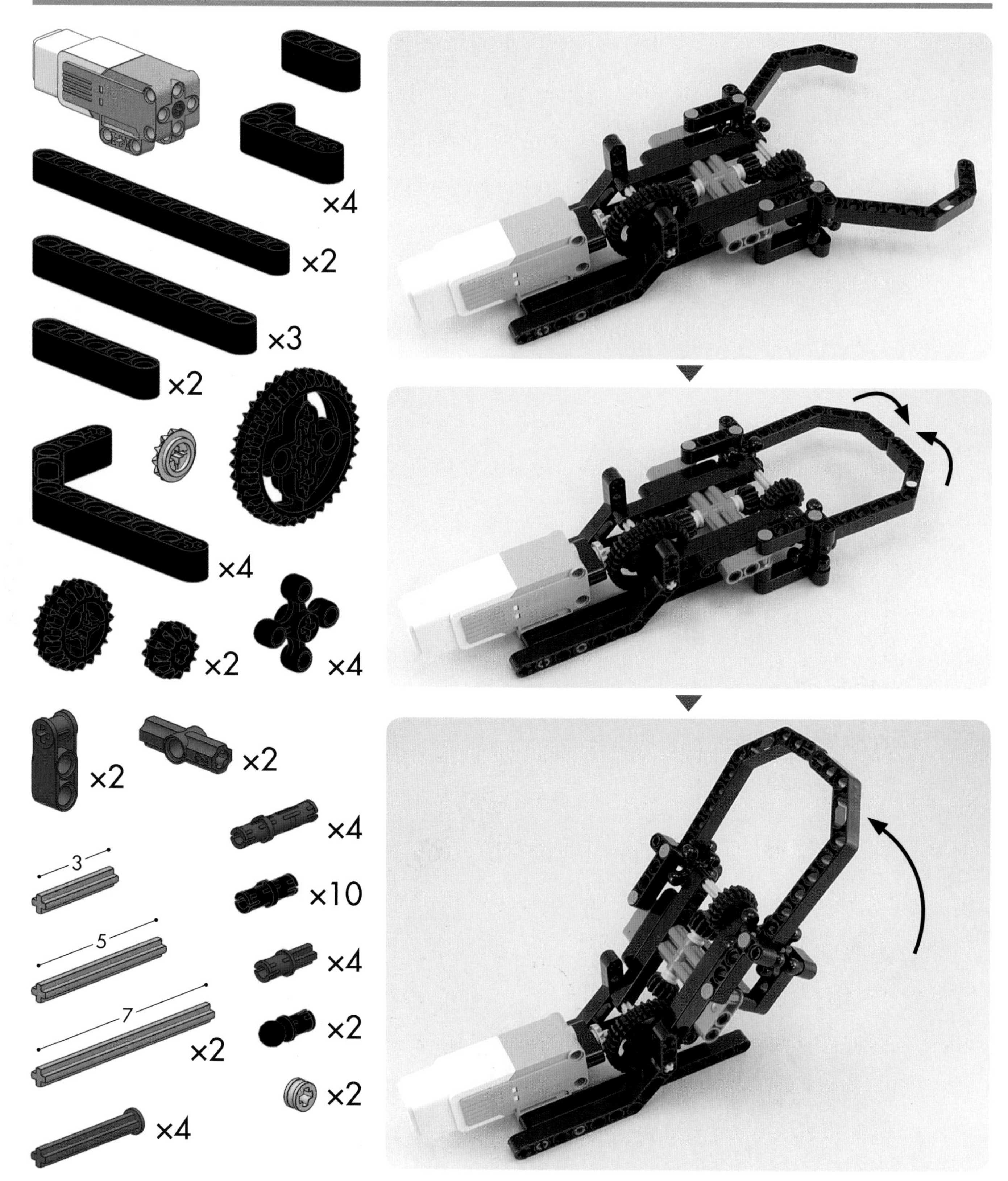
×4
×2
×3
×2
×4
×2
×4
×2
×2
×4
3
×10
5
×4
7
×2
×2
×2
×4

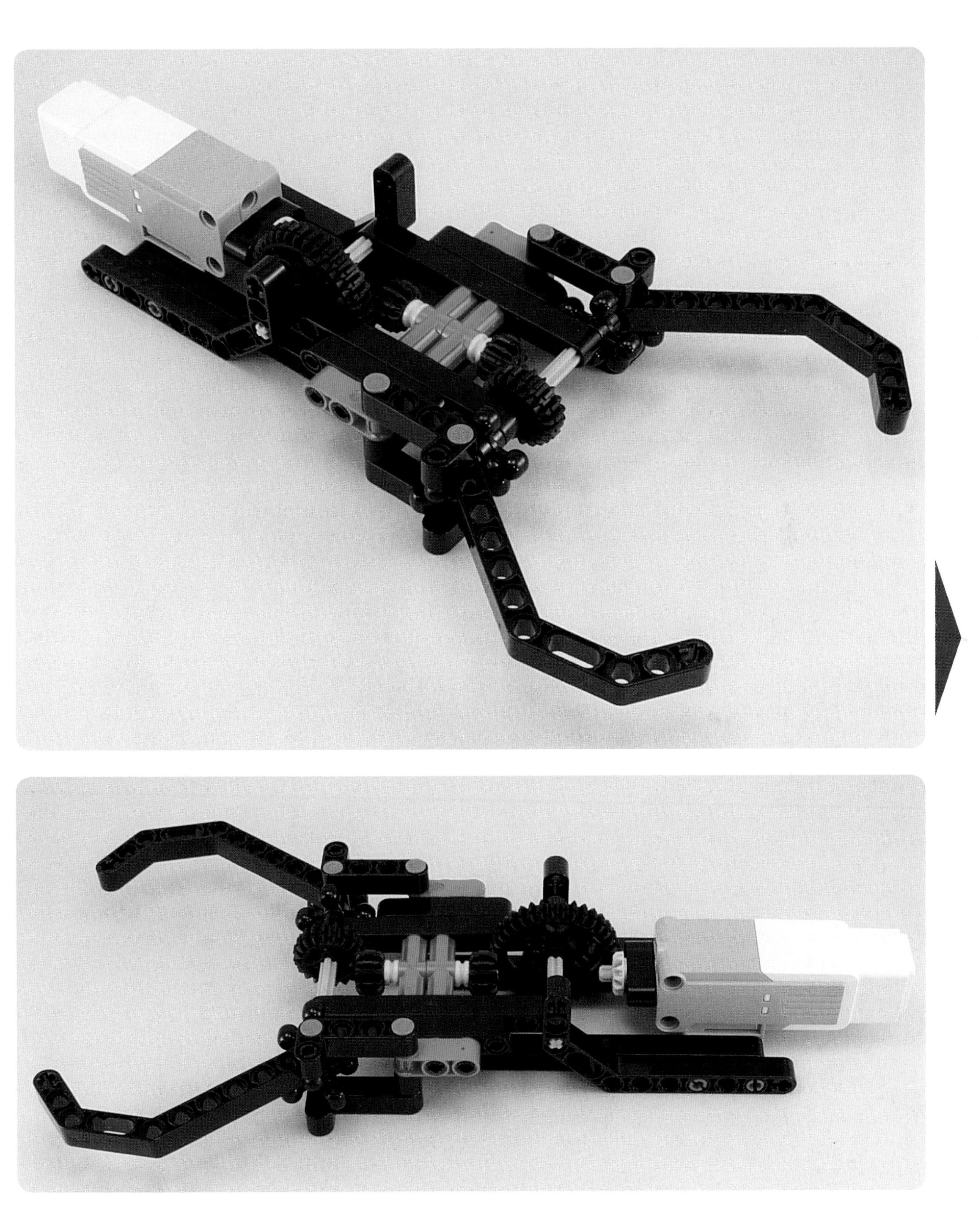

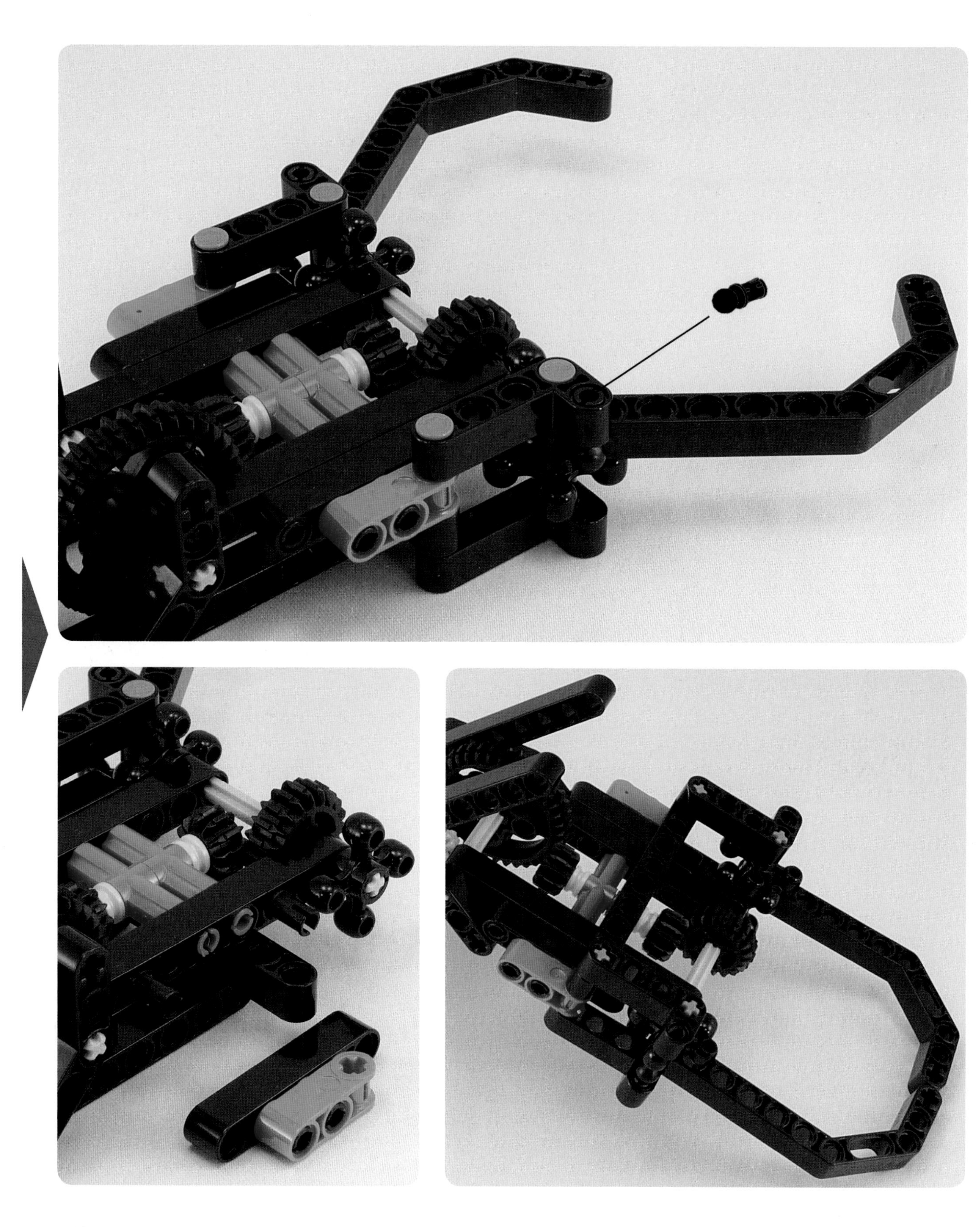

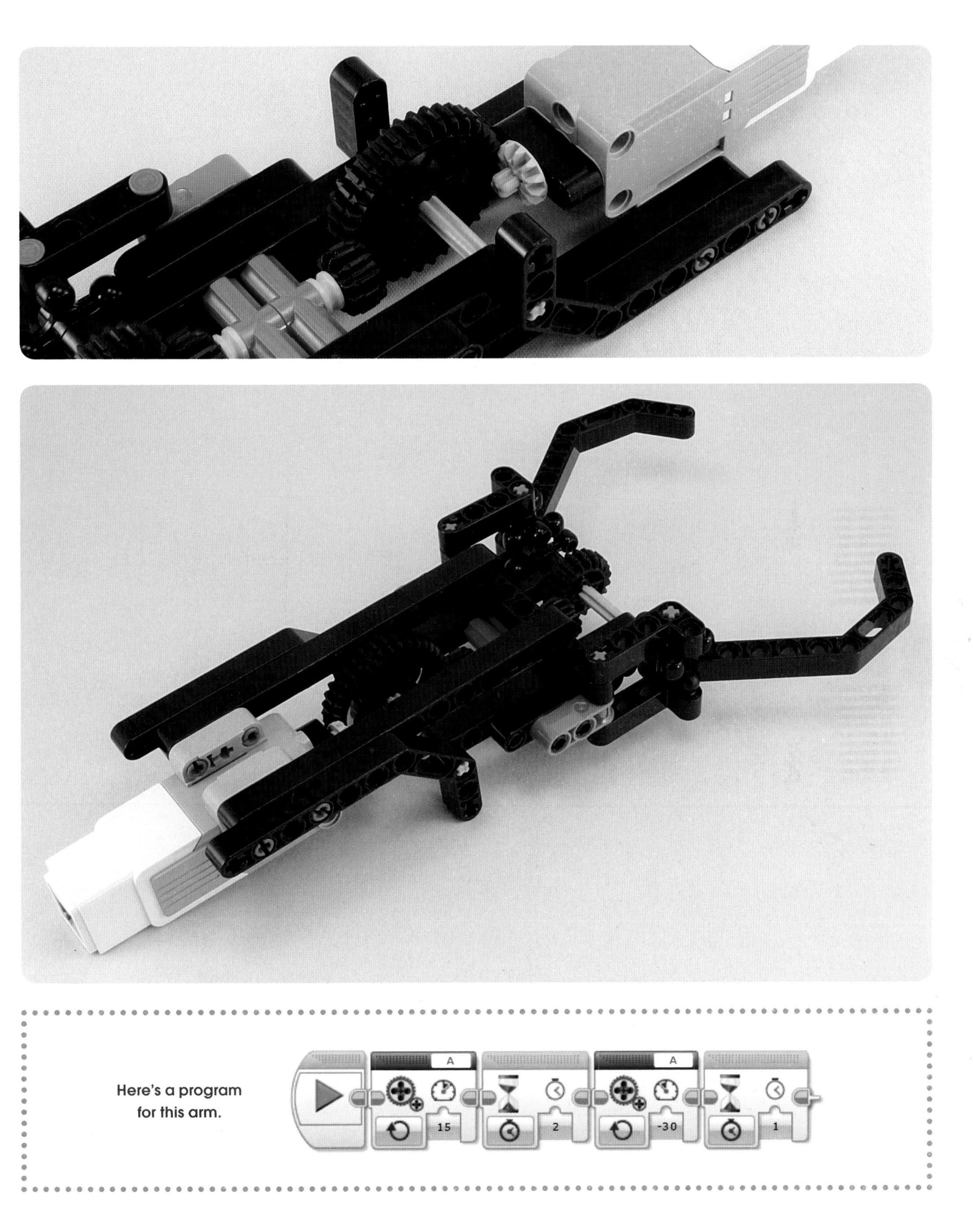
A
15
2
A
-30
1

Here's a program for this arm.

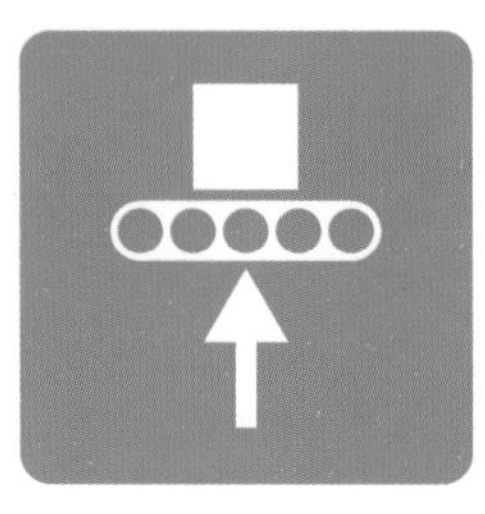

Lifting things

#140

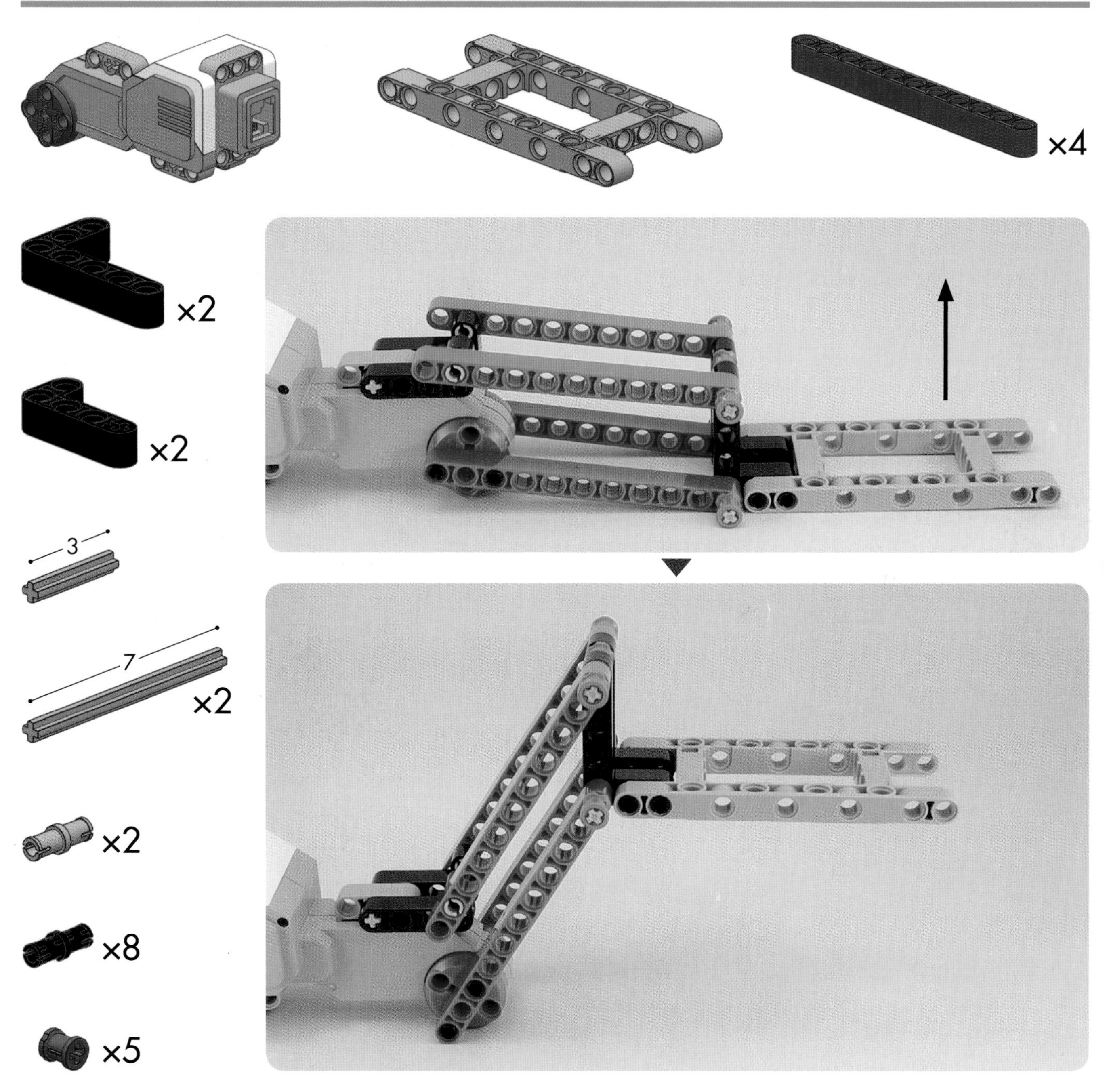
×4
×2
×2
3
7
×2
×2
×8
×5

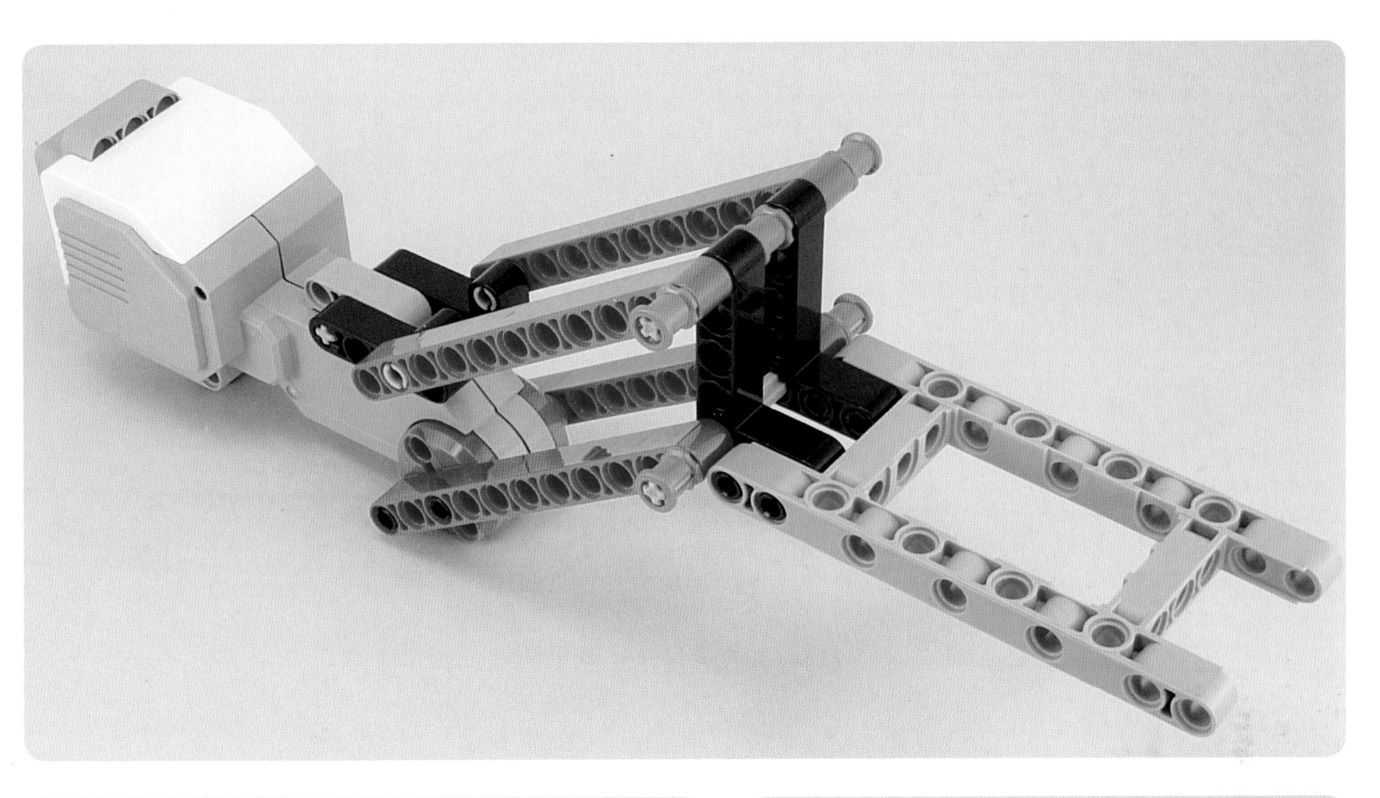

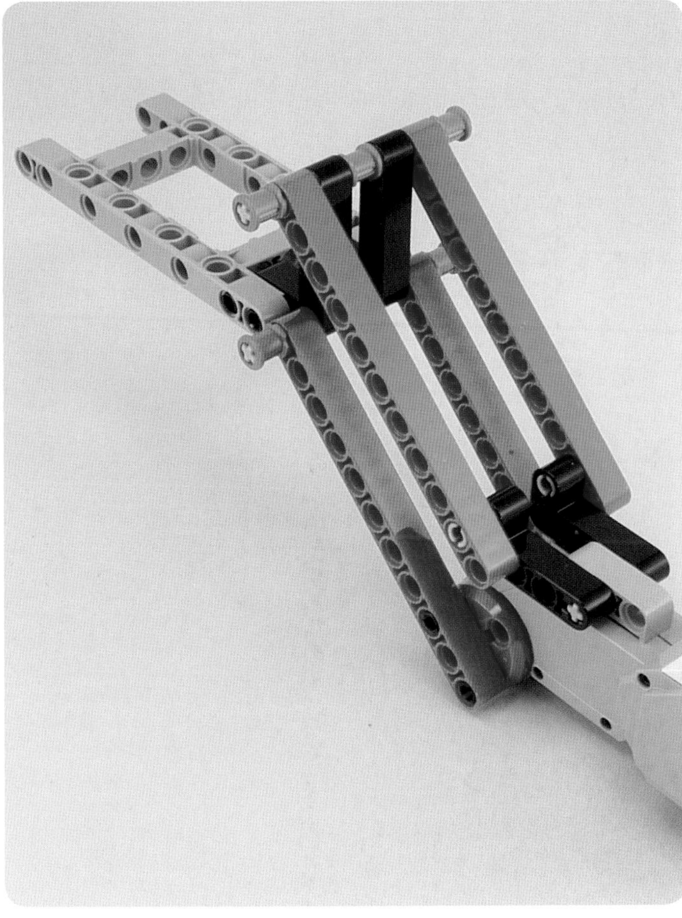

#141

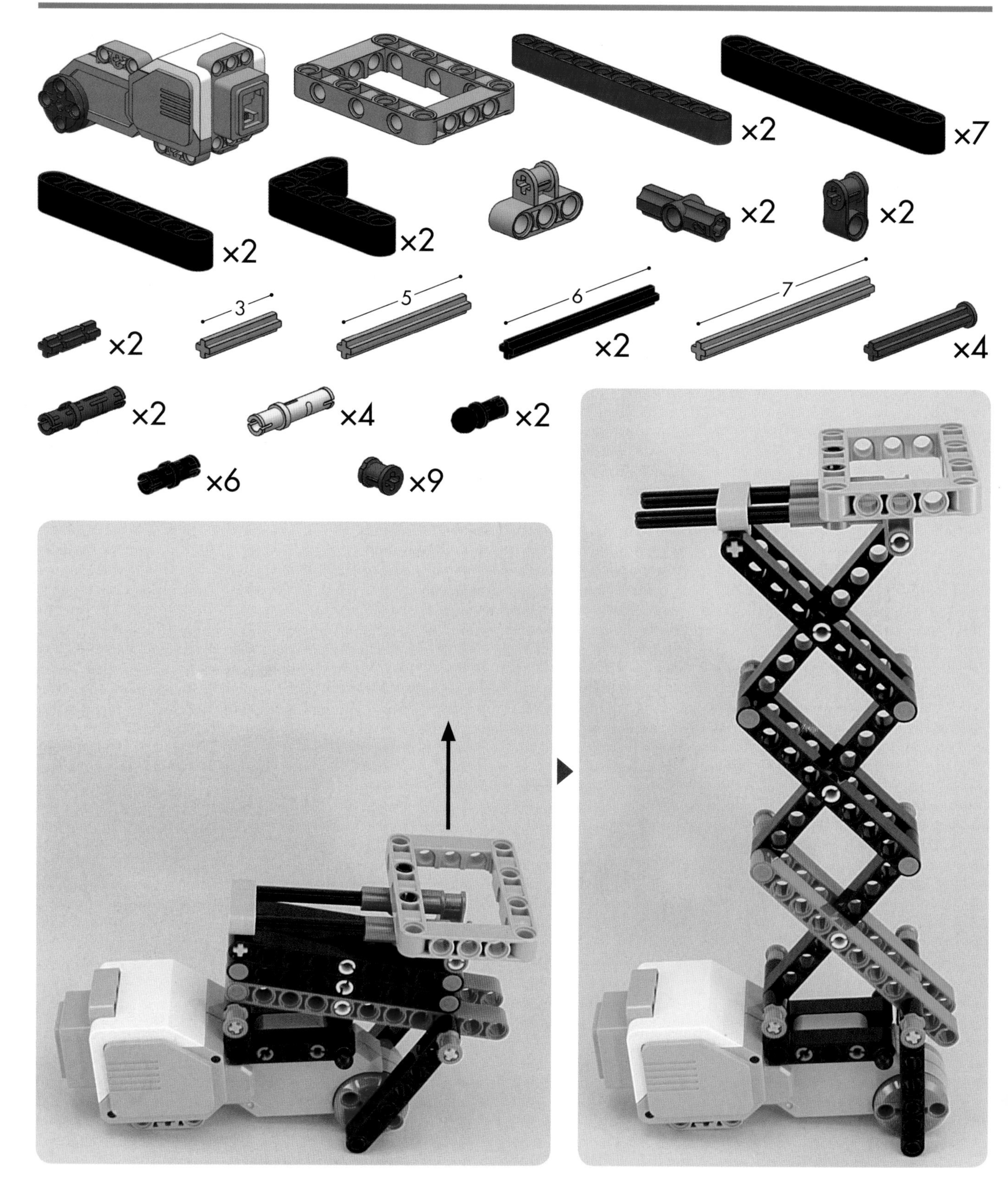

×2
×7
×2
×2
×2
×2
3
5
6
7
×2
×2
×4
×2
×4
×2
×6
×9

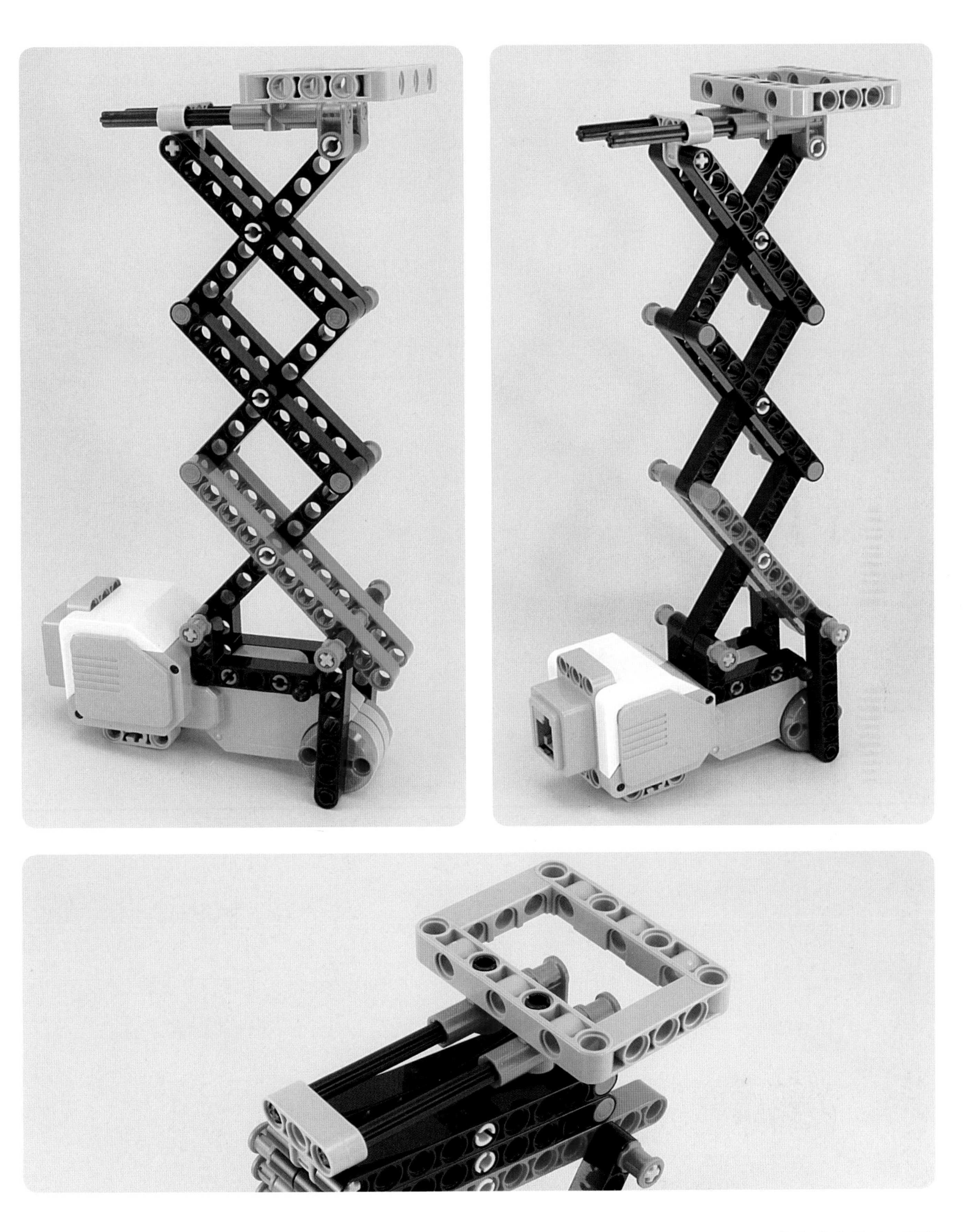

#142

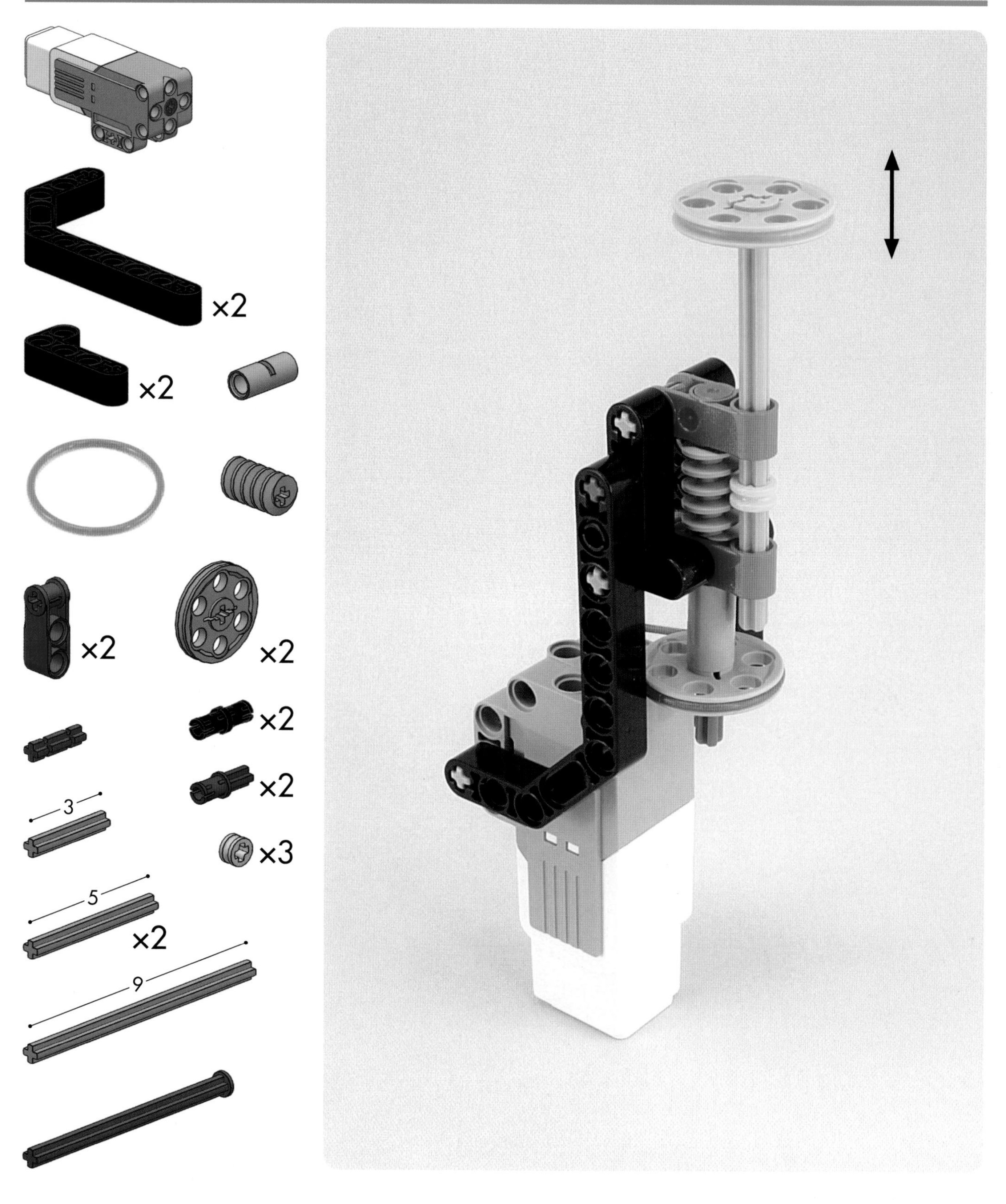
×2
×2
×2
×2
×2
×2
3
×3
5
×2
9

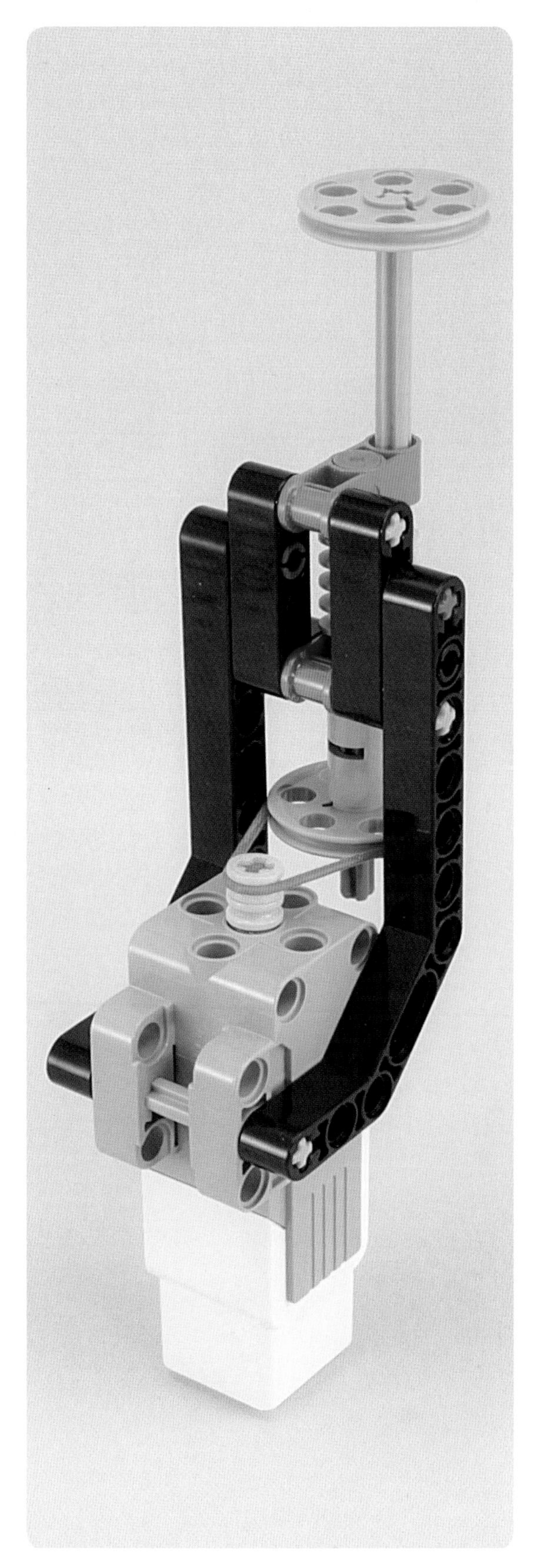

Shooting things

#143

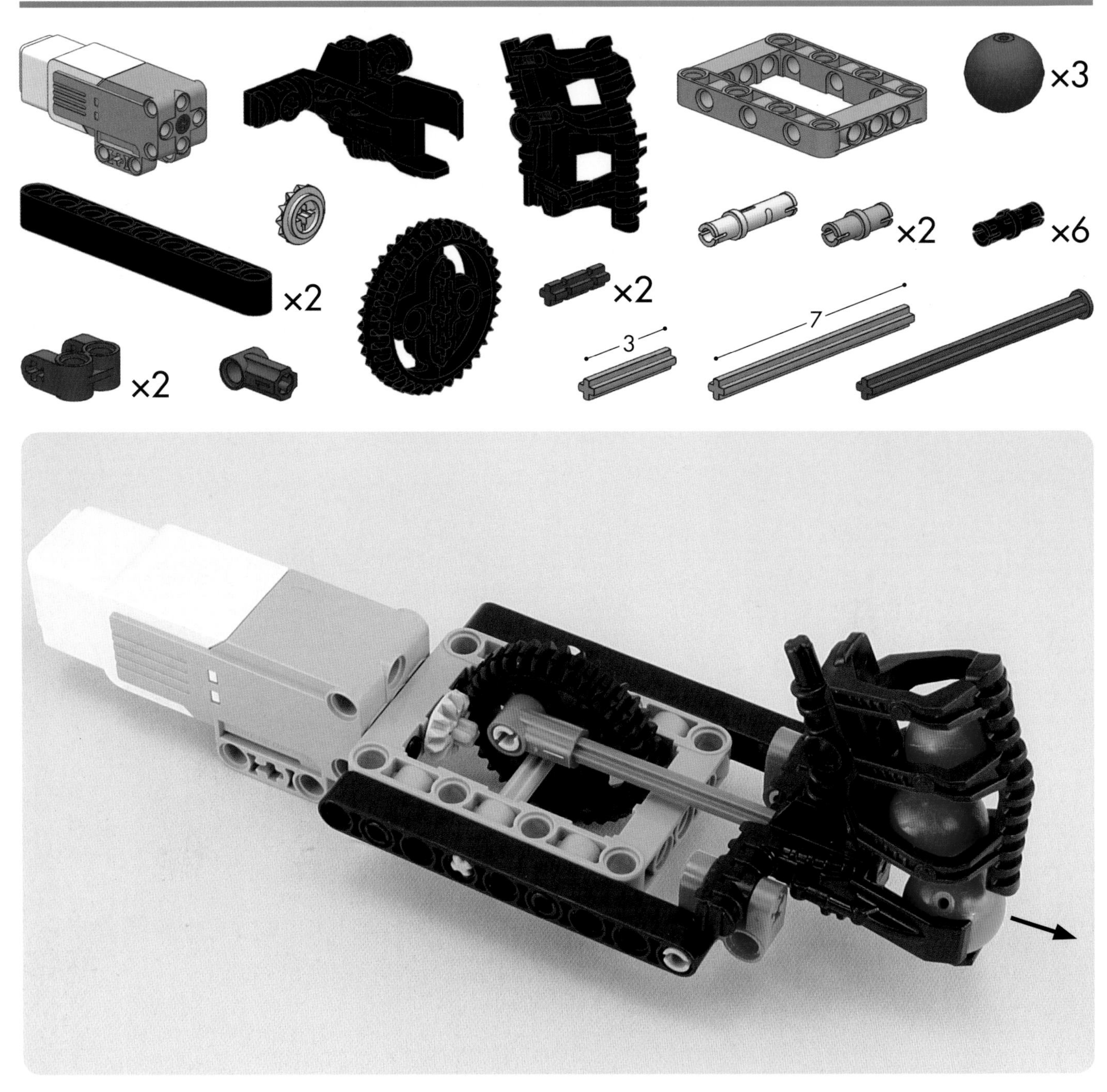

×3
×2
×2
×6
×2
×2
3
7

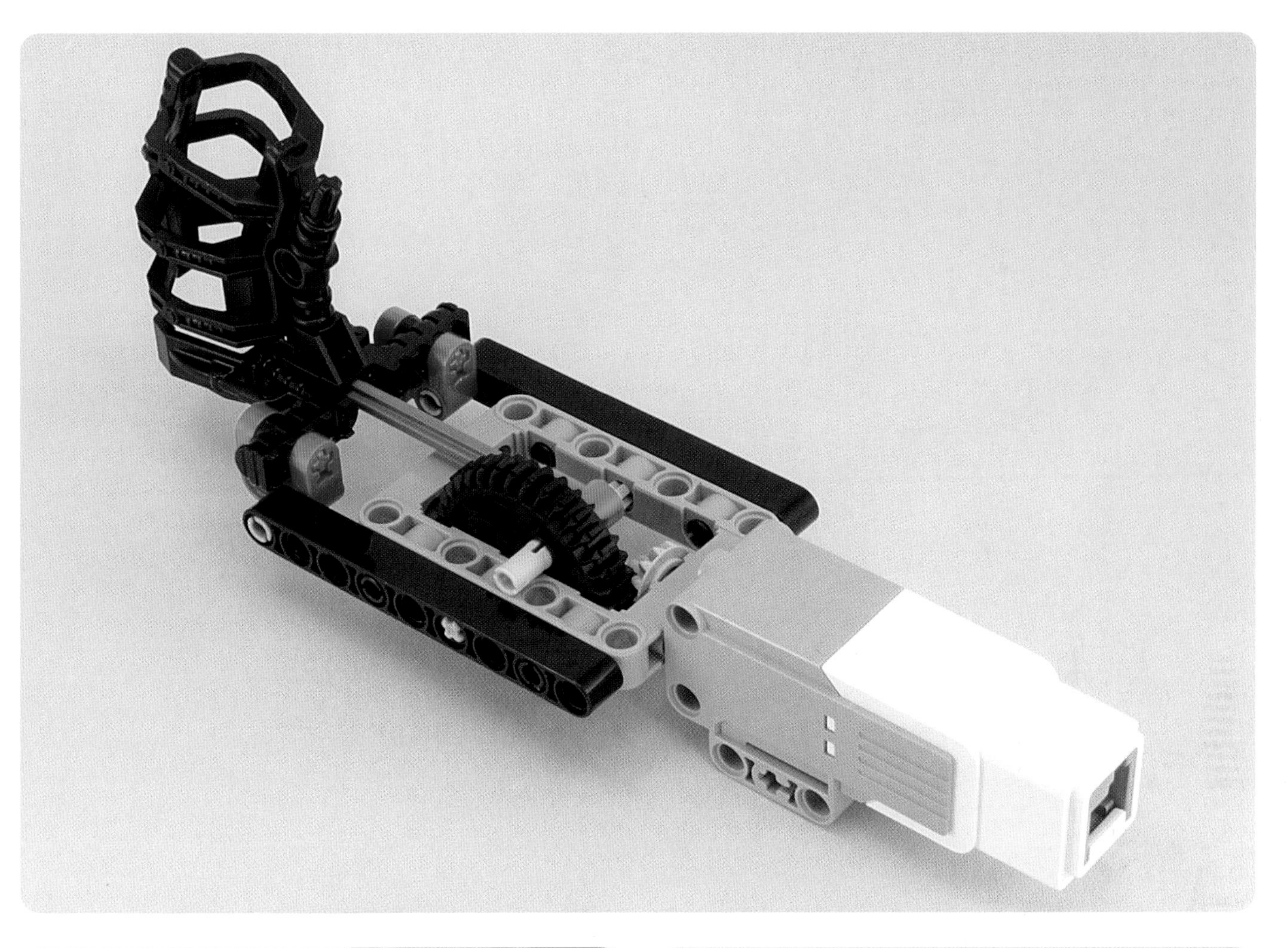

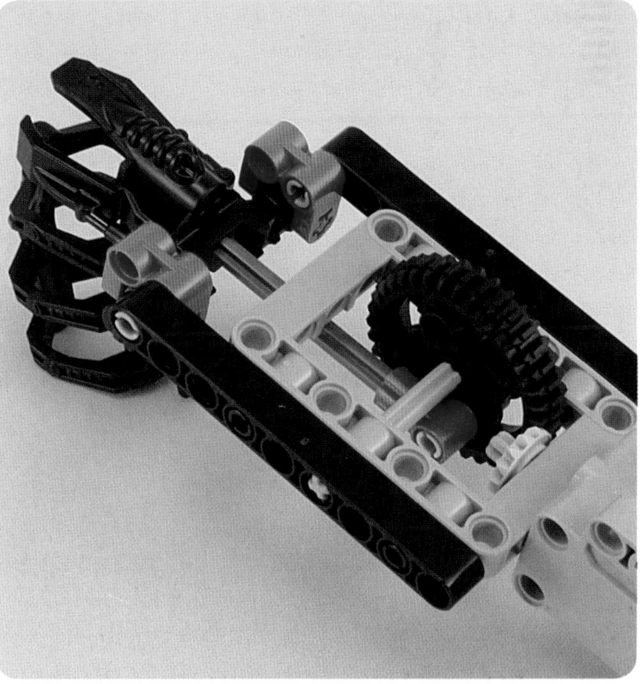

#144

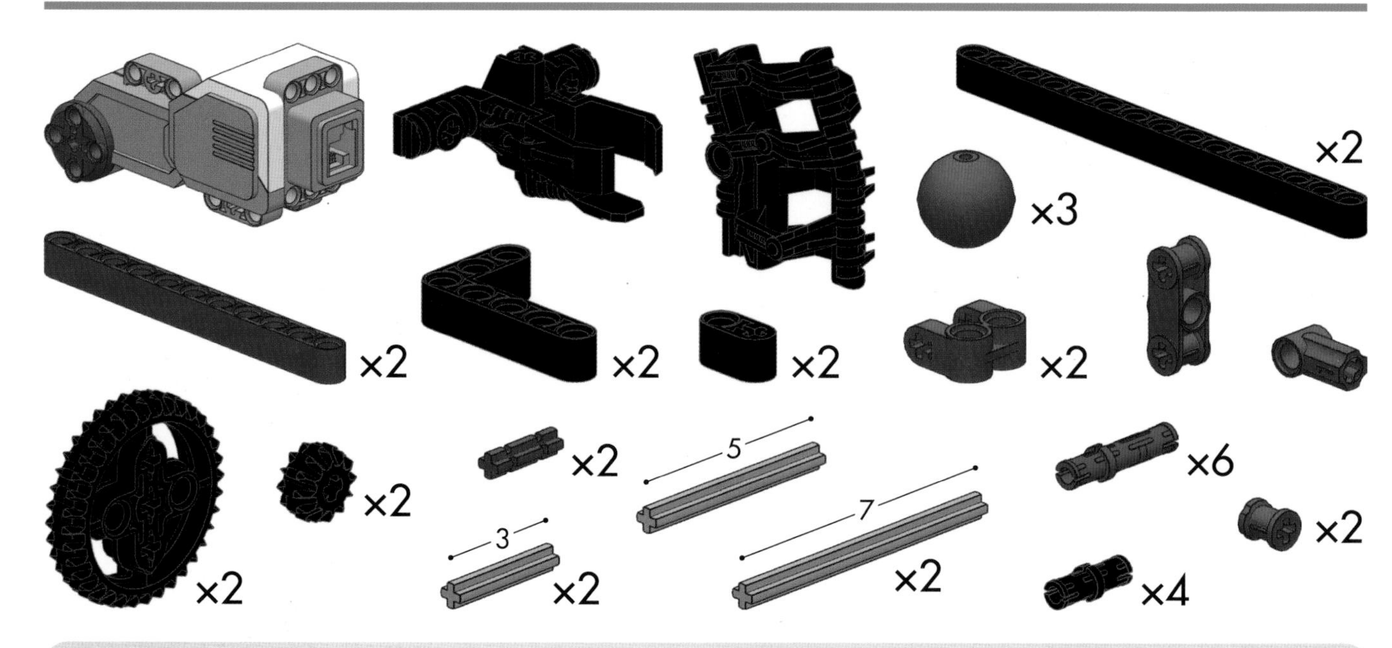
×2
×3
×2
×2
×2
×2
×2
×2
5
×2
×6
×2
7
3
×2
×2
×2
×4

#145

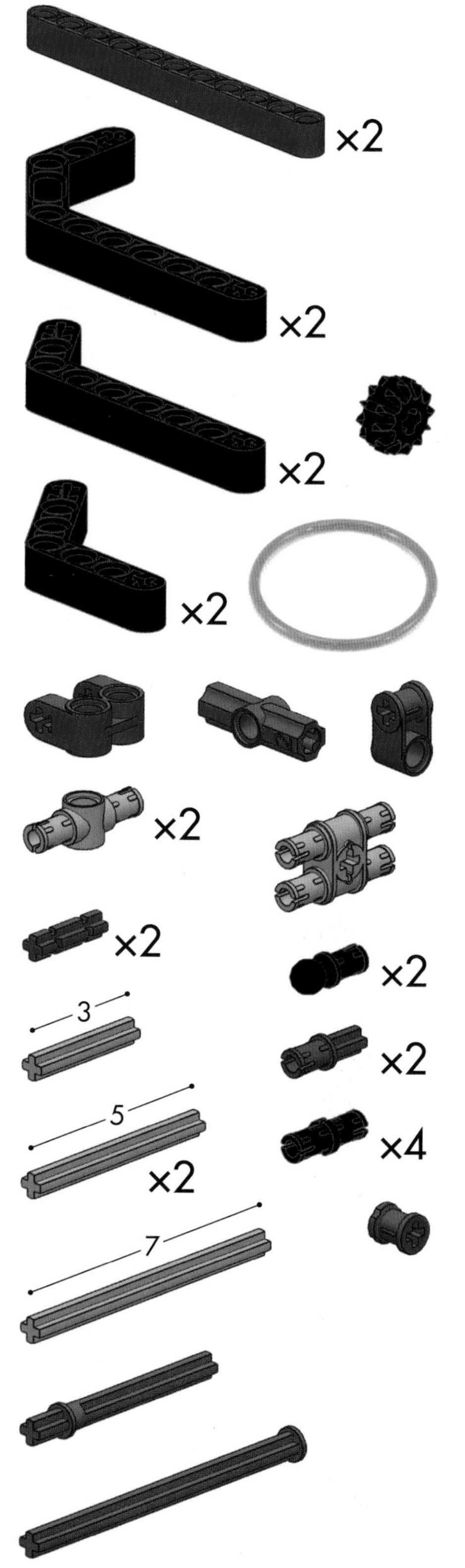
×2
×2
×2
×2
×2
×2
×2
×2
×2
×4
3
5
×2
7

#146

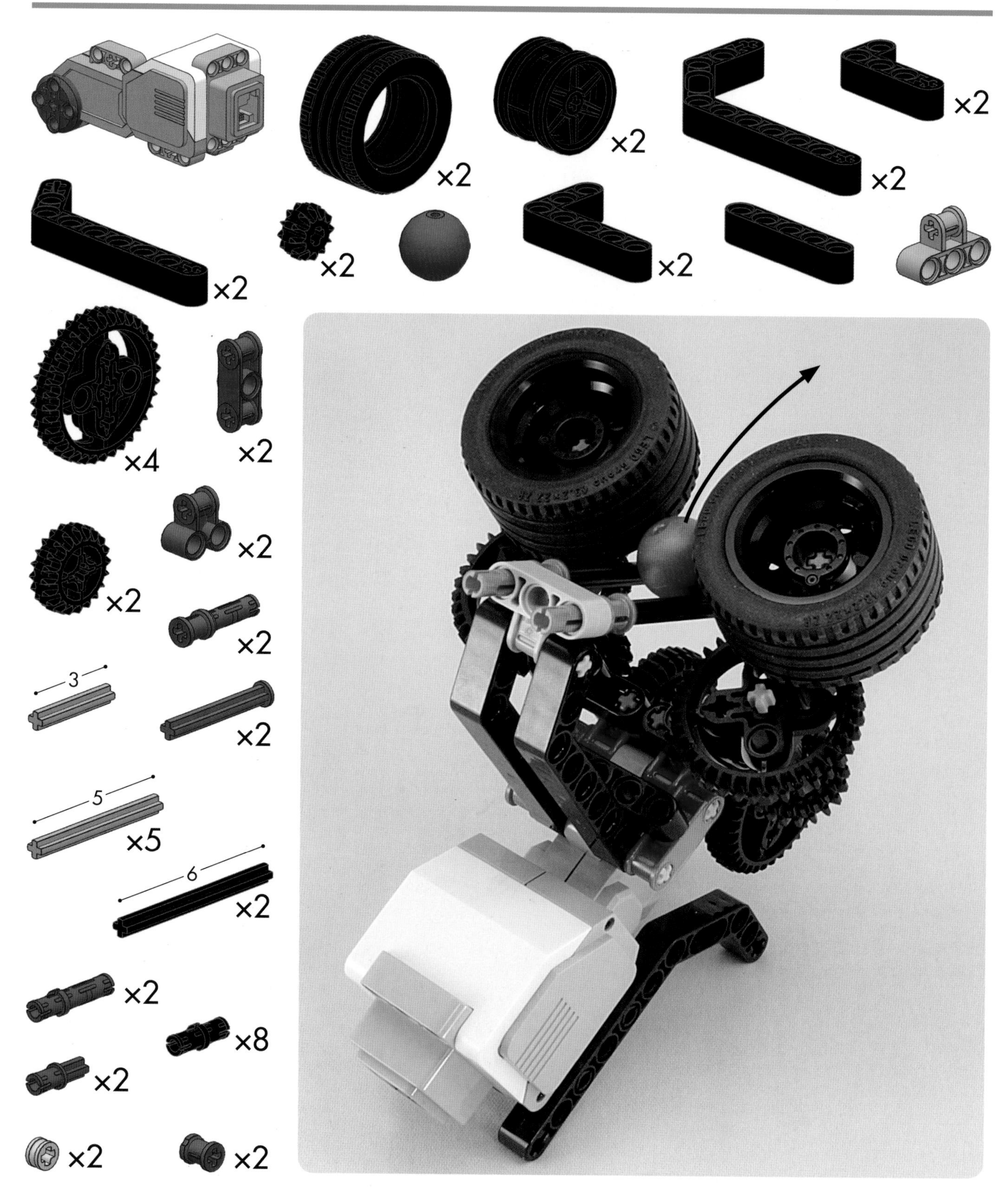

×2
×2
×2
×2
×2
×2
×2
×4
×2
×2
×2
×2
3
×2
5
×5
6
×2
×2
×8
×2
×2
×2

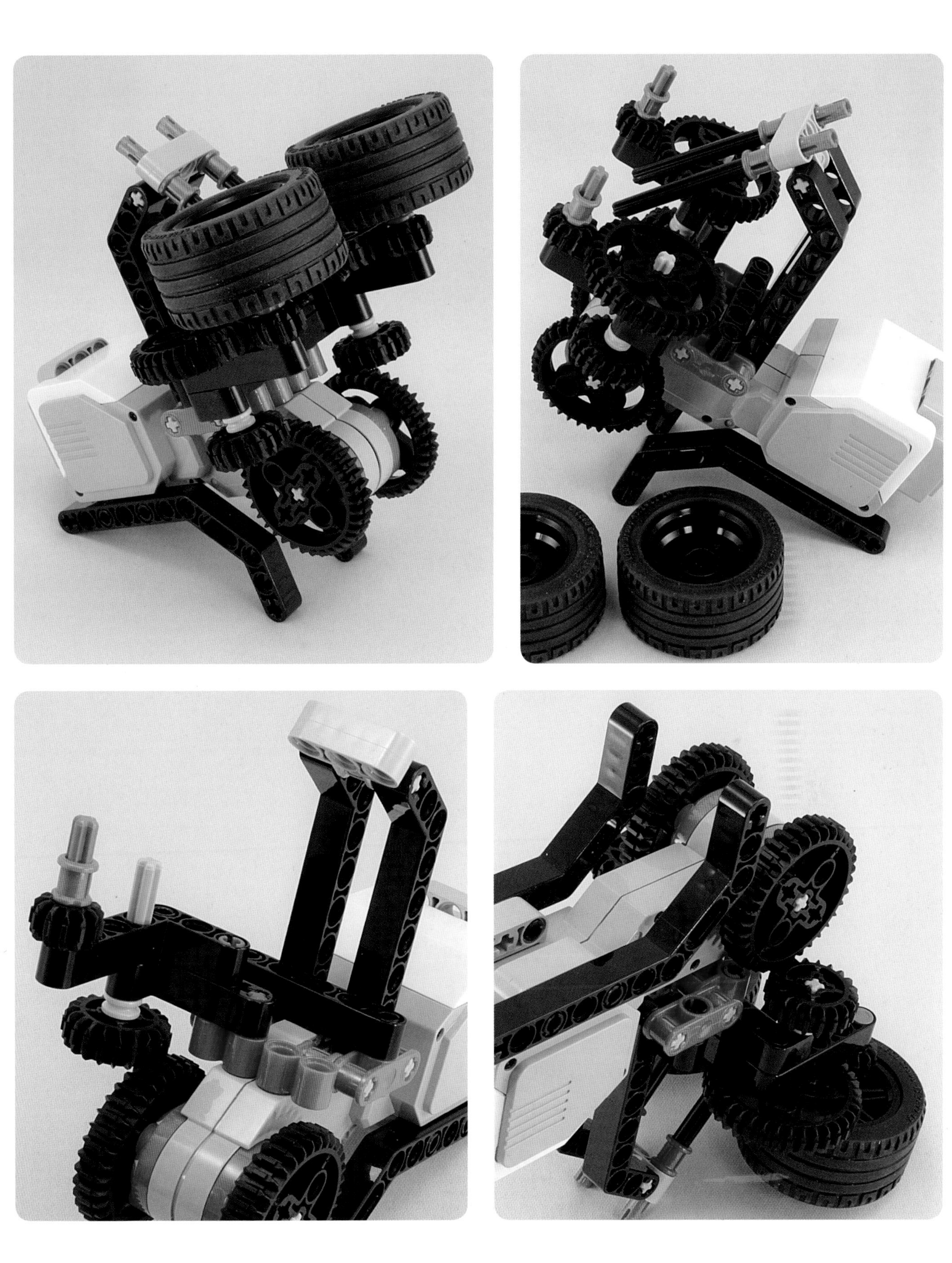

#147

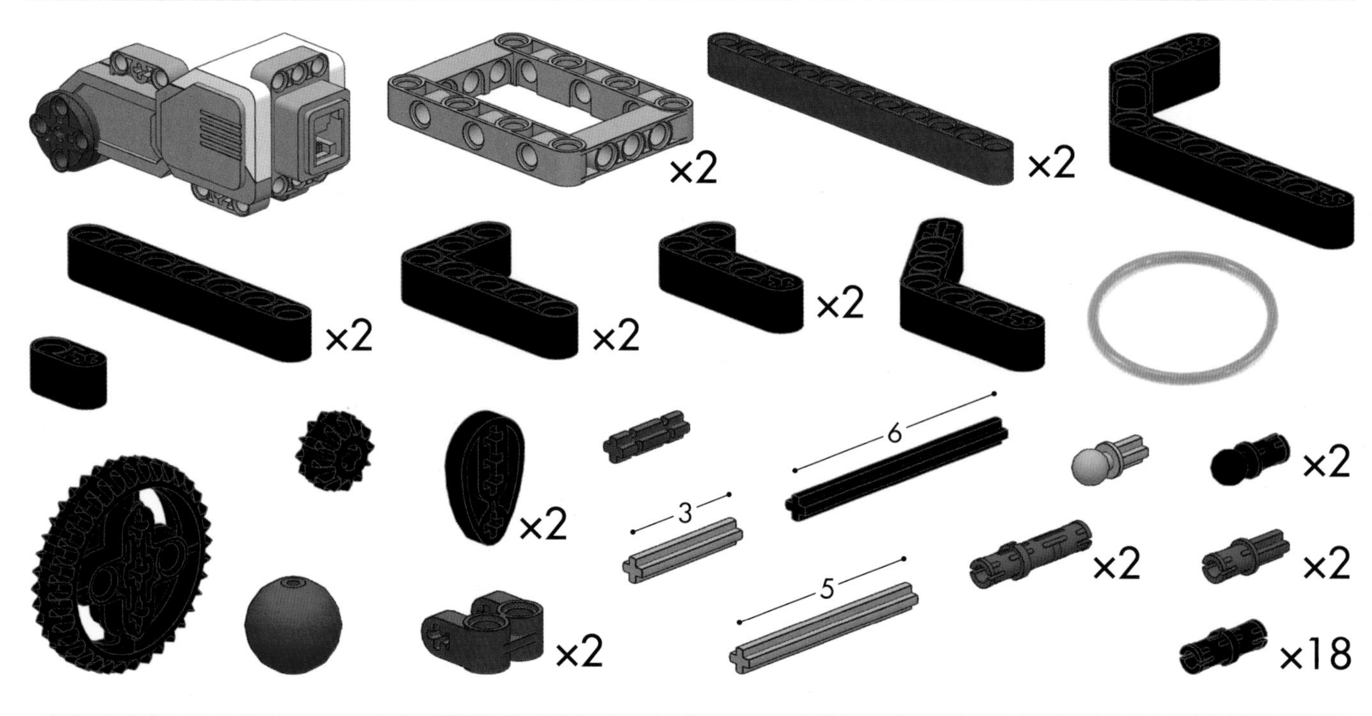
×2
×2
×2
×2
×2
×2
6
3
×2
5
×2
×2
×2
×2
×18

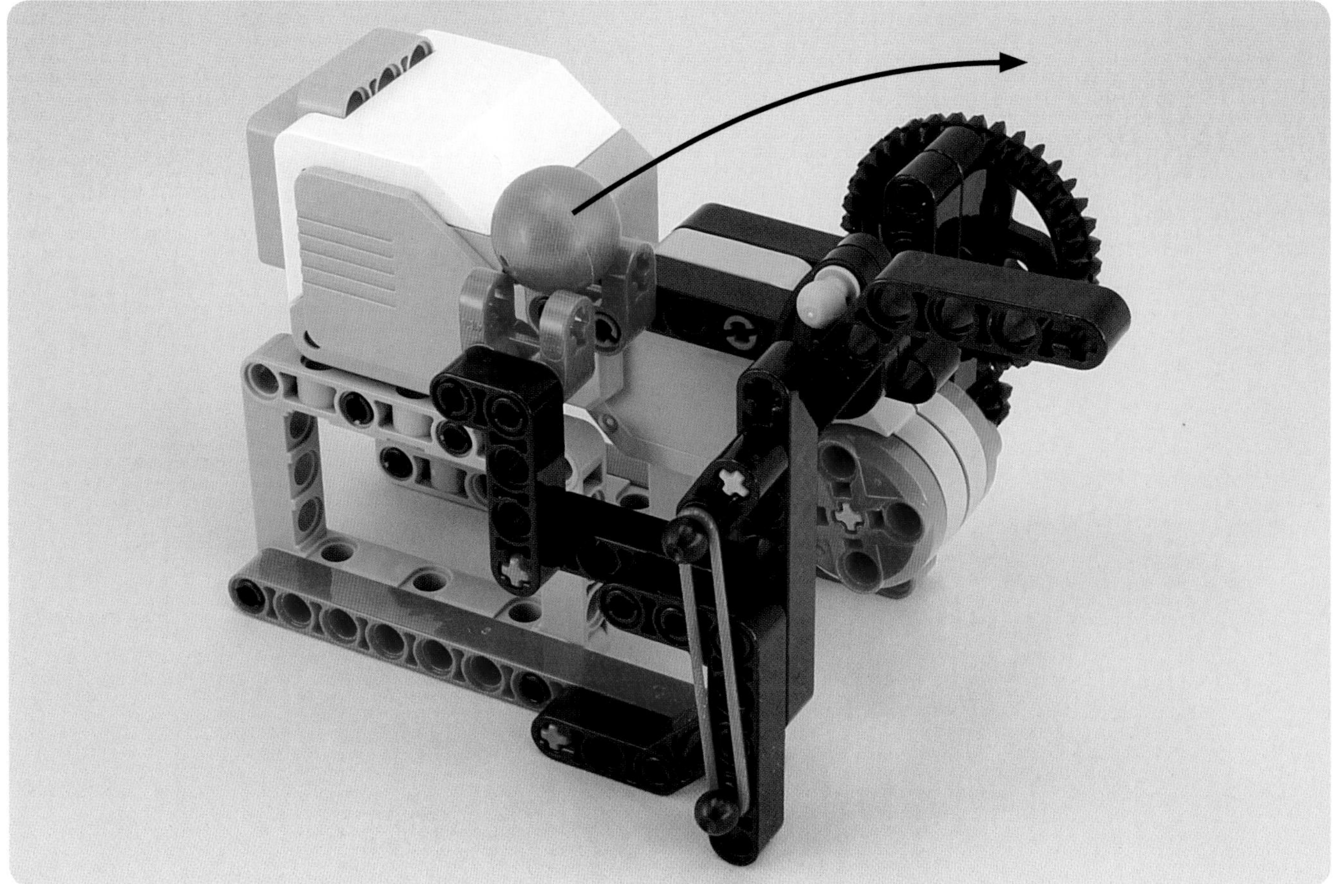

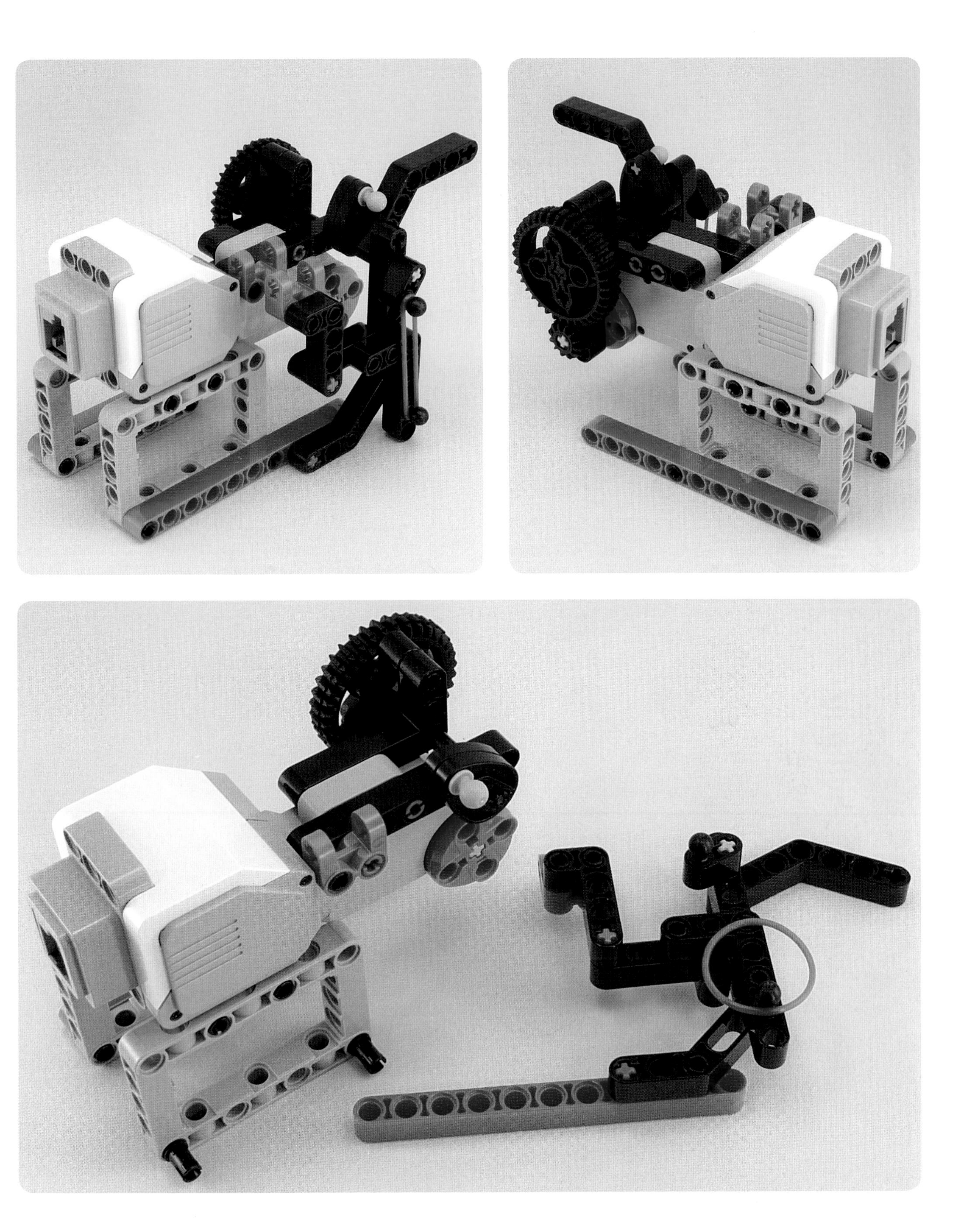

Automatic doors

#148

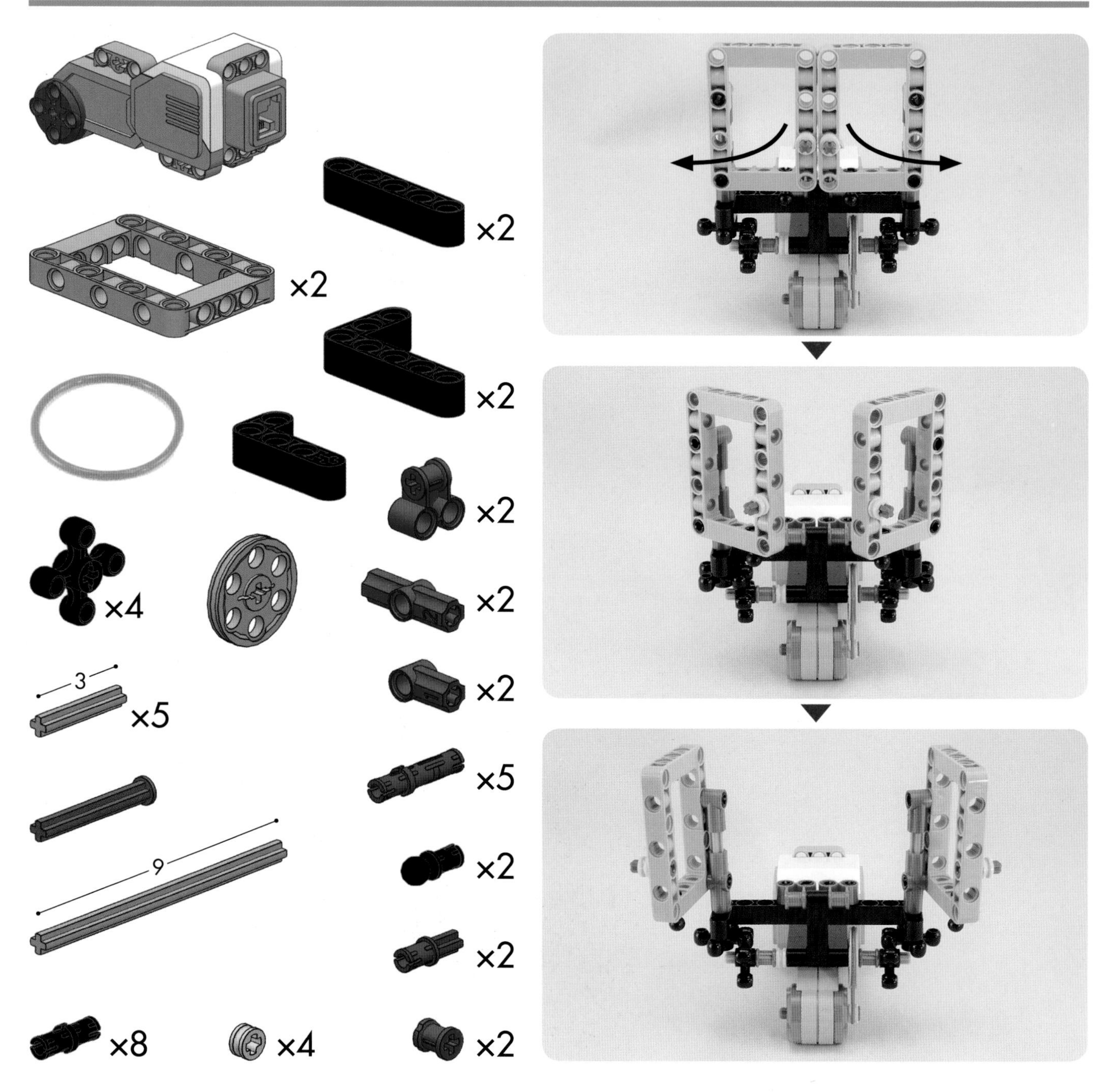

×2
×2
×2
×2
×4
×2
3
×5
×2
×5
9
×2
×2
×8
×4
×2

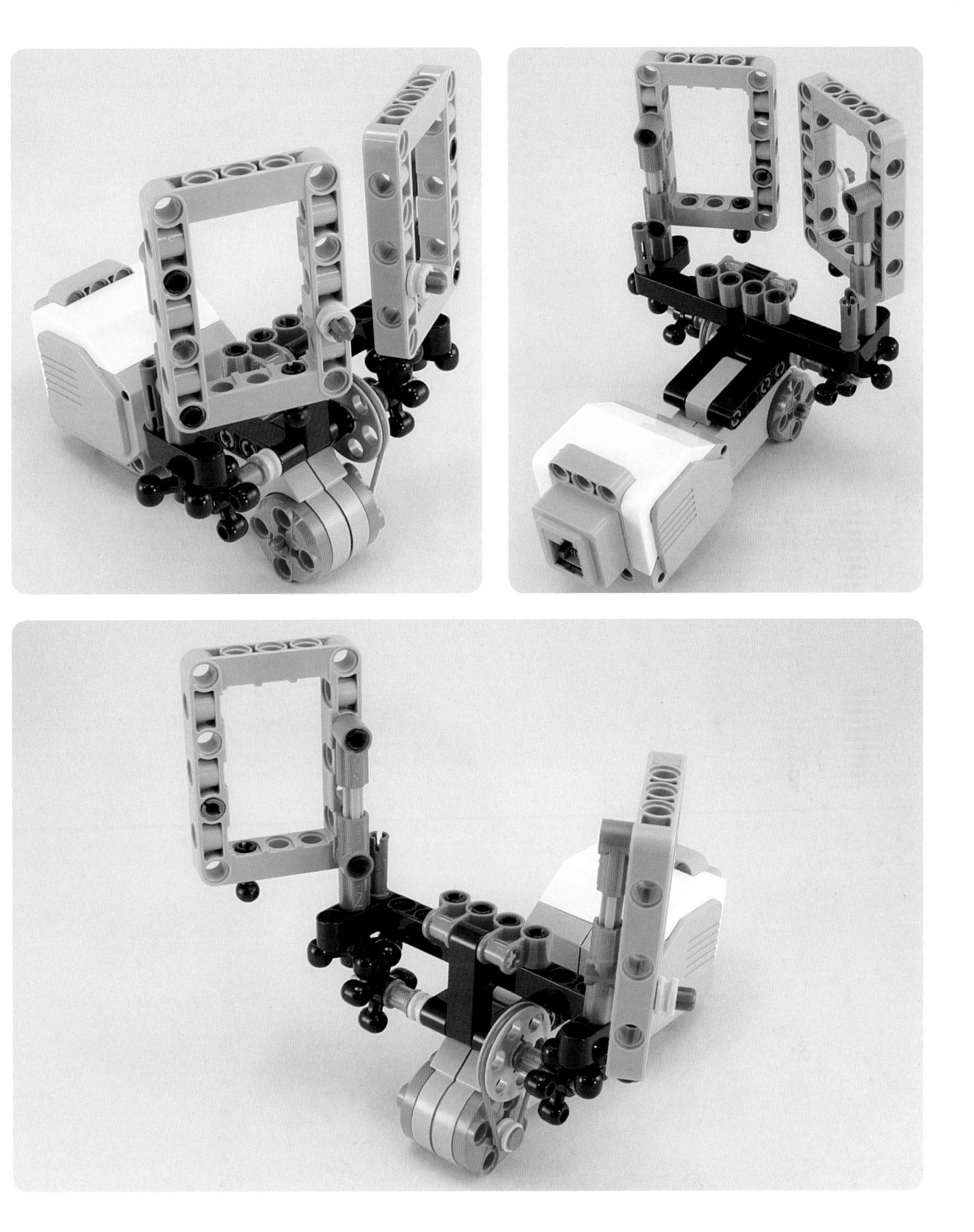

#149

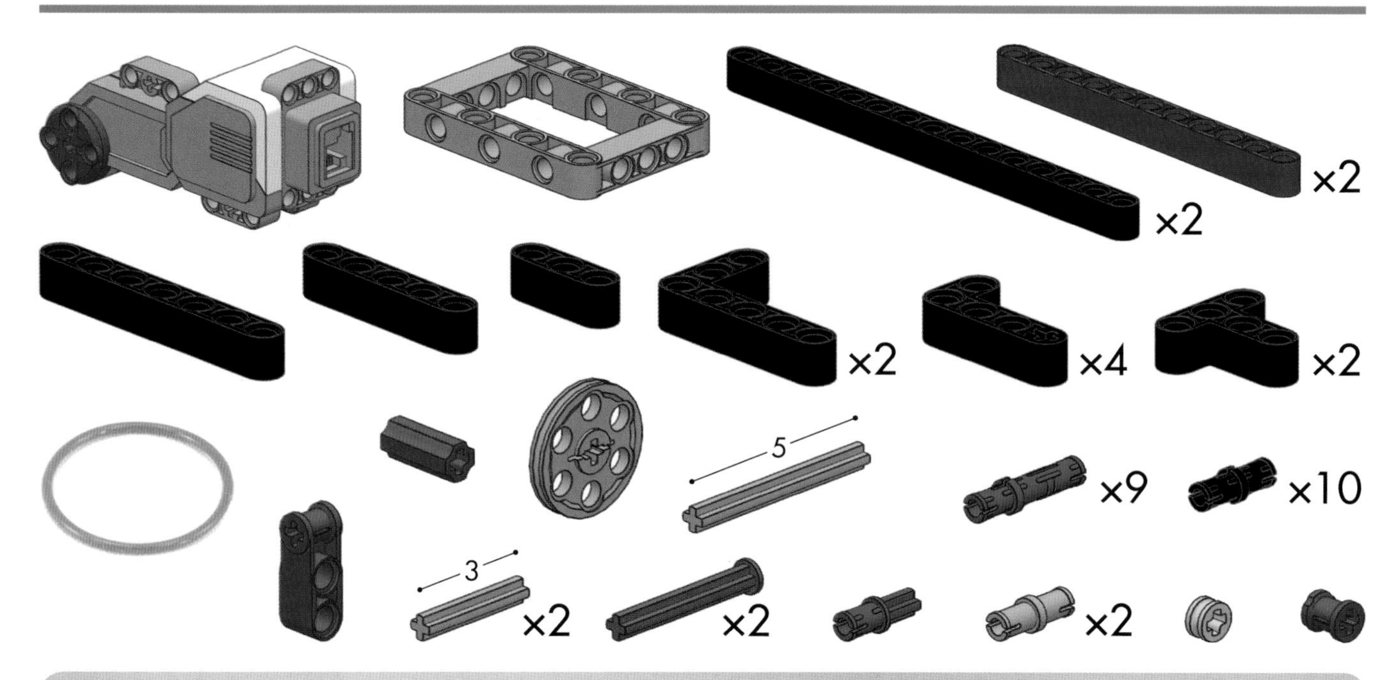
×2
×2
×2
×4
×2
5
×9
×10
3
×2
×2
×2

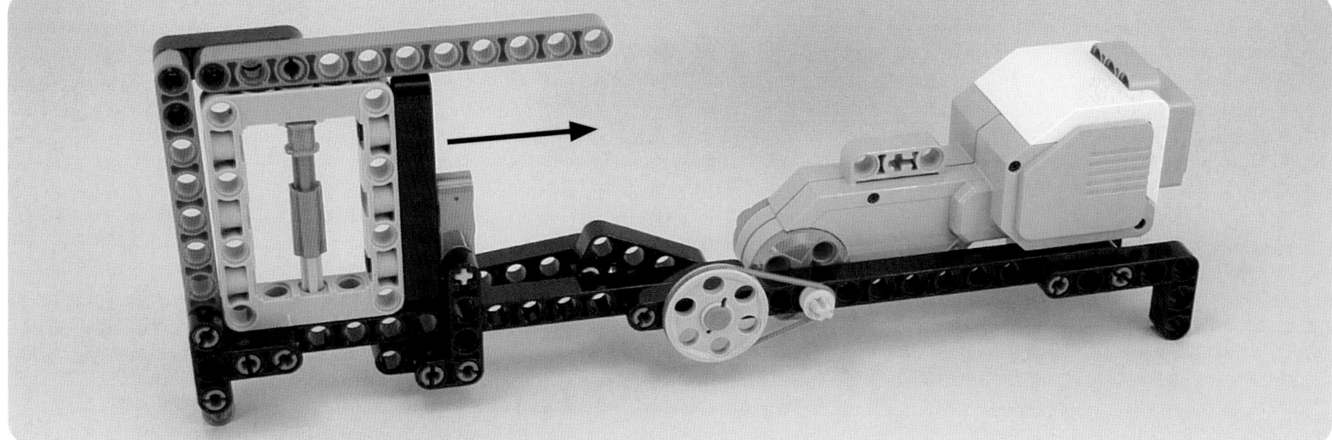

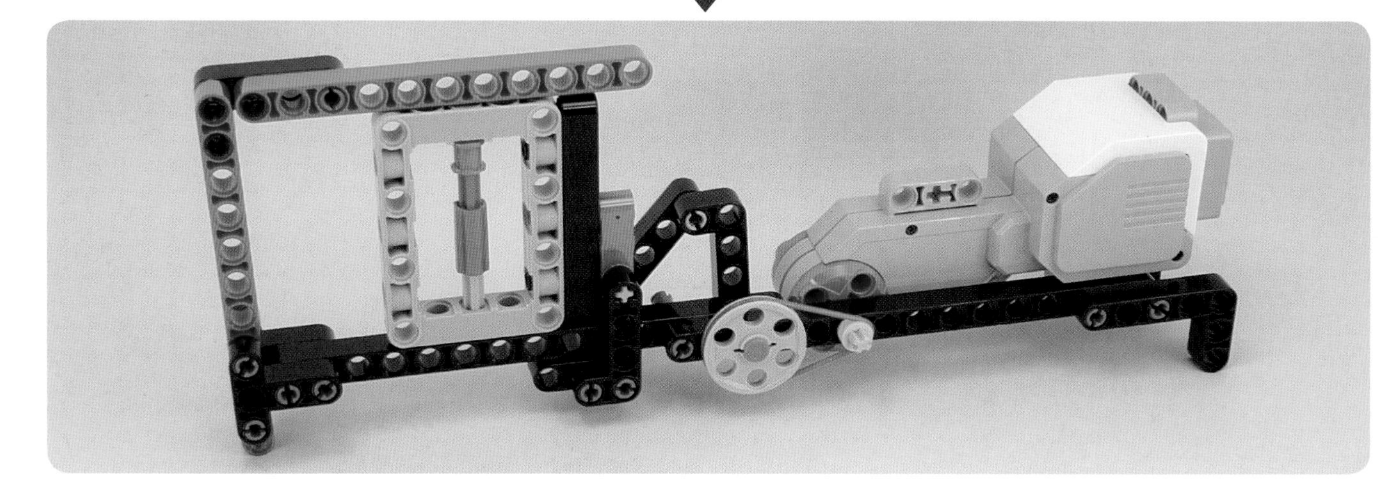

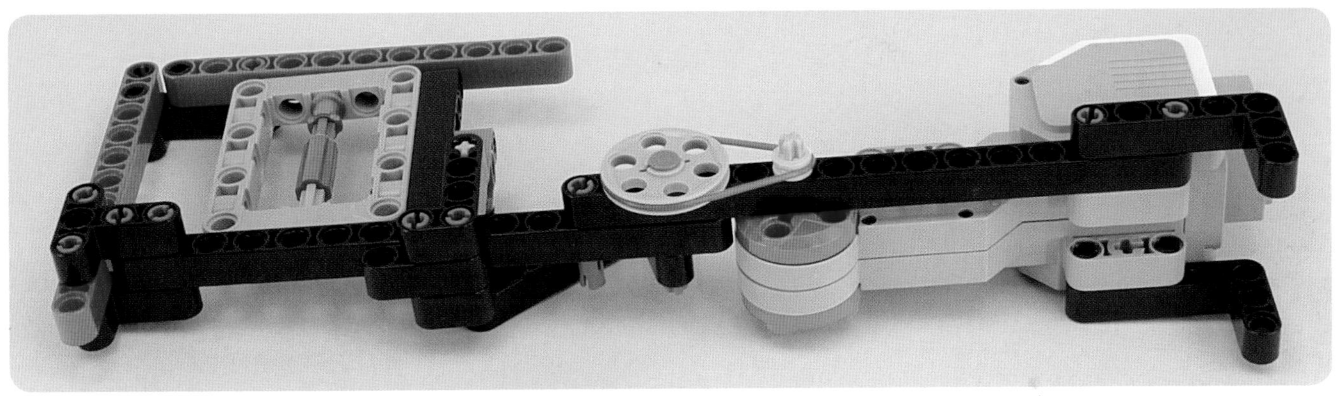

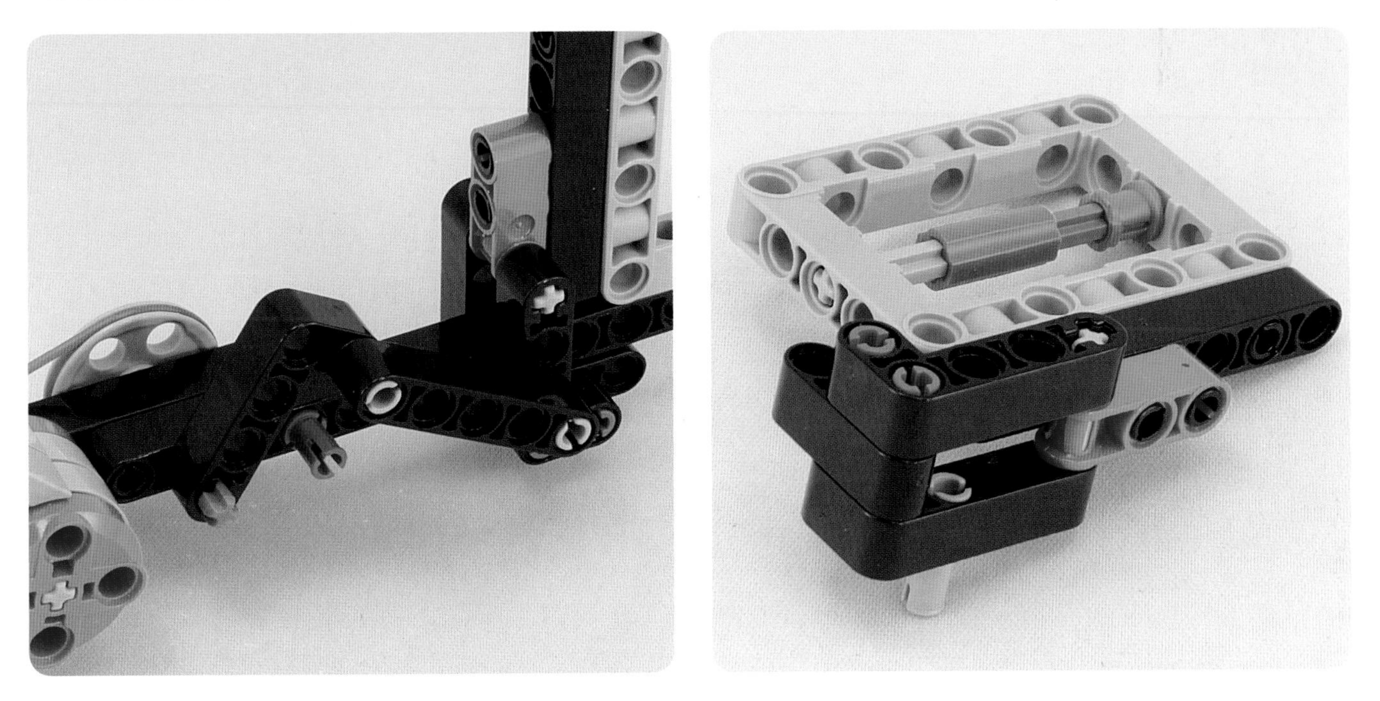

#150

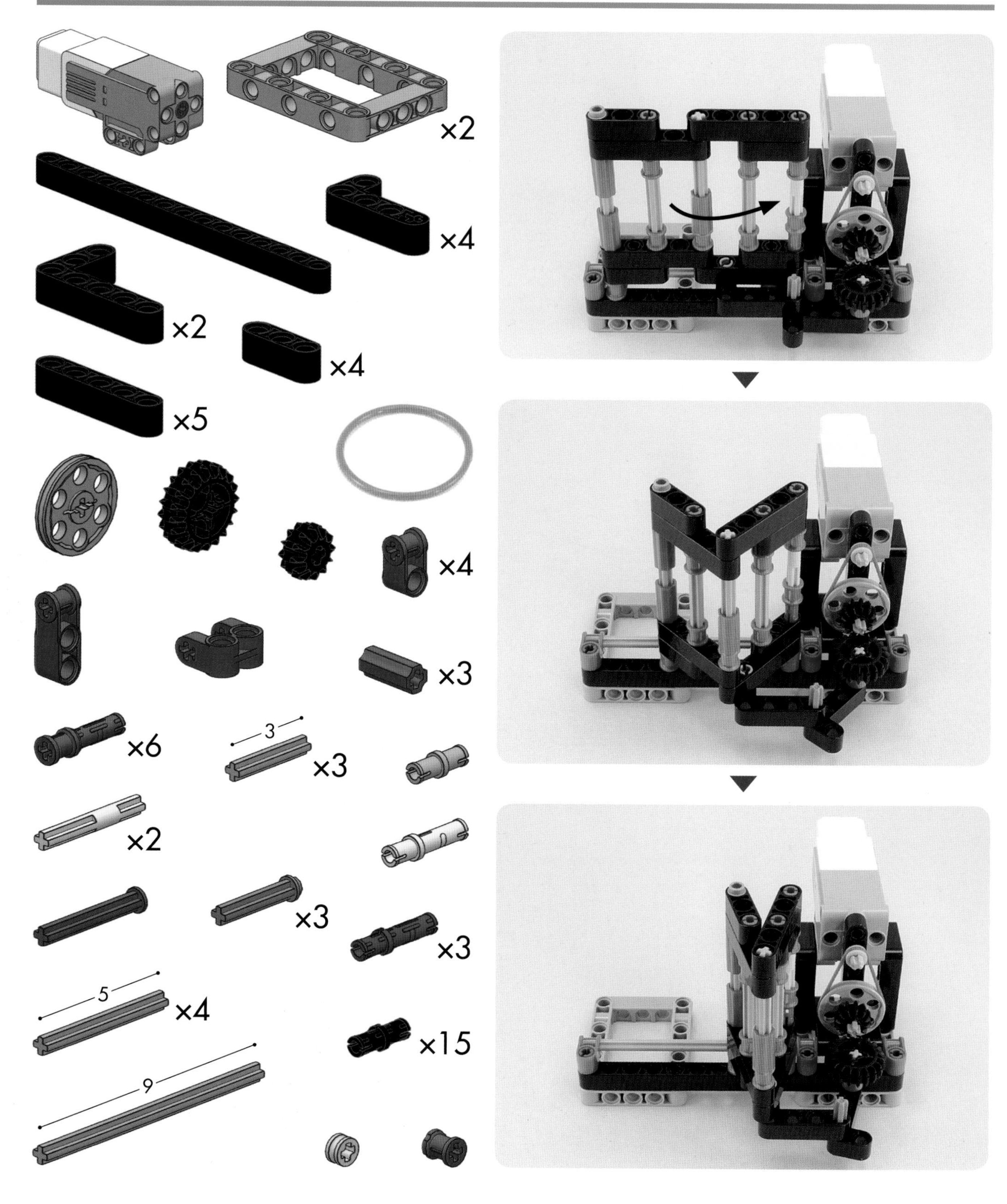
×2
×4
×2
×4
×5
×4
×3
×6
3
×3
×2
×3
×3
5
×4
×15
9

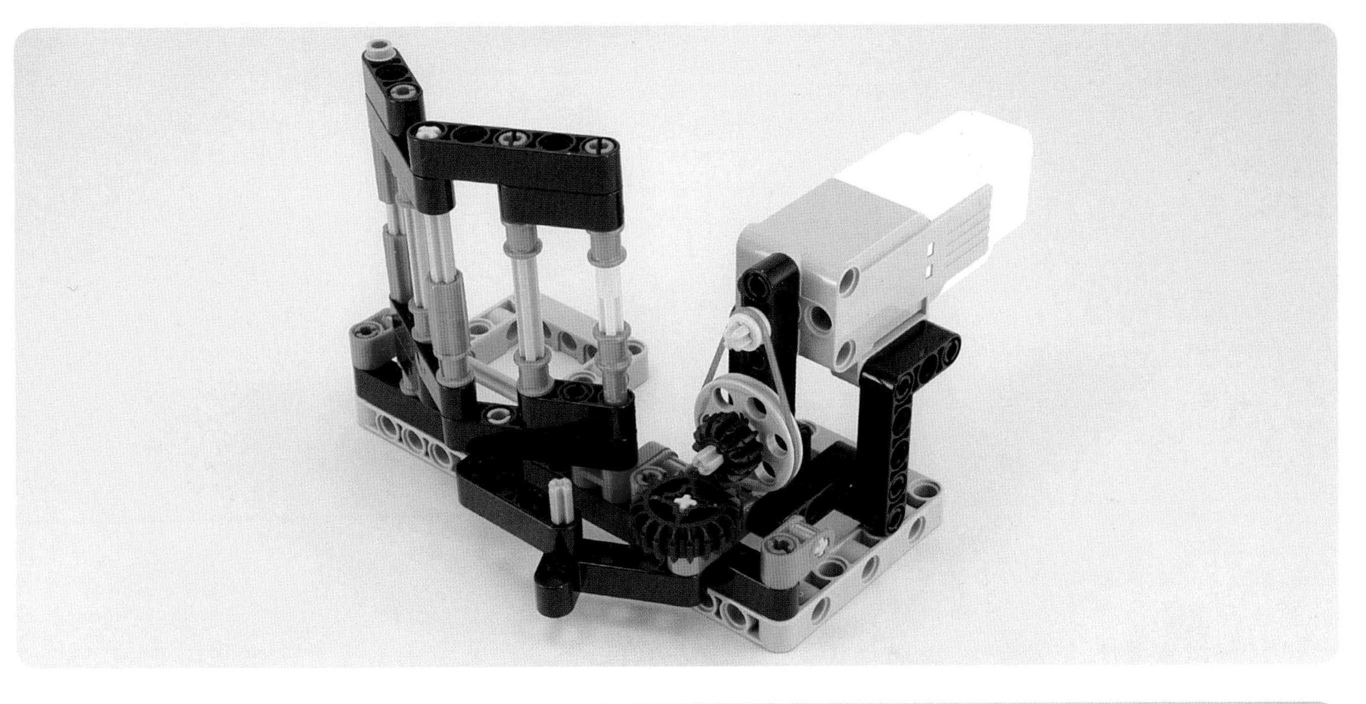

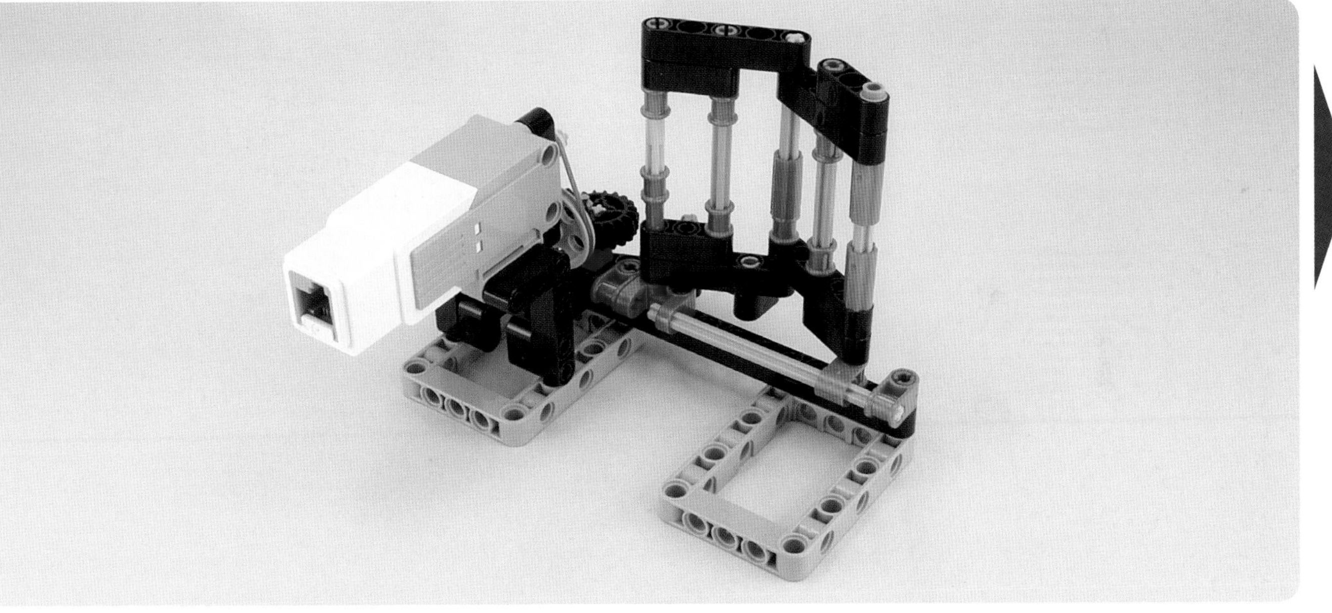

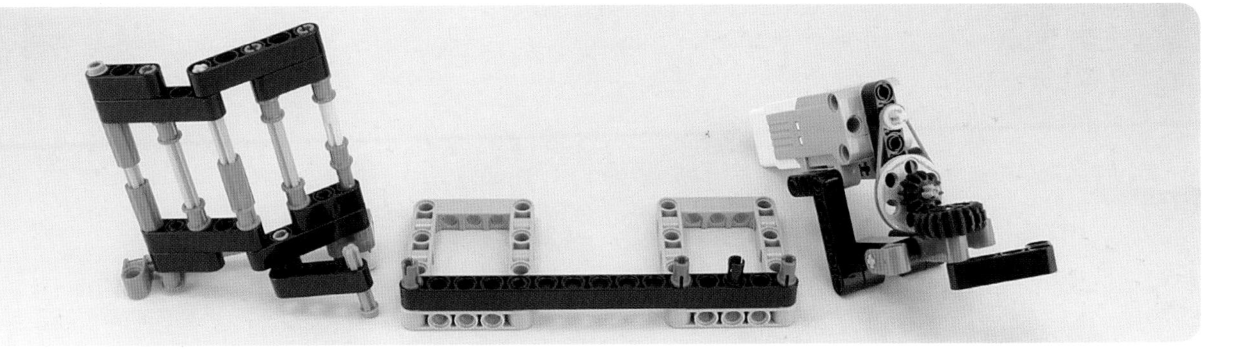

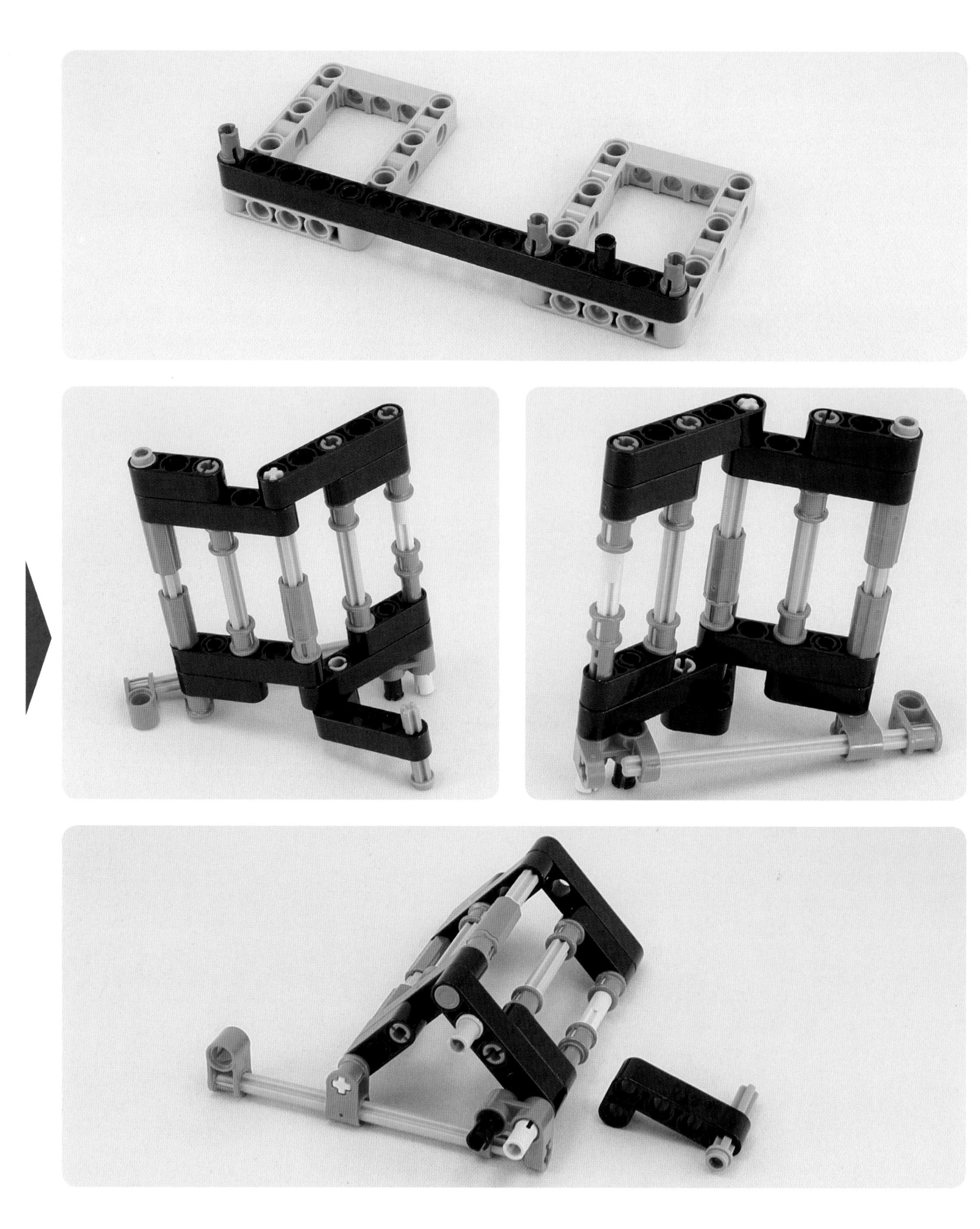

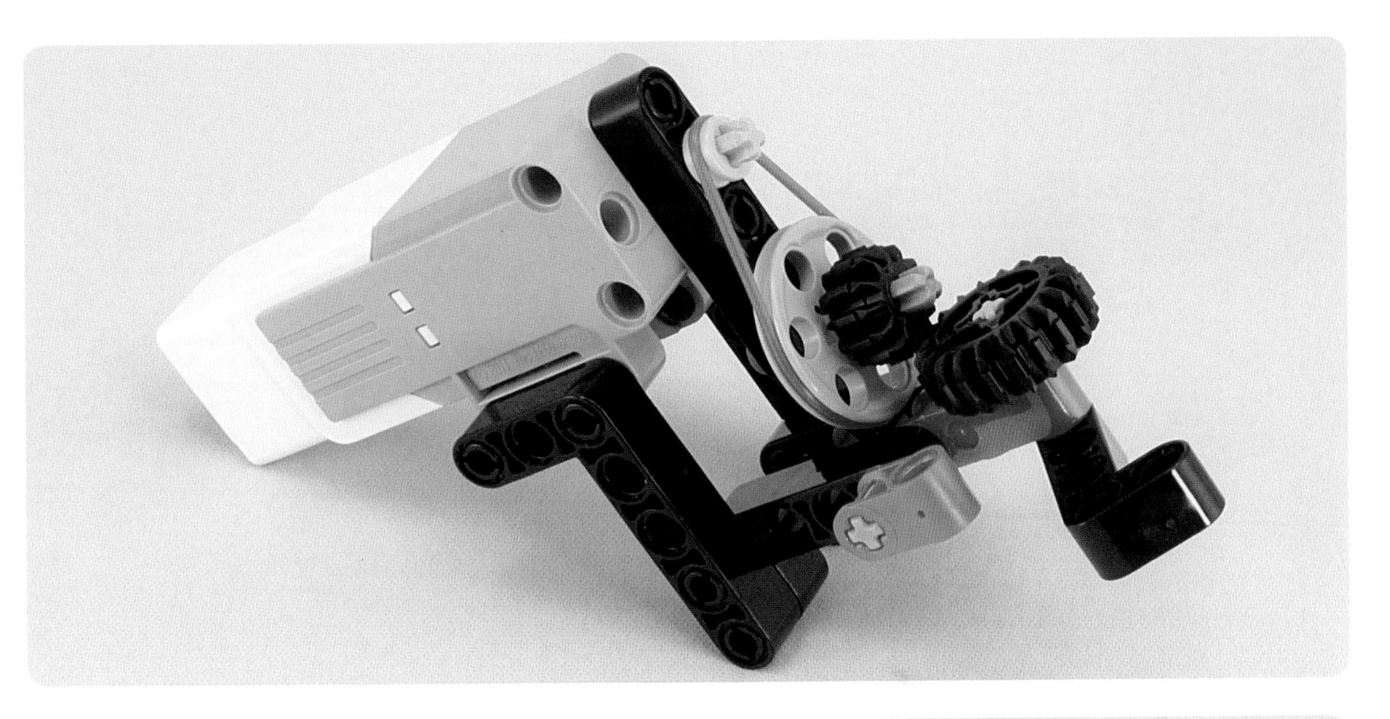

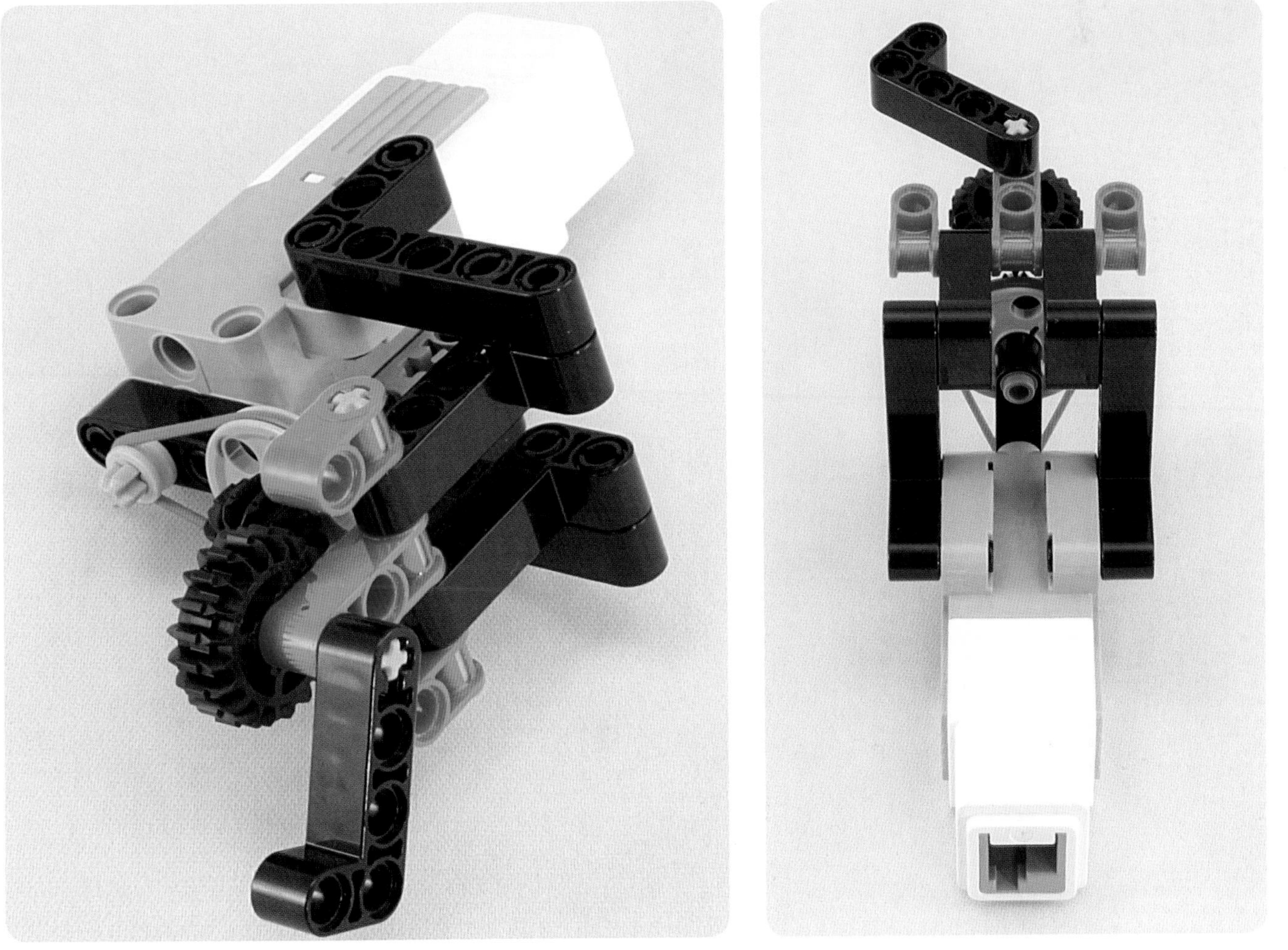

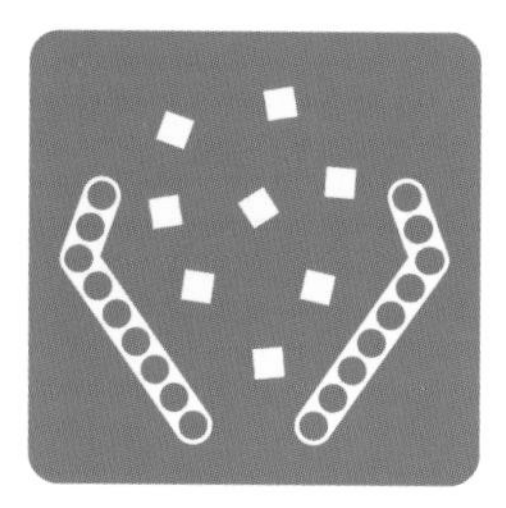

Raking up or out

#151

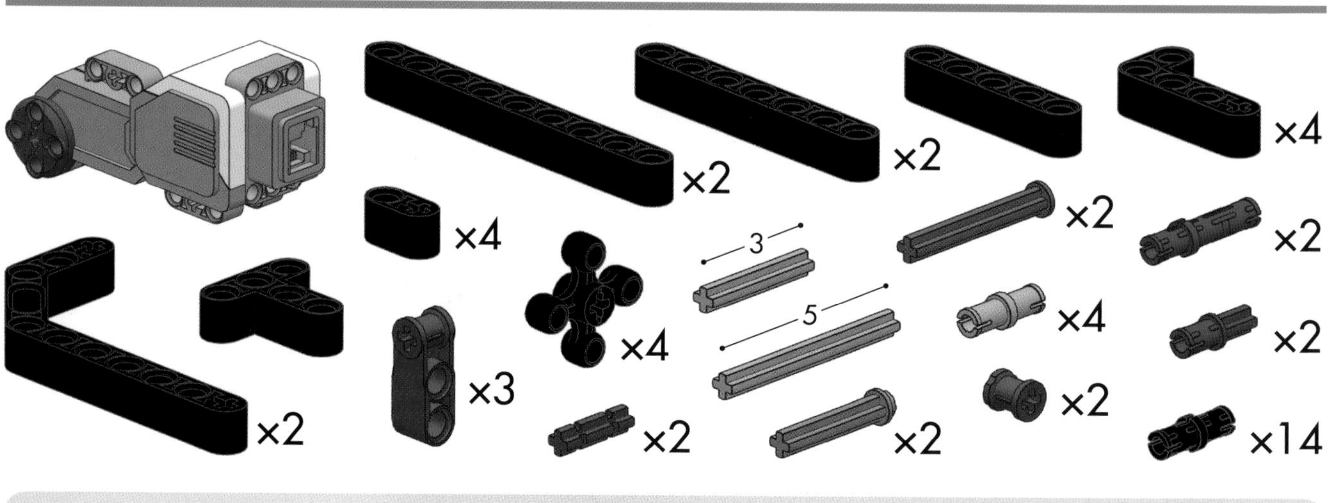
×2
×2
×4
×4
×2
×2
3
5
×2
×2
×3
×4
×4
×2
×2
×2
×2
×2
×14

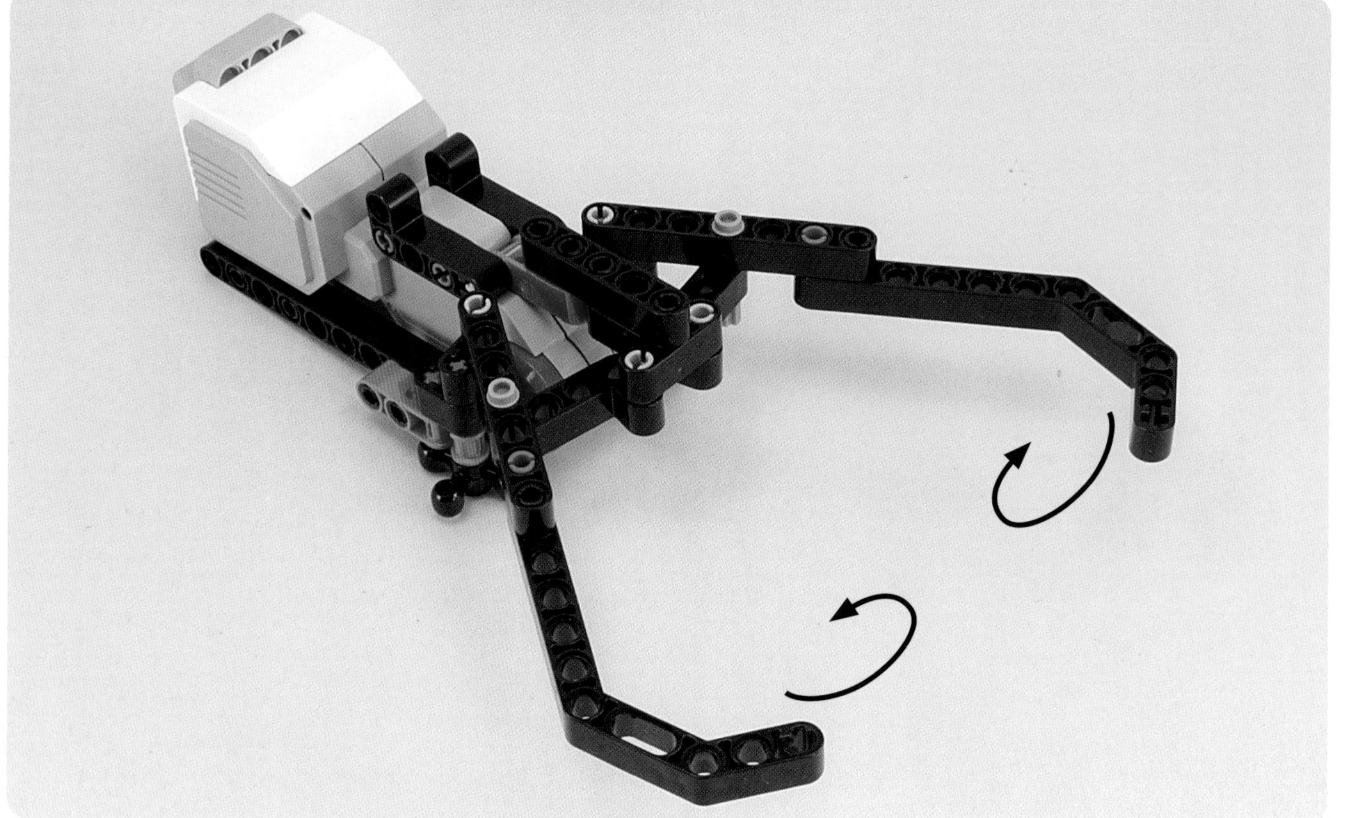

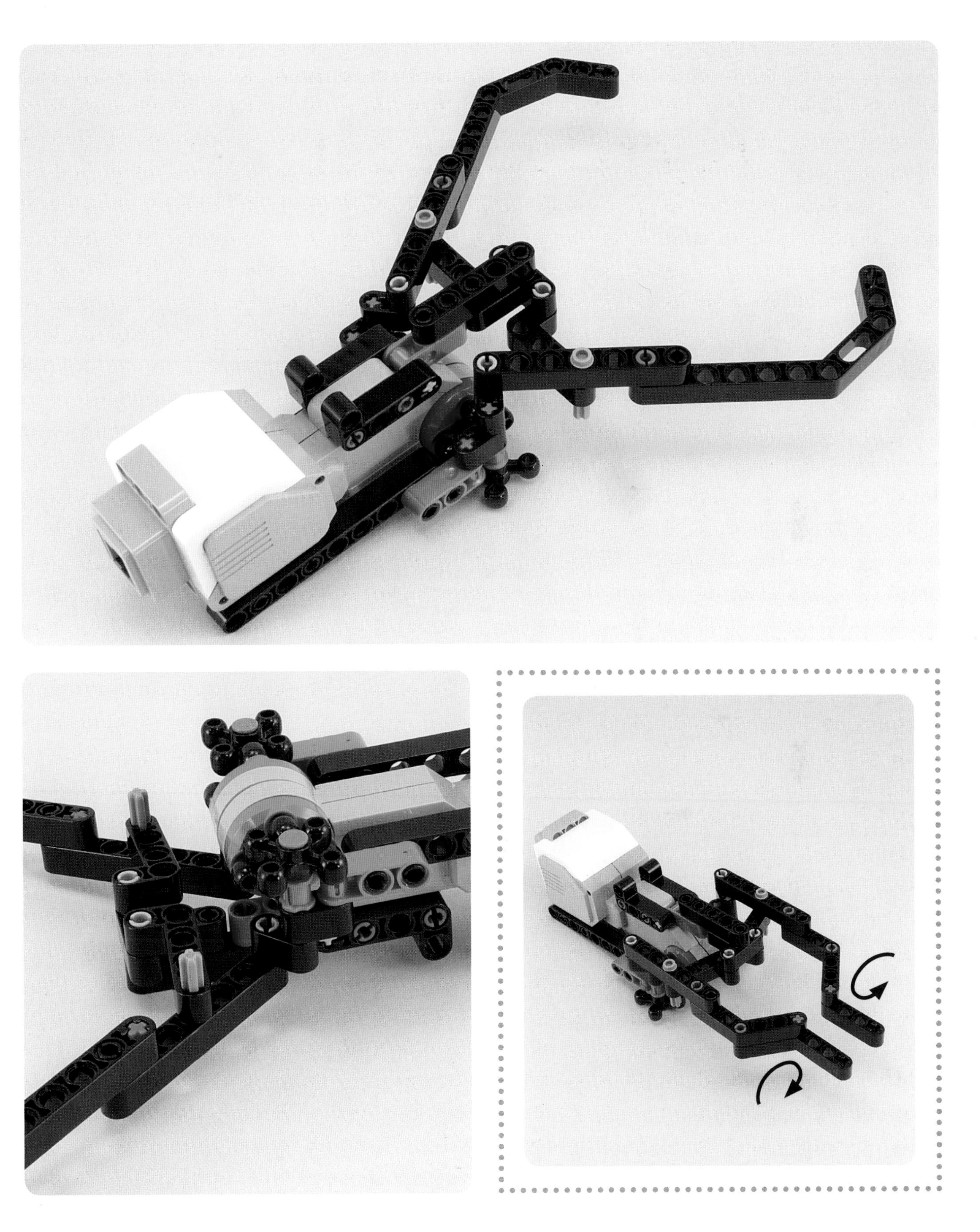

#152

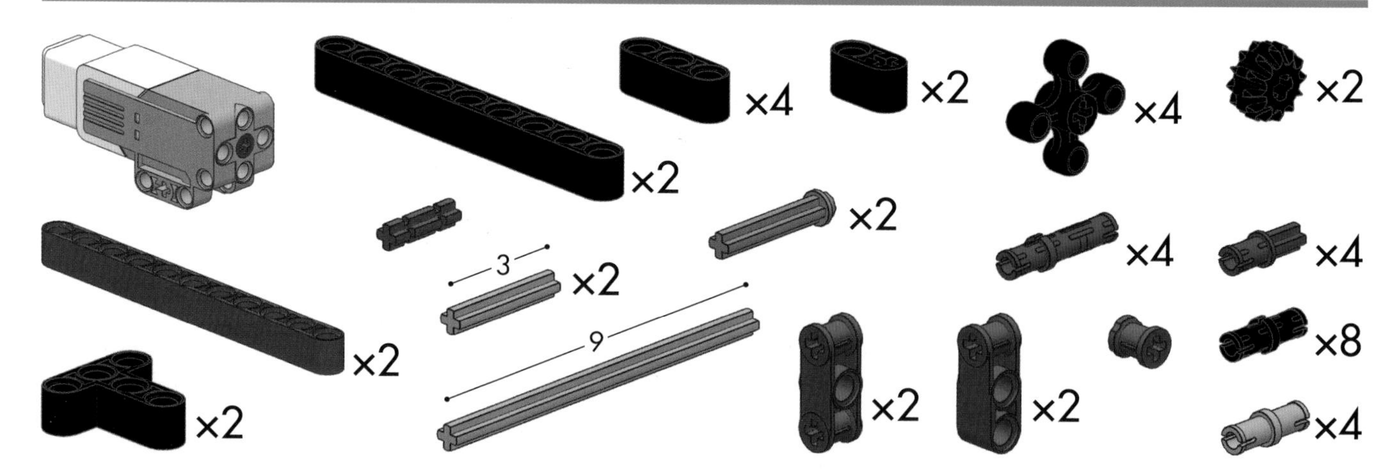
×4
×2
×4
×2
×2
×2
×2
×4
×4
3
×2
×2
9
×2
×2
×8
×4

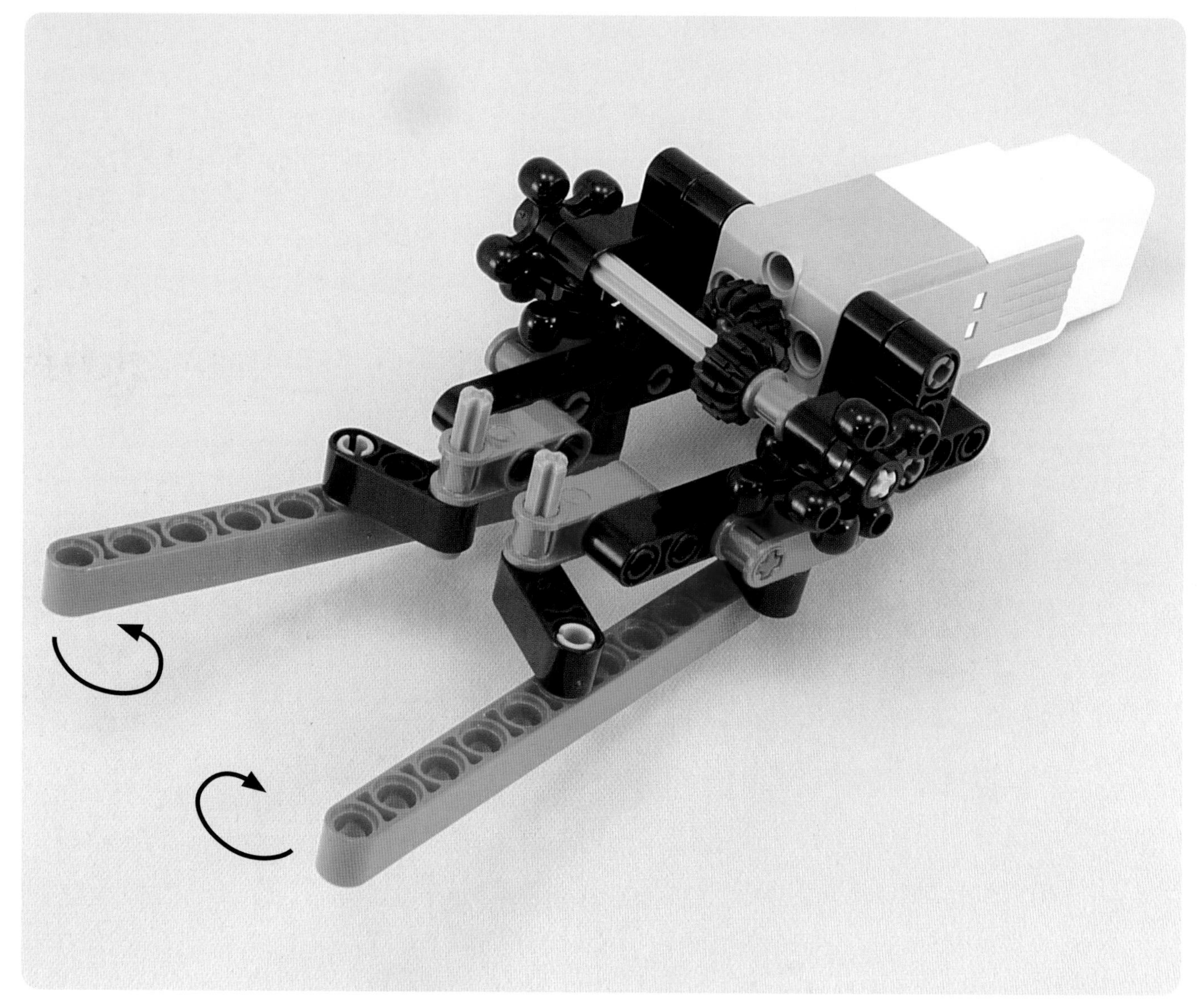

Creating wind

#153

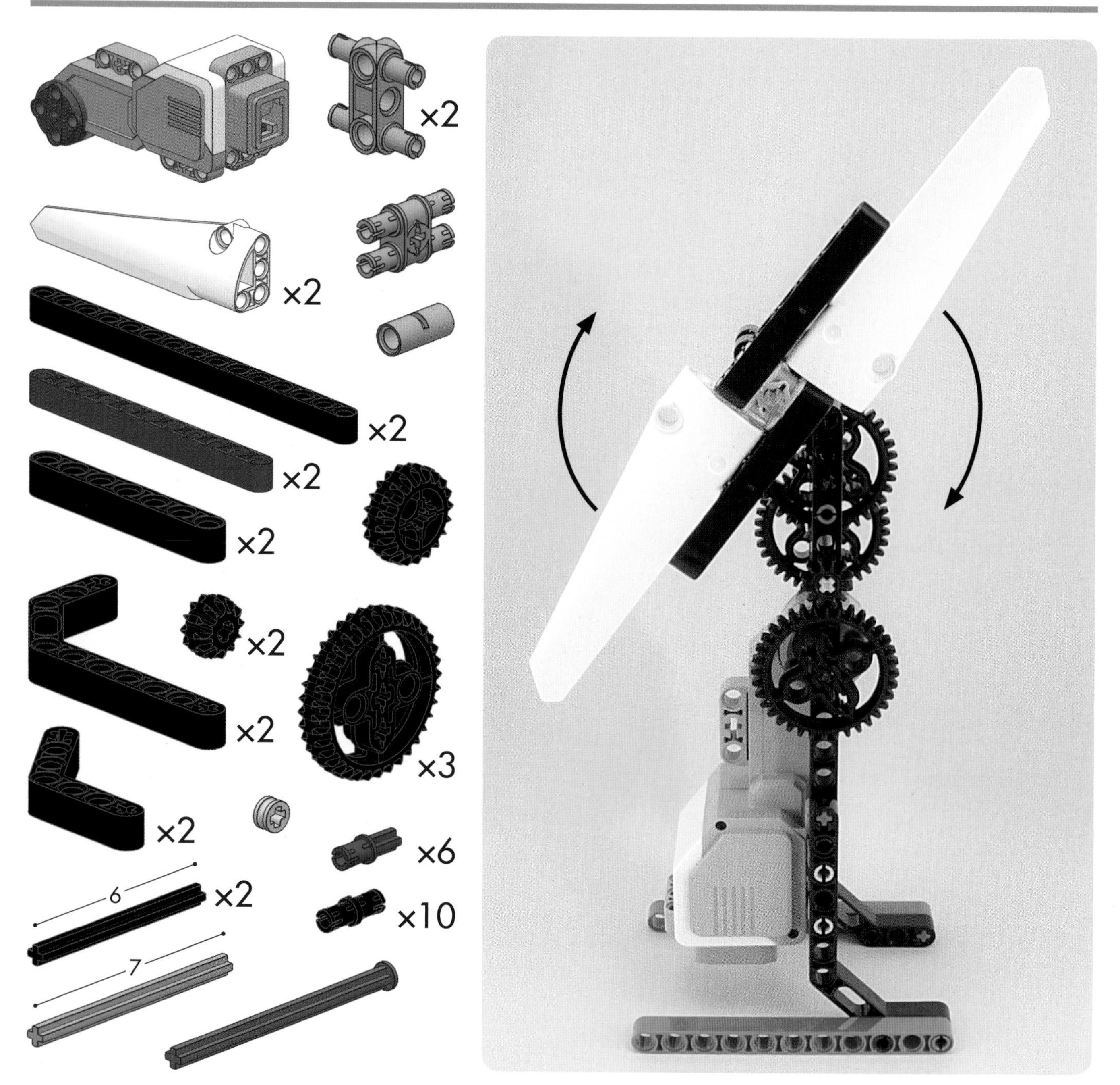
×2
×2
×2
×2
×2
×2
×2
×3
×2
×6
6
×2
×10
7

#154

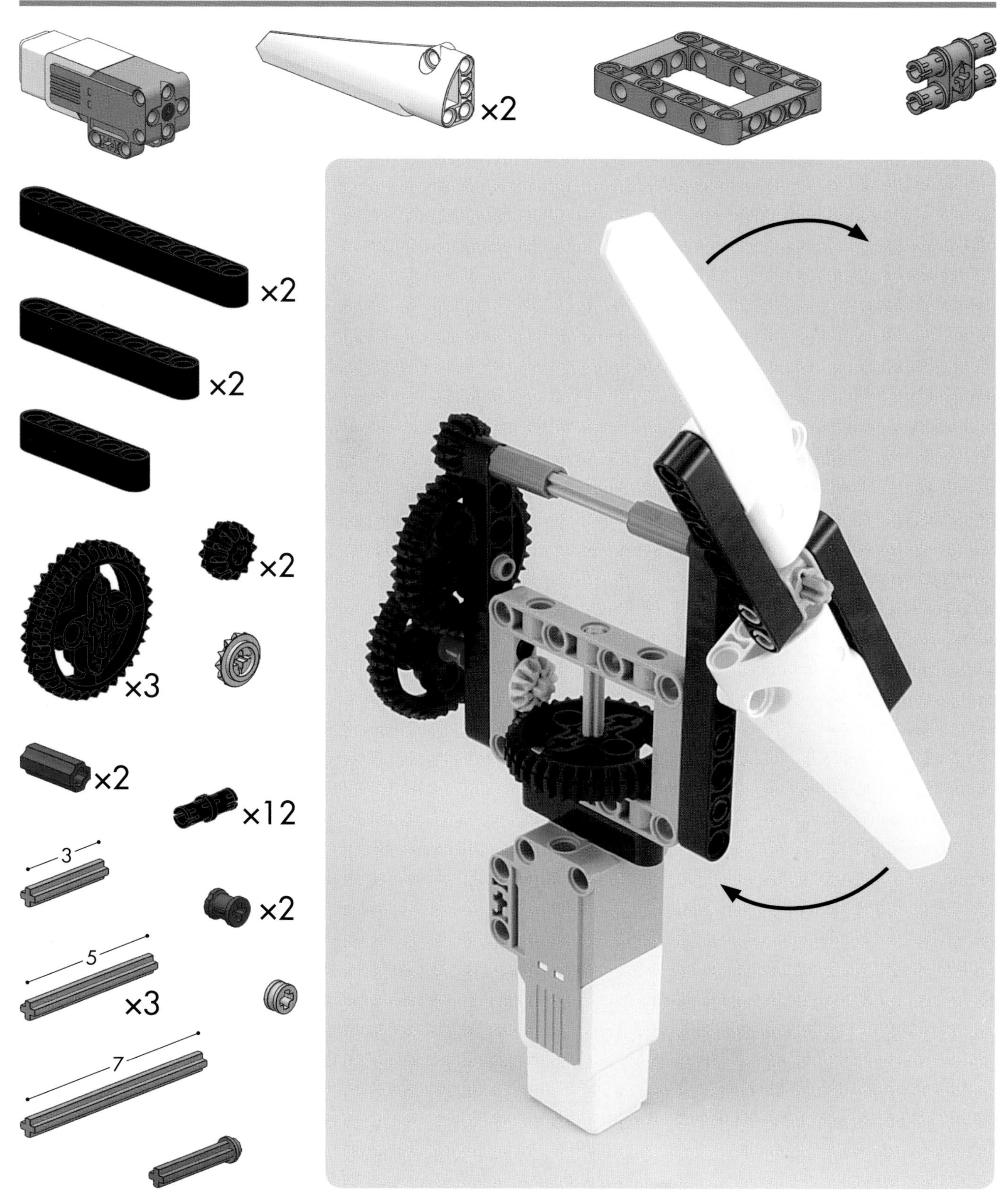
×2
×2
×2
×2
×3
×2
×2
×12
3
×2
5
×3
7

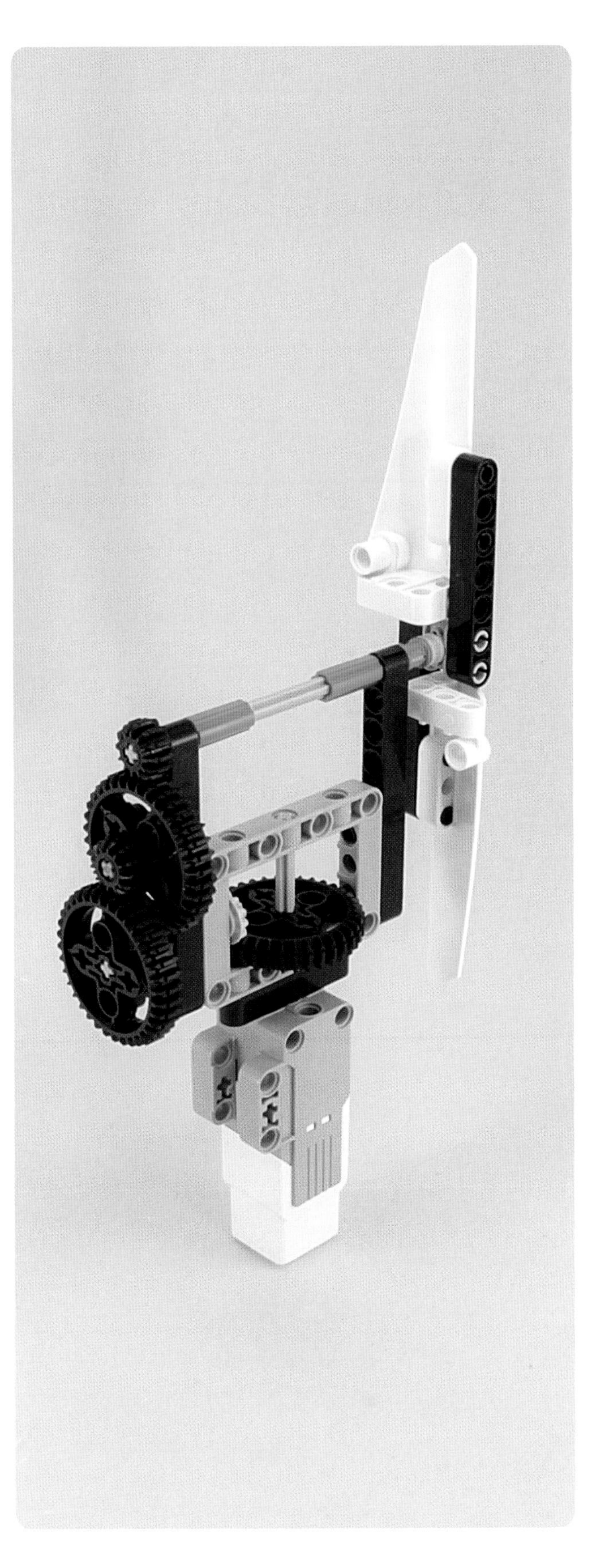

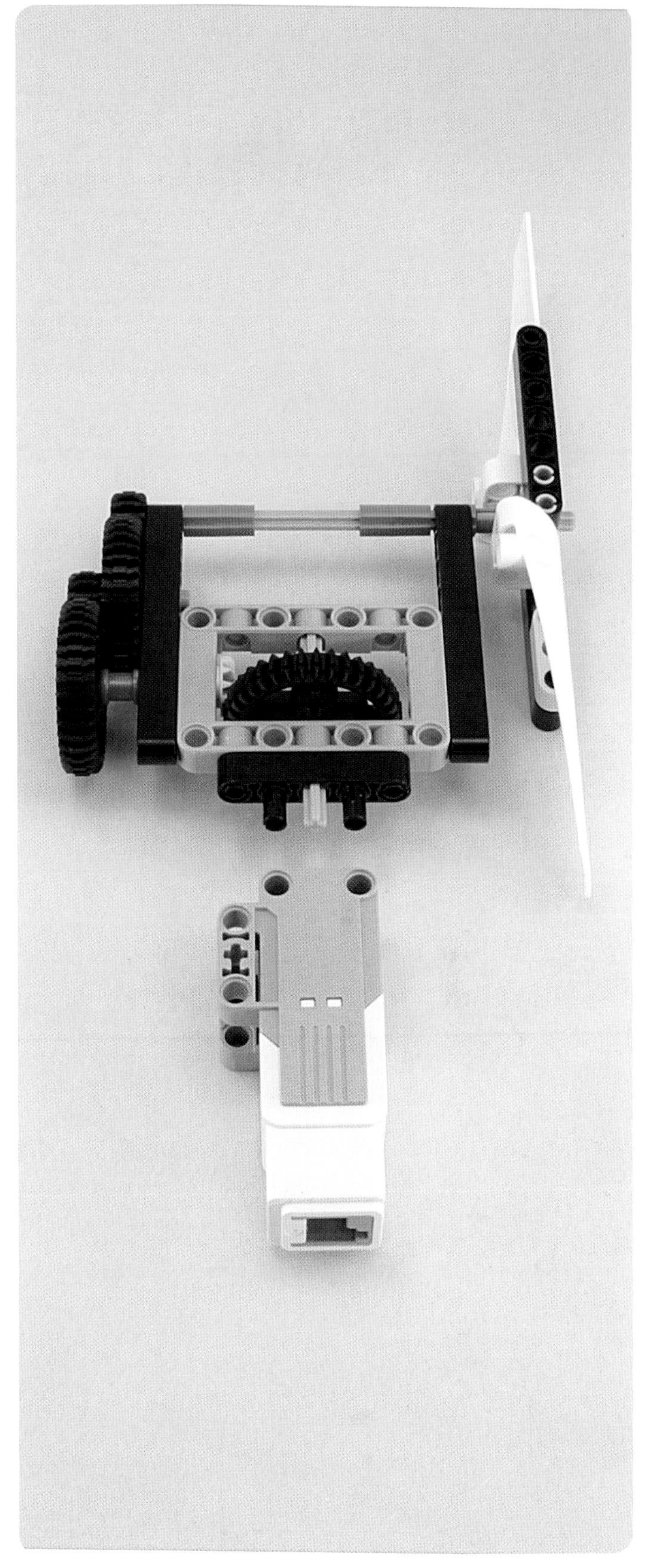

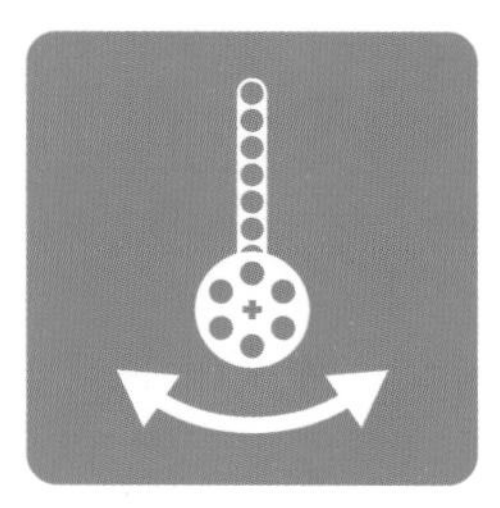

Swinging a pendulum

#155

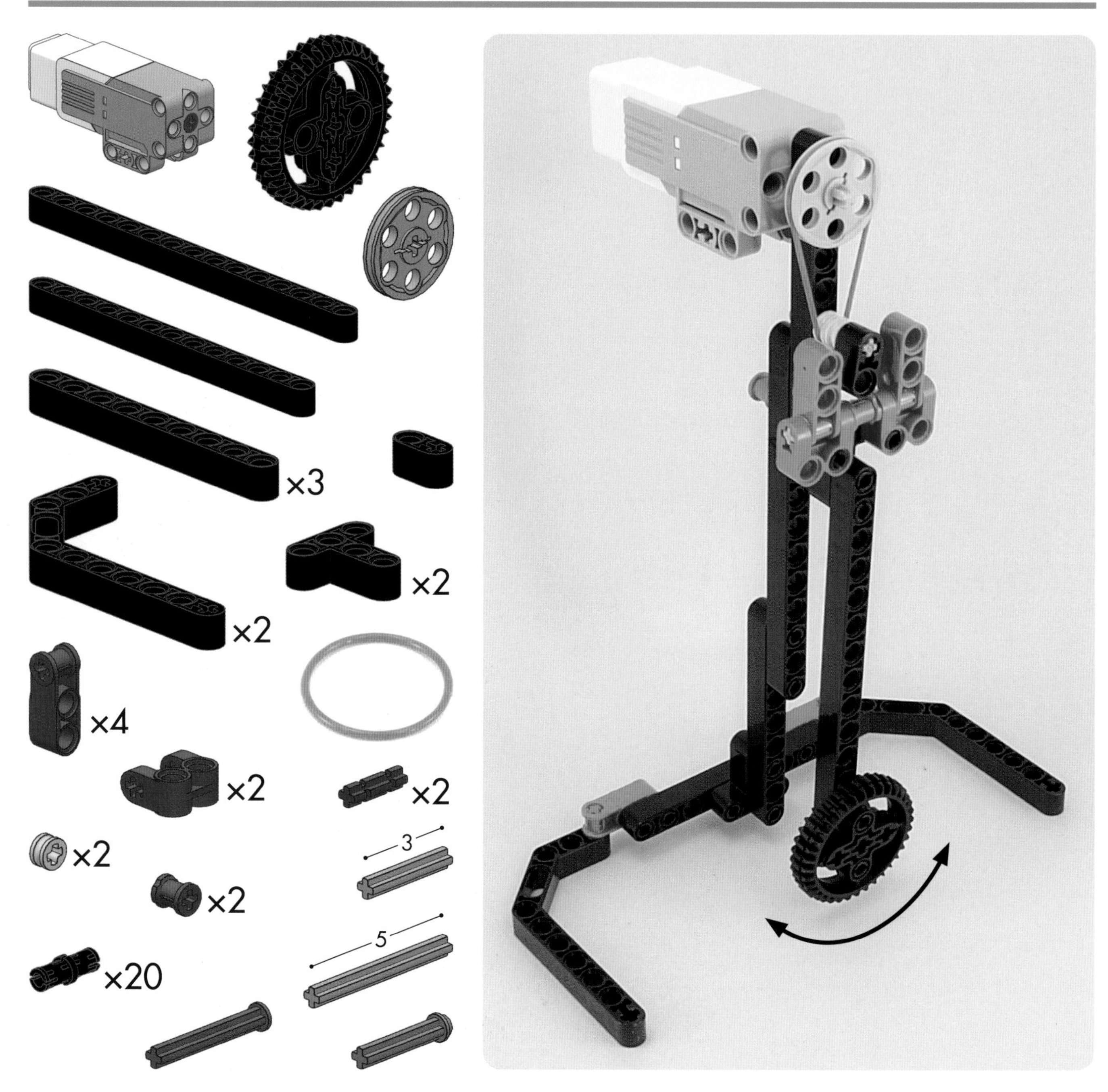
×3
×2
×2
×4
×2
×2
×2
×2
3
×20
5

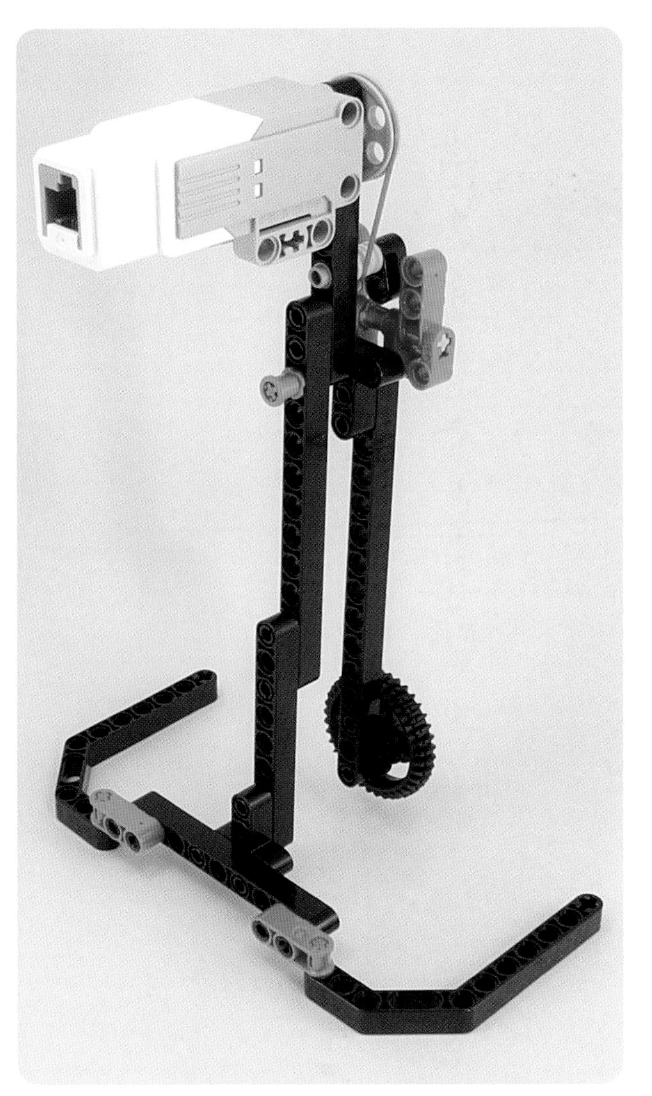

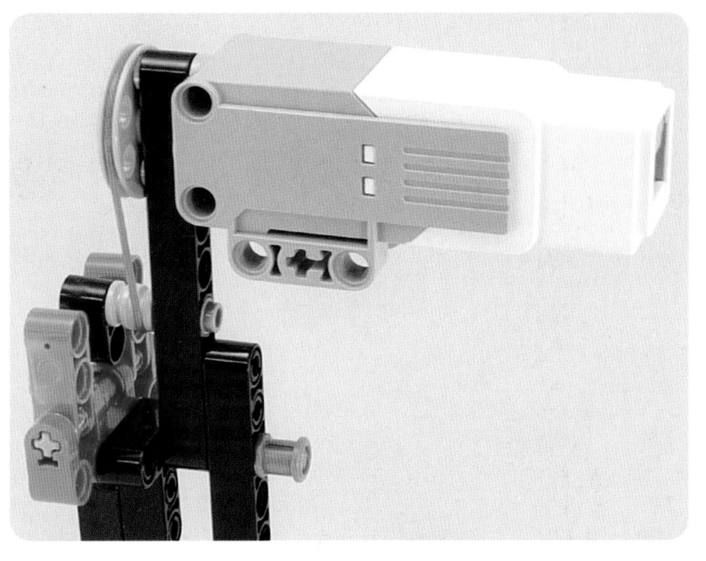

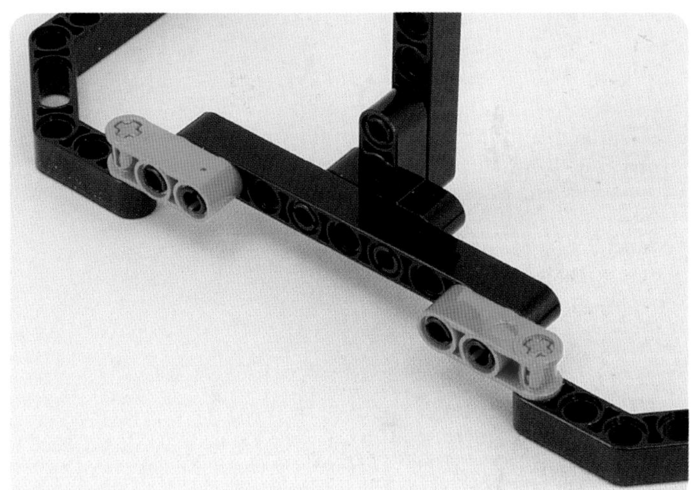

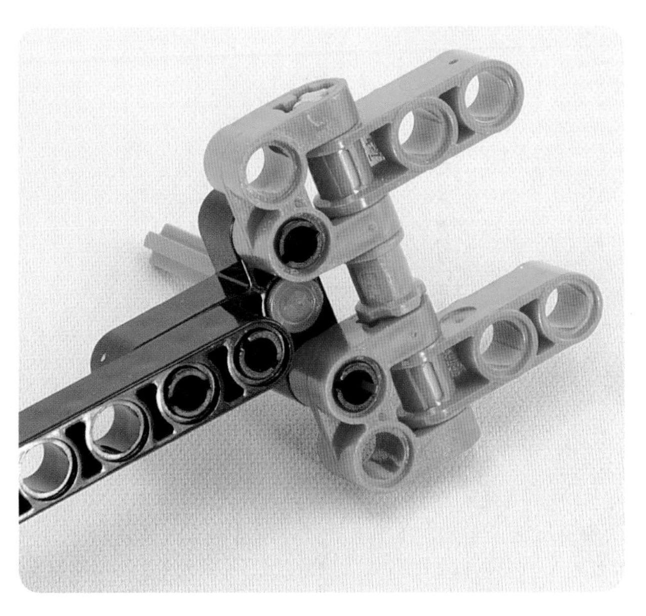

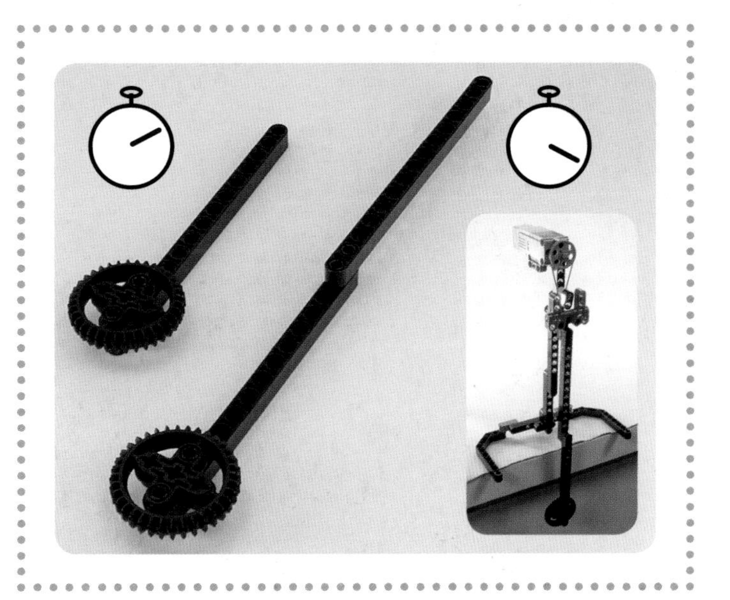

#156

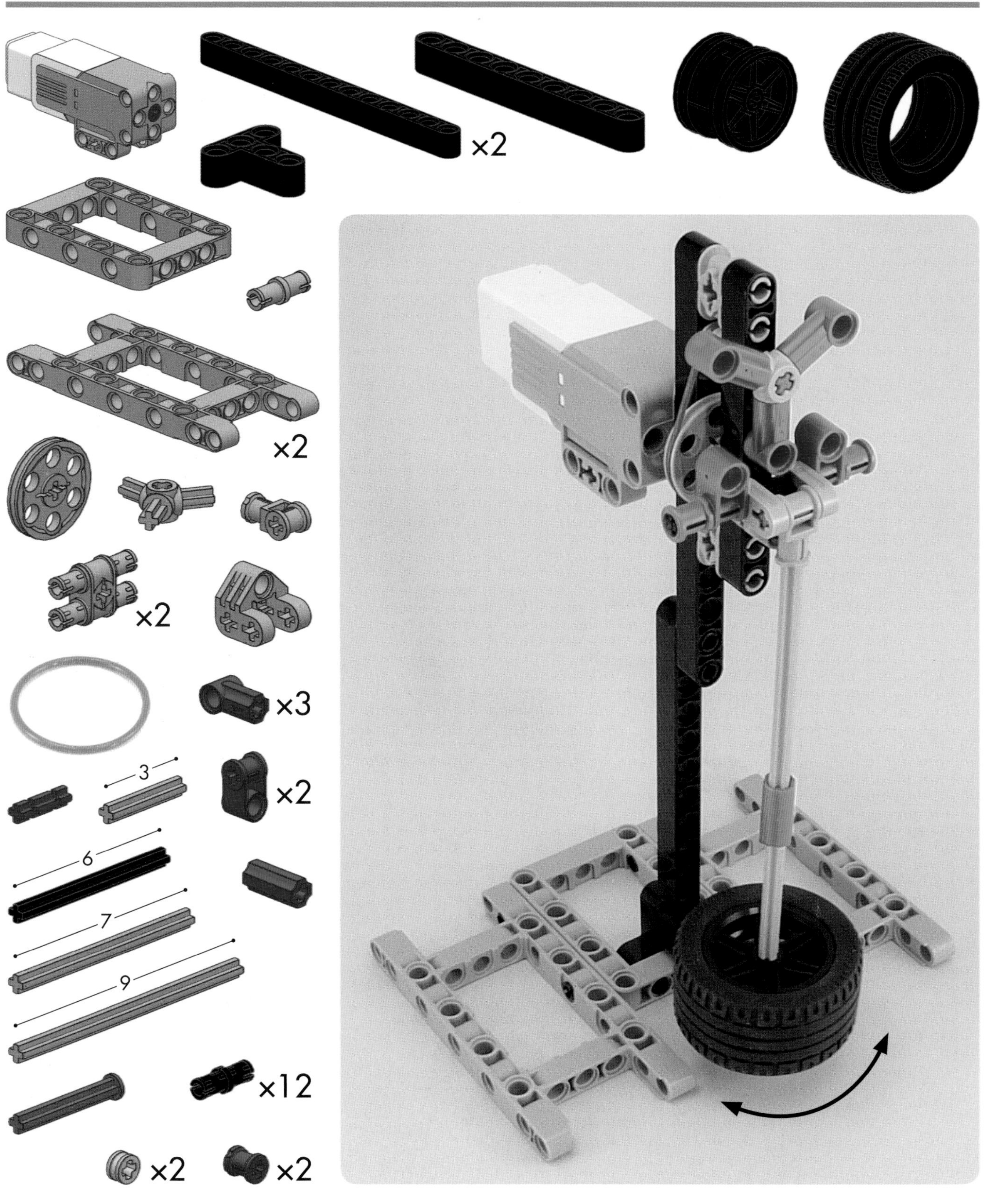
×2
×2
×2
×3
3
×2
6
7
9
×12
×2
×2

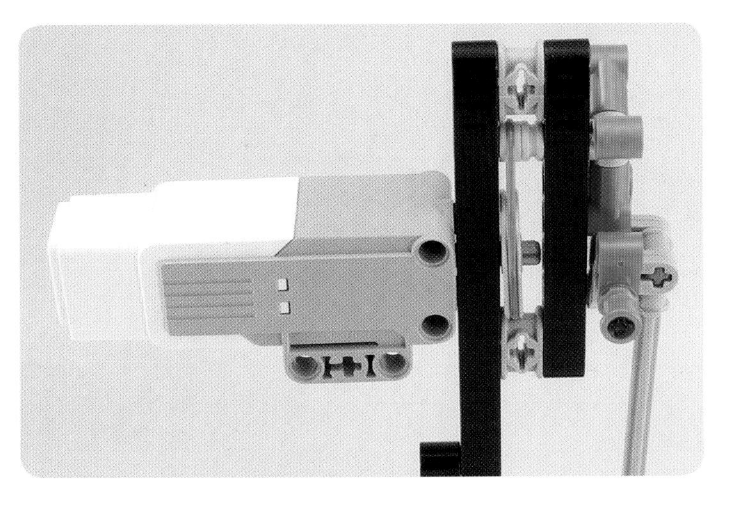

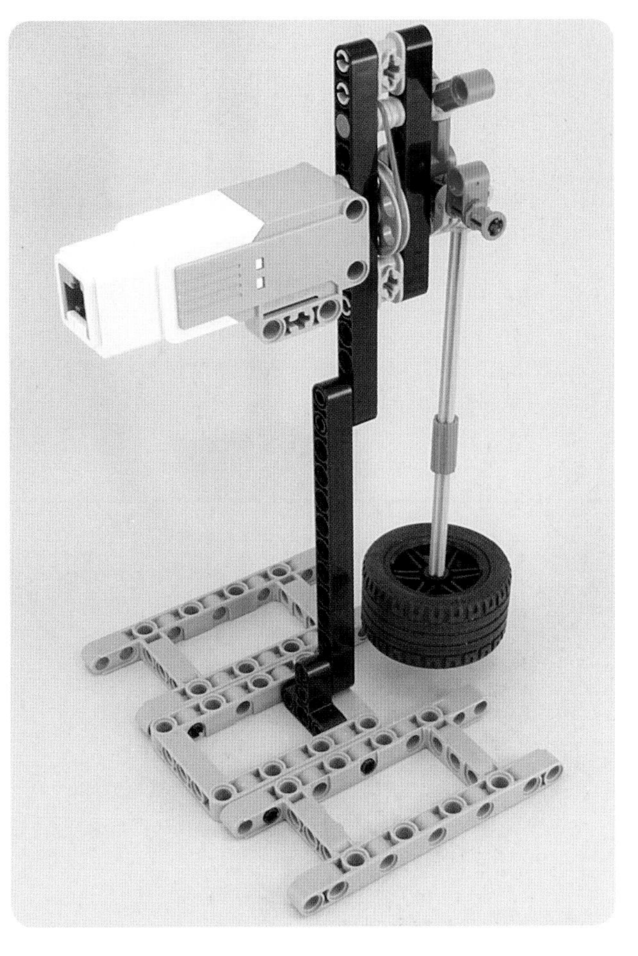

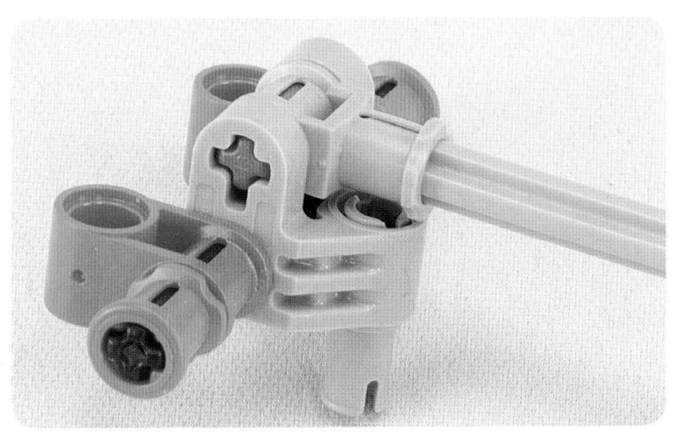

Using attachments to change motion

#157

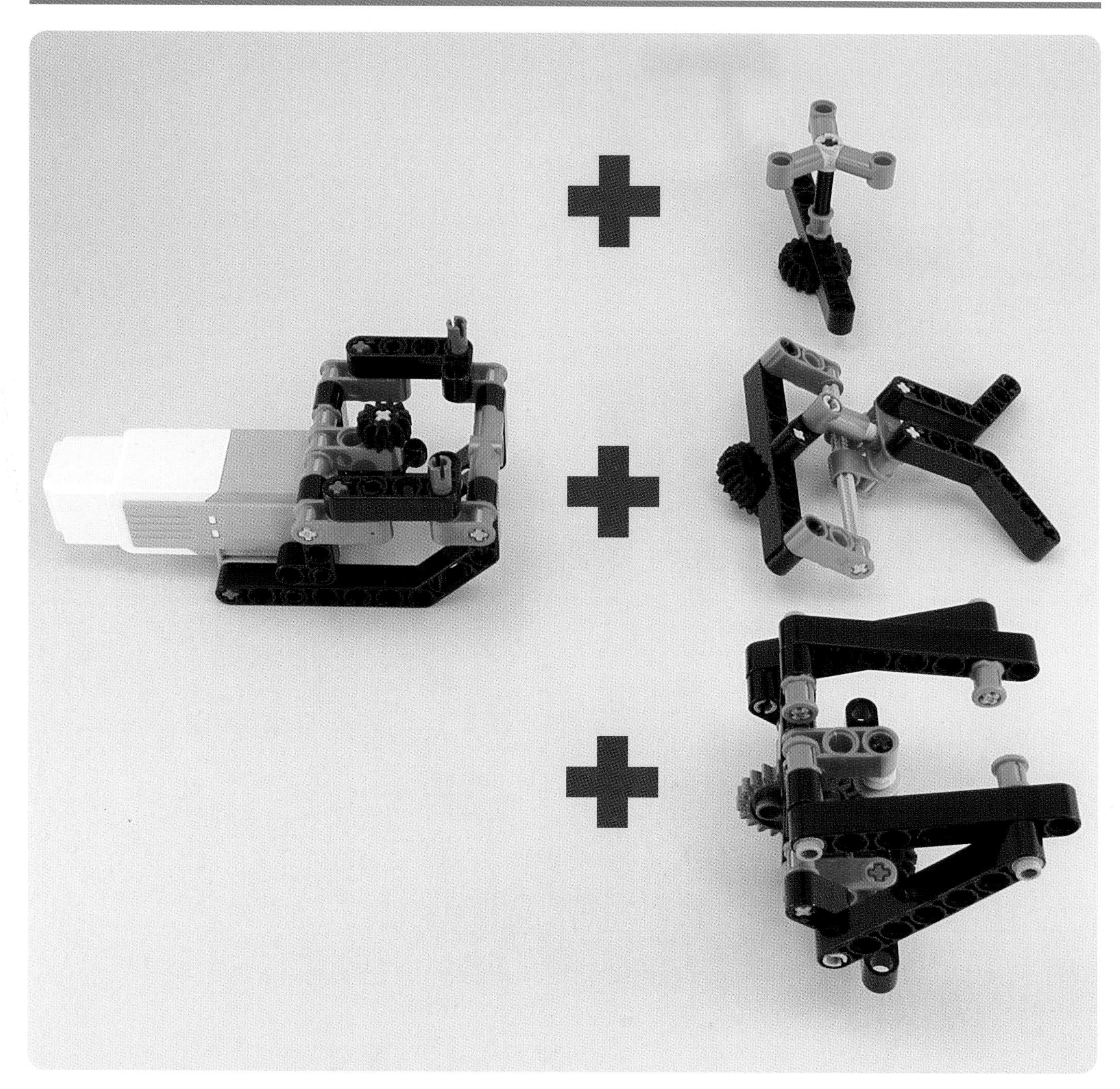

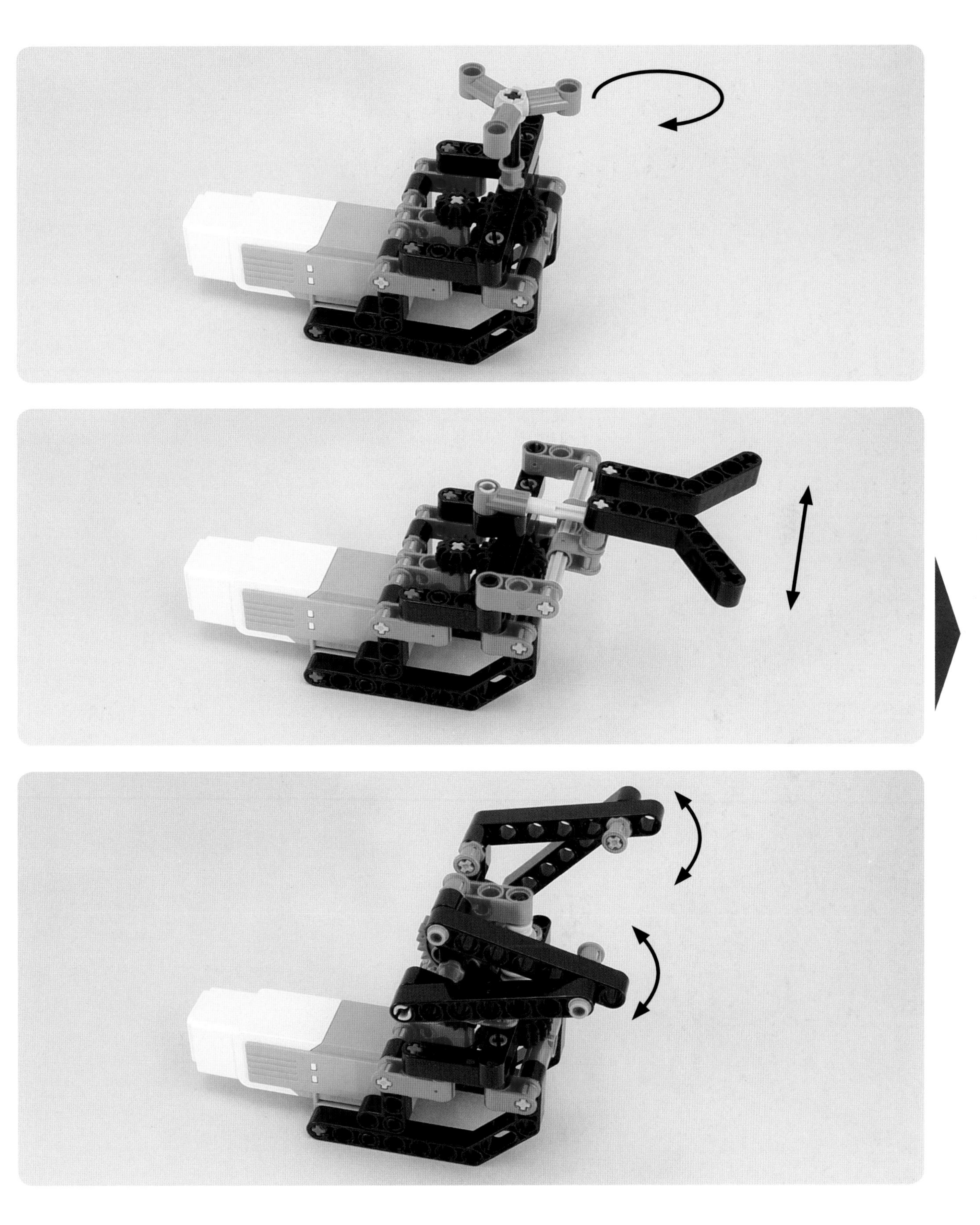

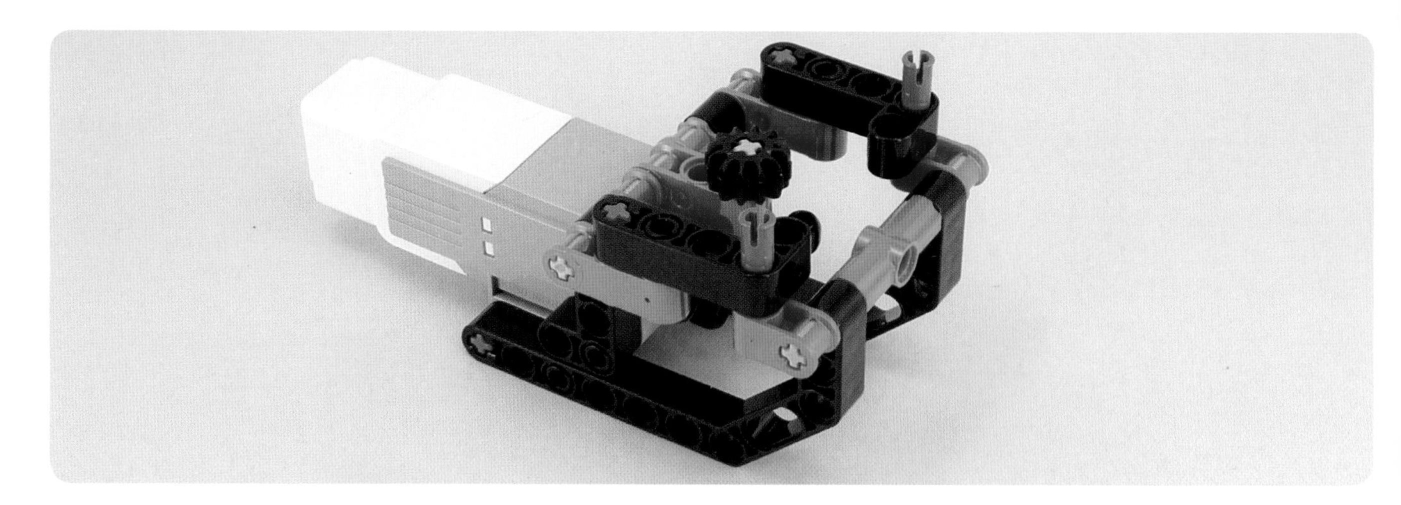

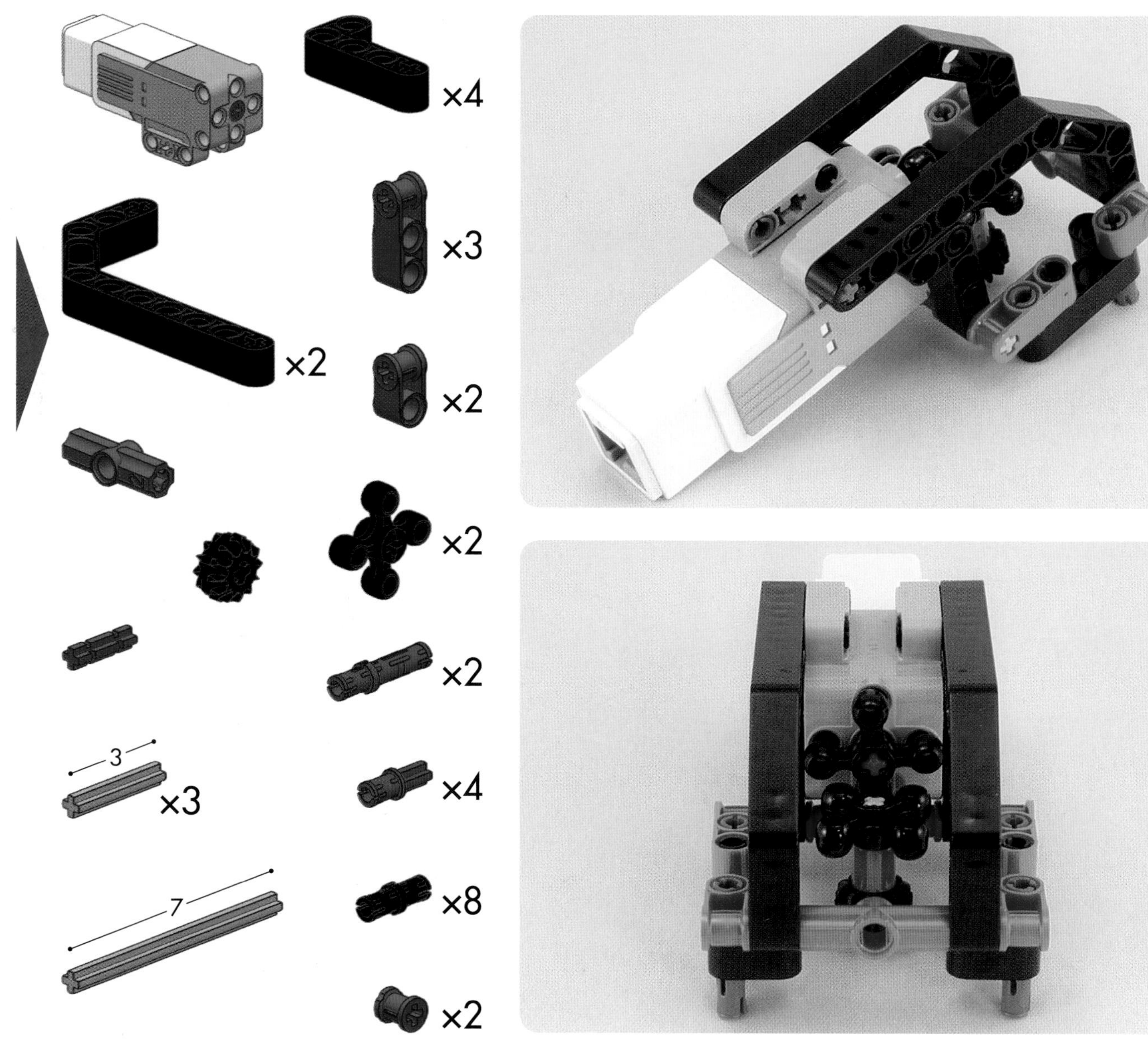
×4
×3
×2
×2
×2
3
×3
×2
×4
7
×8
×2

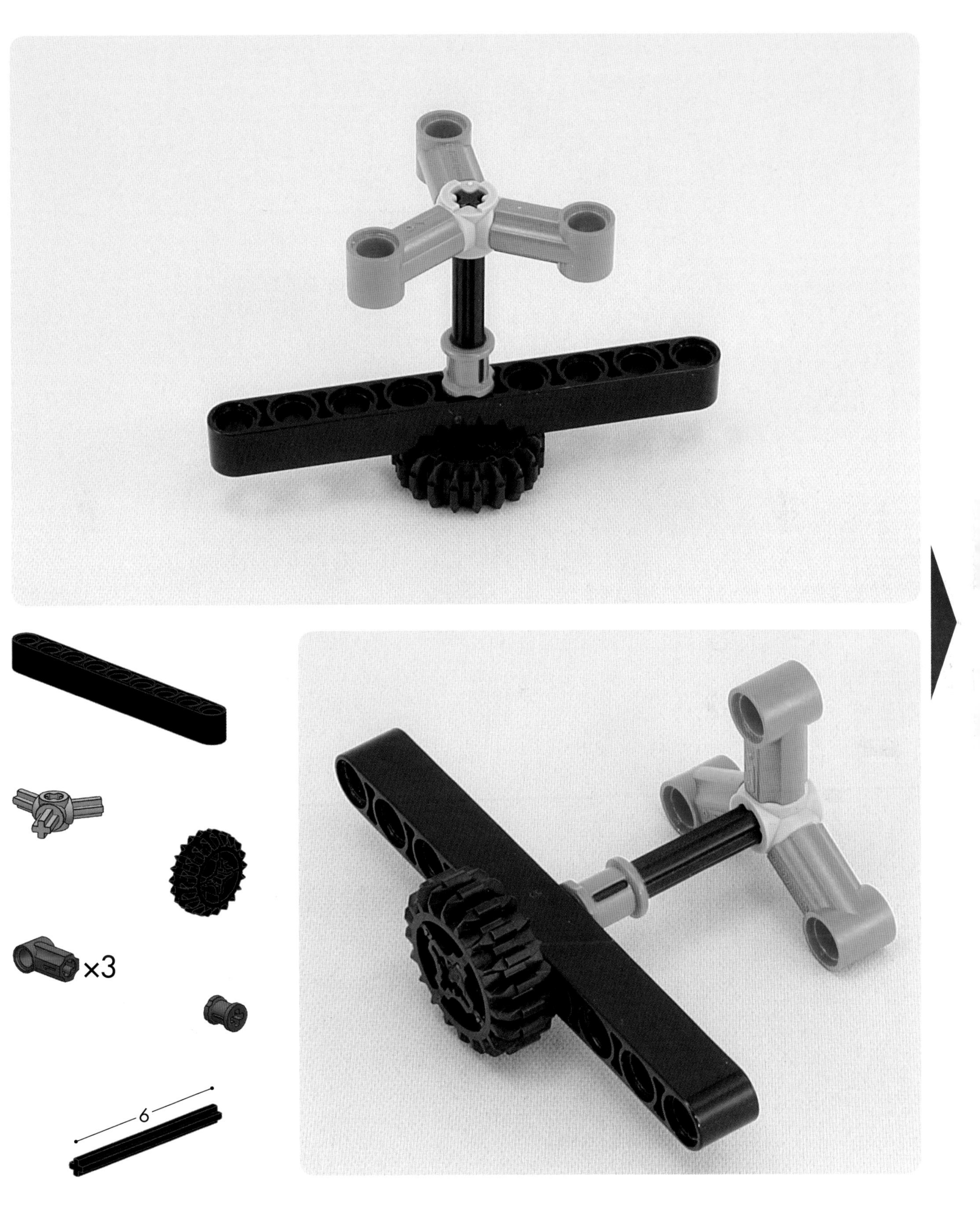
×3
6

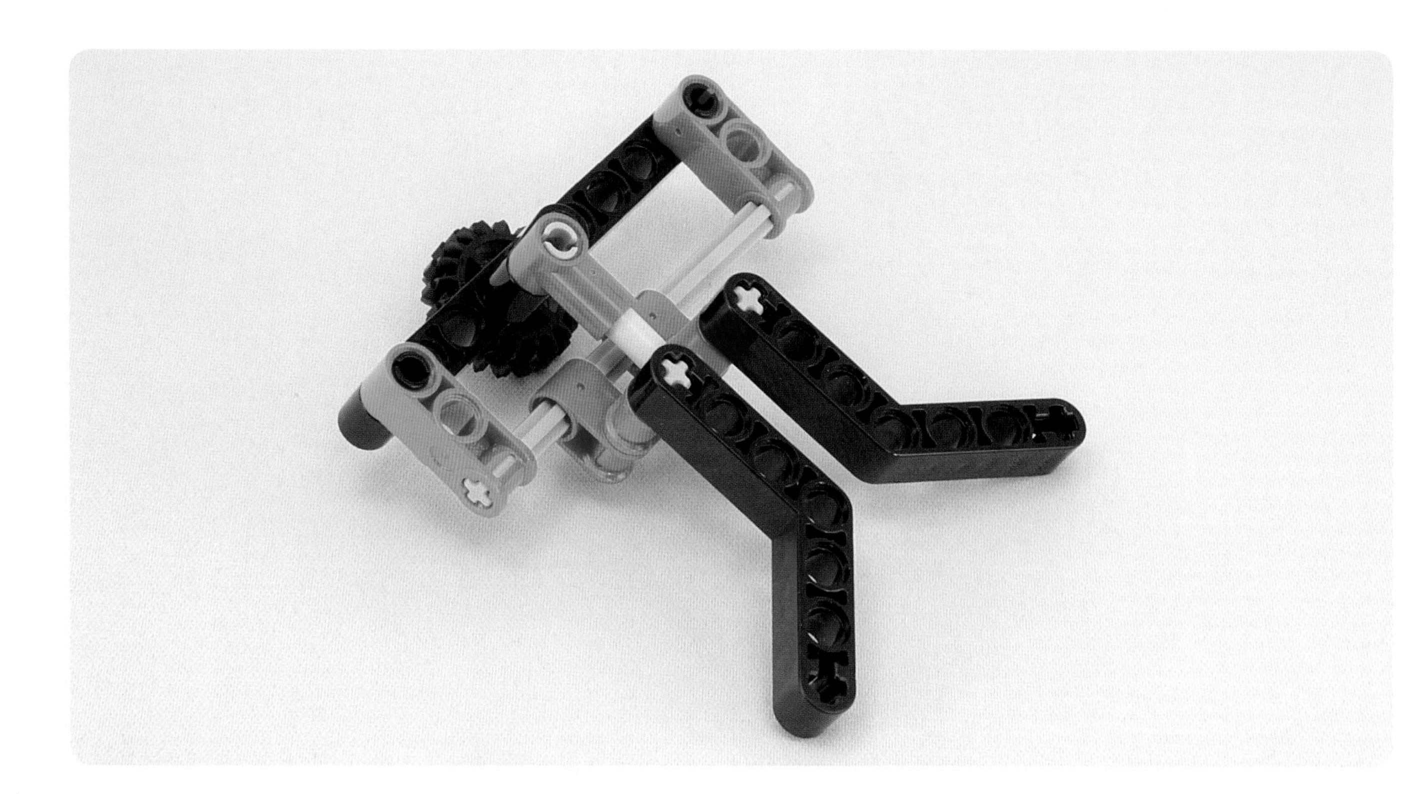

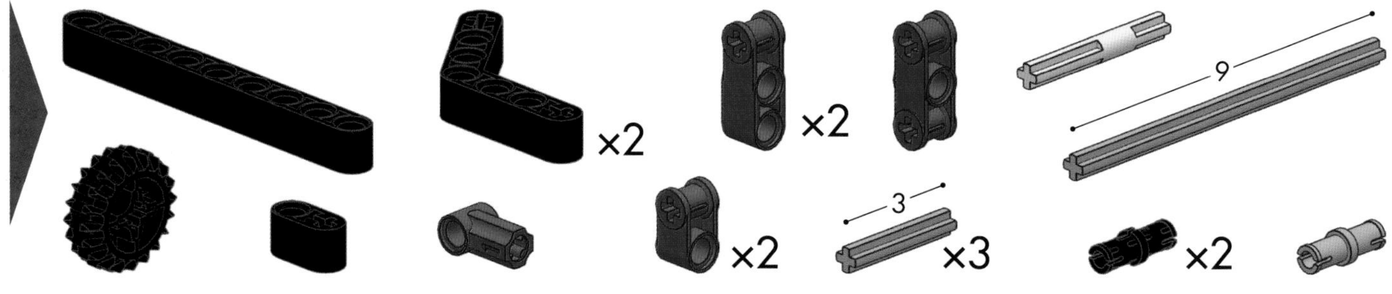
×2
×2
9
3
×2
×3
×2

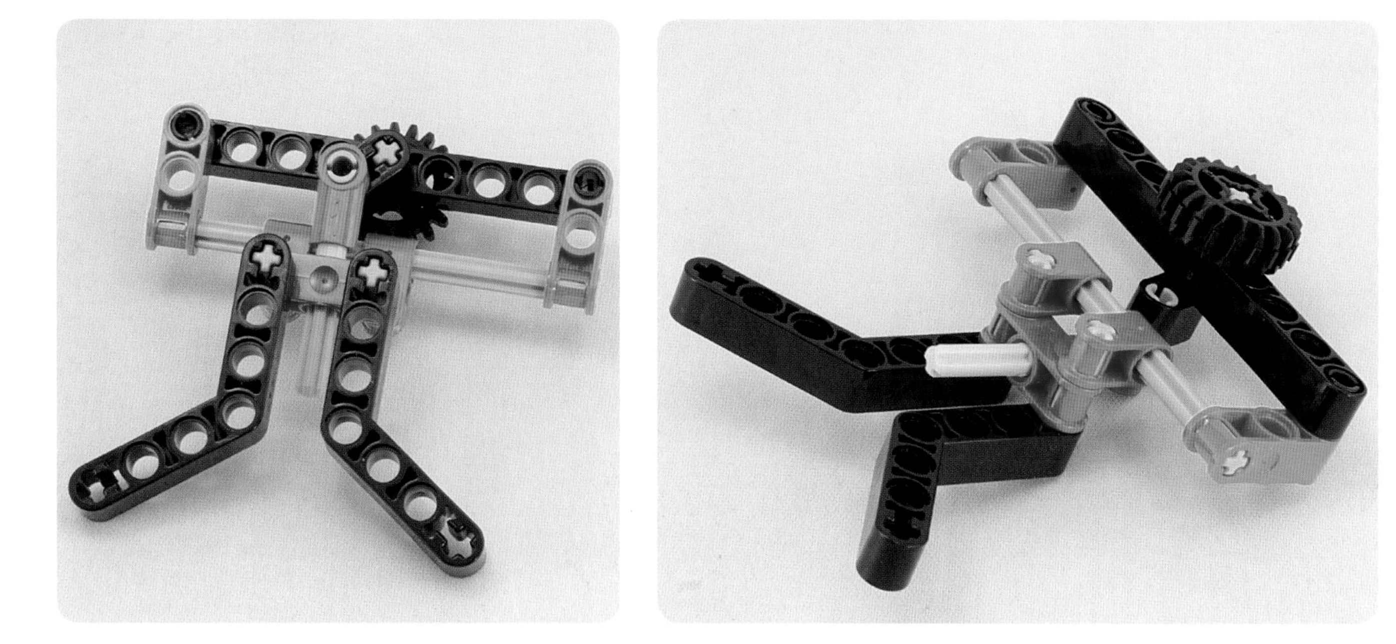

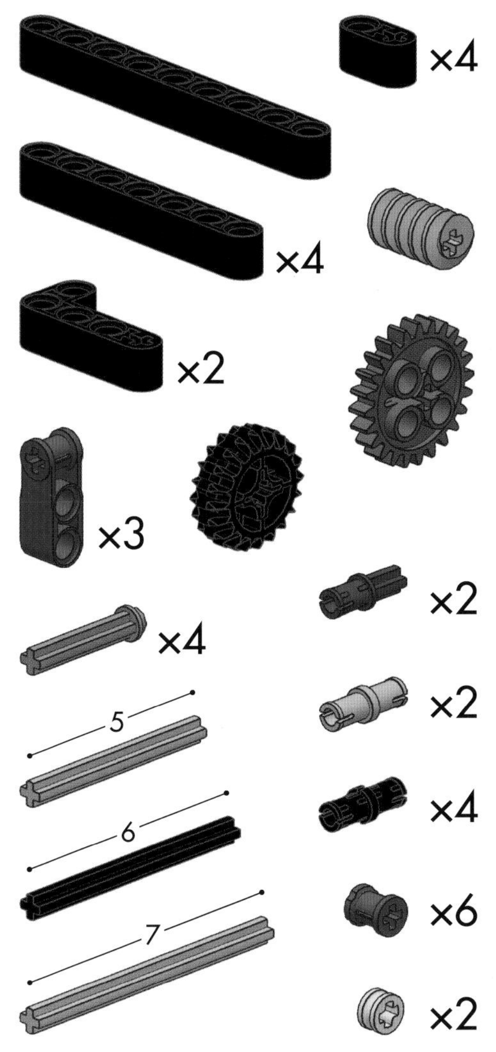
×4
×4
×2
×3
×2
×4
5
×2
6
×4
7
×6
×2

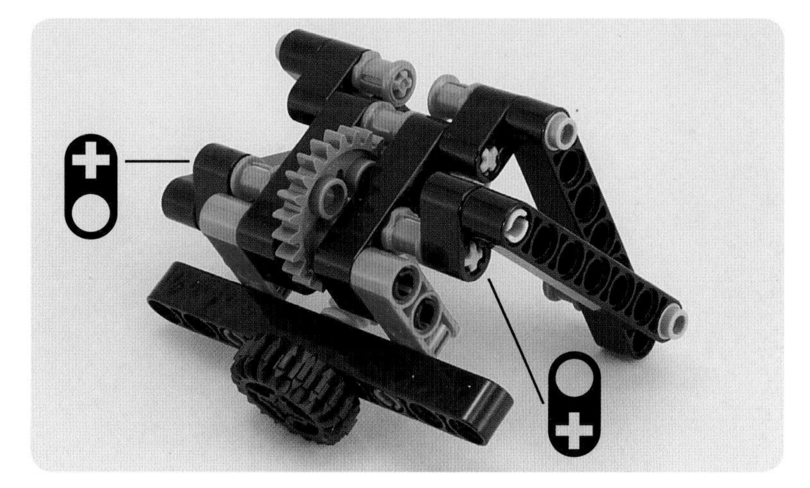

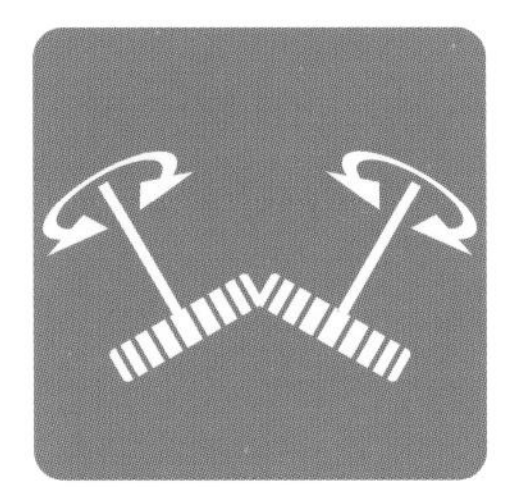

Meshing gears diagonally

#158

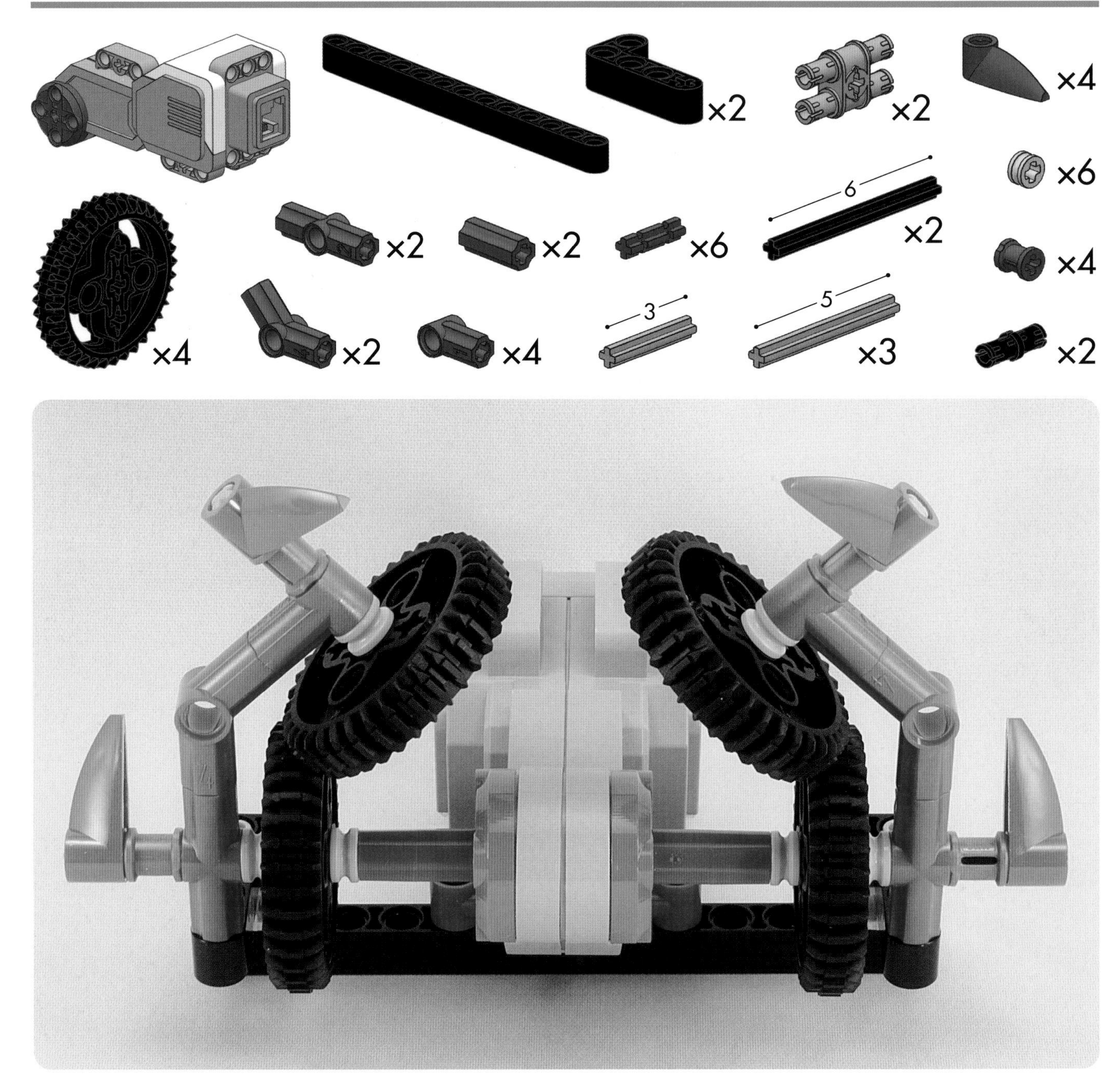
×2
×2
×4
6
×6
×2
×2
×6
×2
×4
×4
×2
3
5
×4
×3
×2

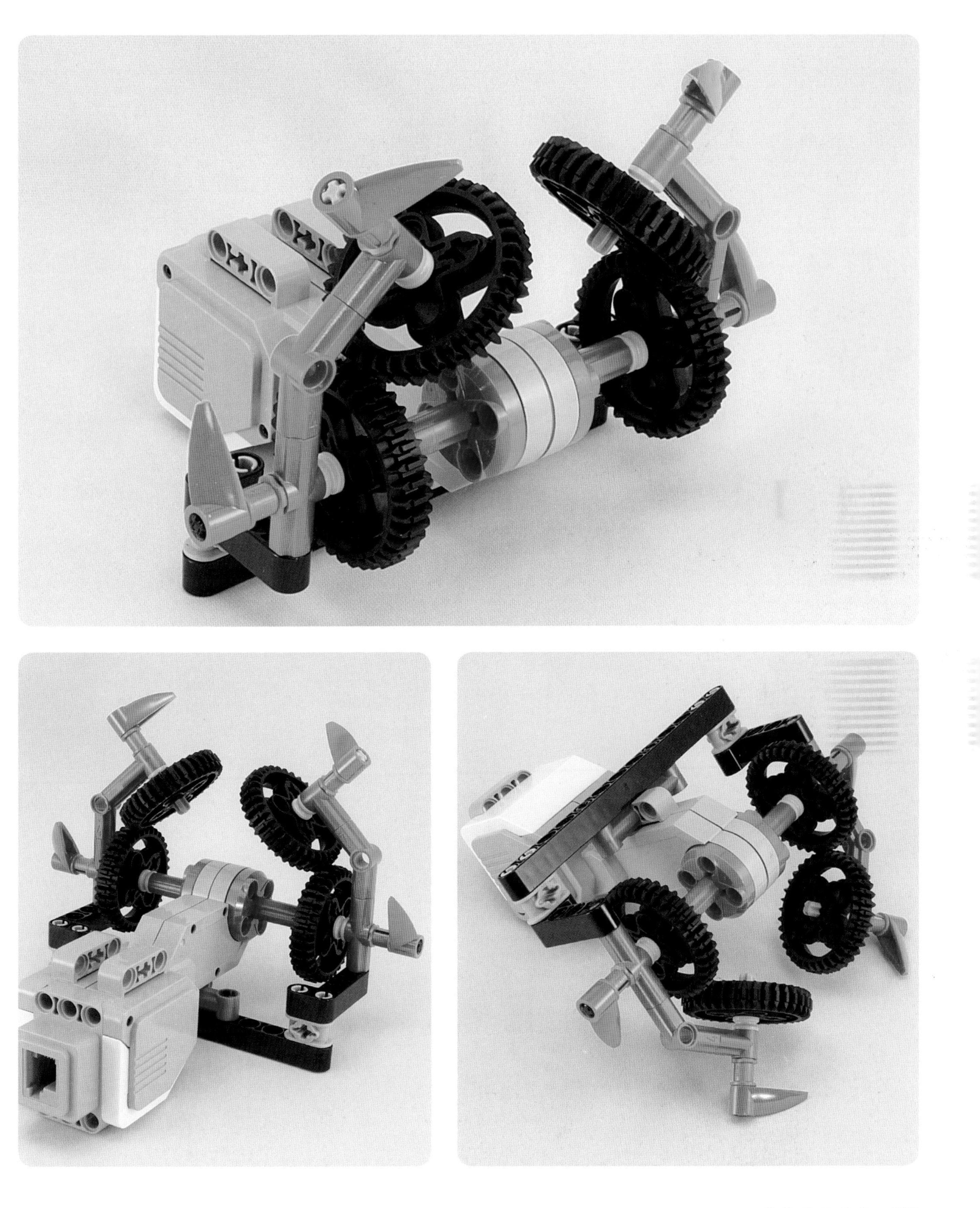

#159

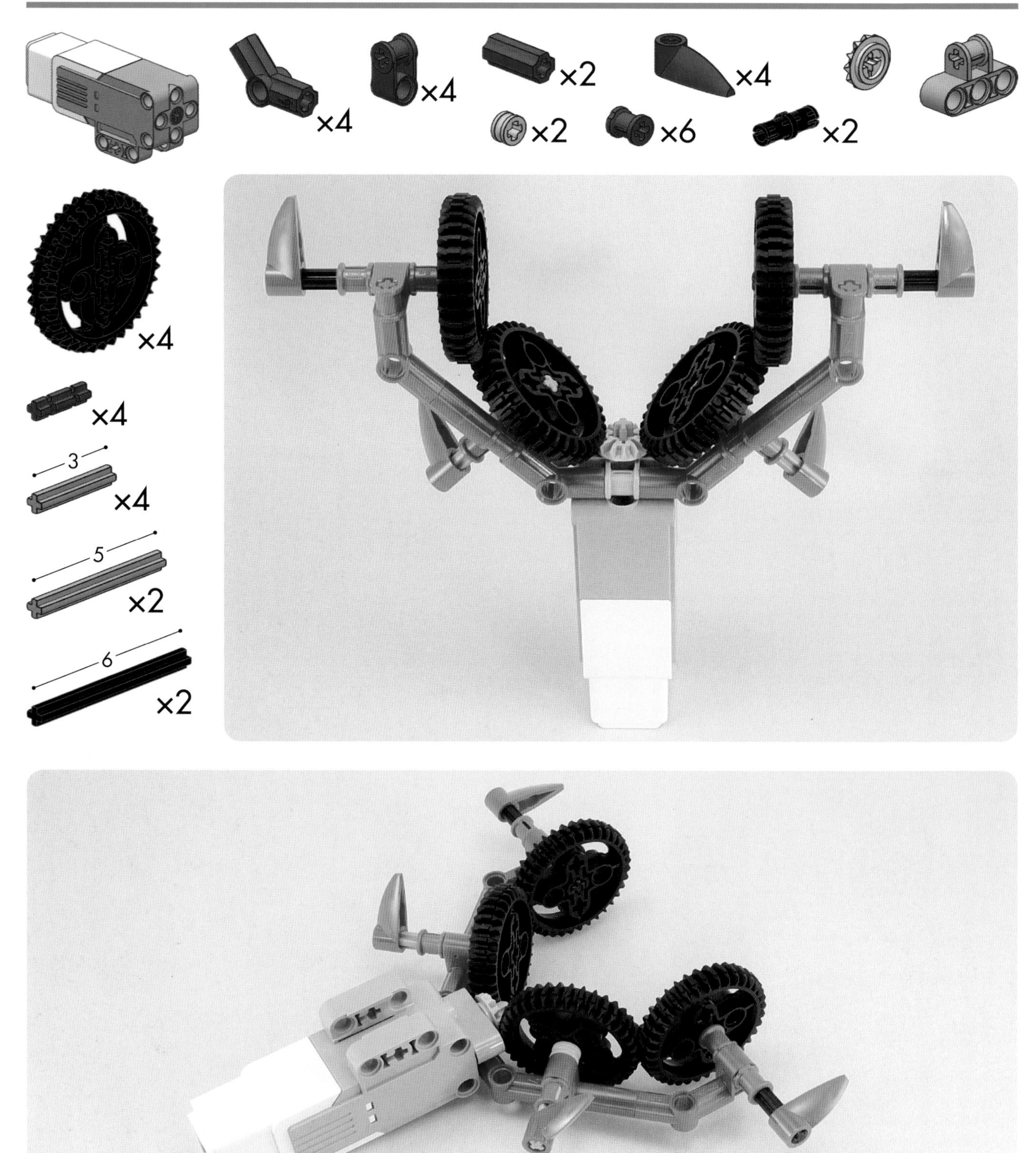

×4
×4
×2
×4
×2
×6
×2
×4
×4
3
×4
5
×2
6
×2

#160

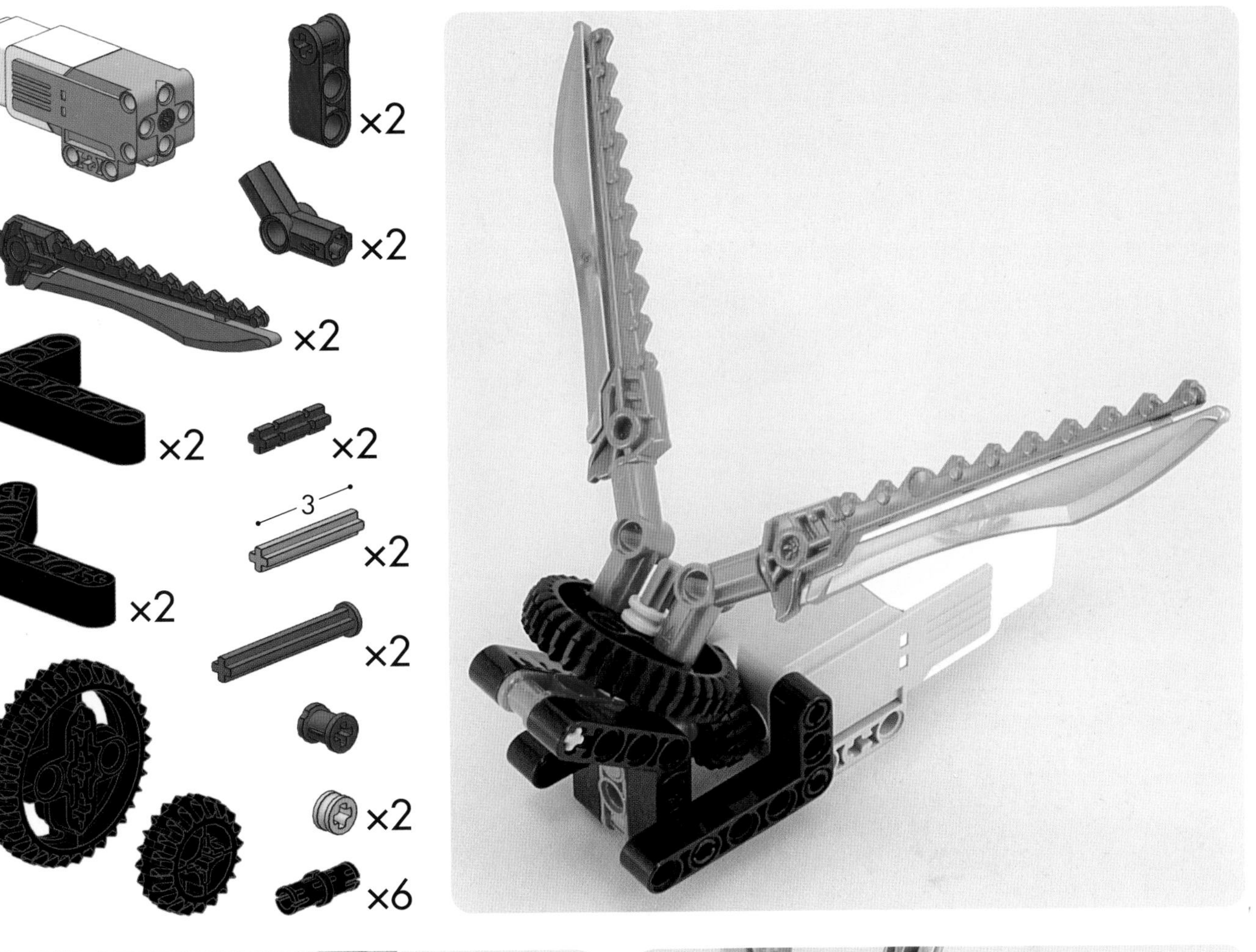
×2
×2
×2
×2
×2
3
×2
×2
×2
×2
×6

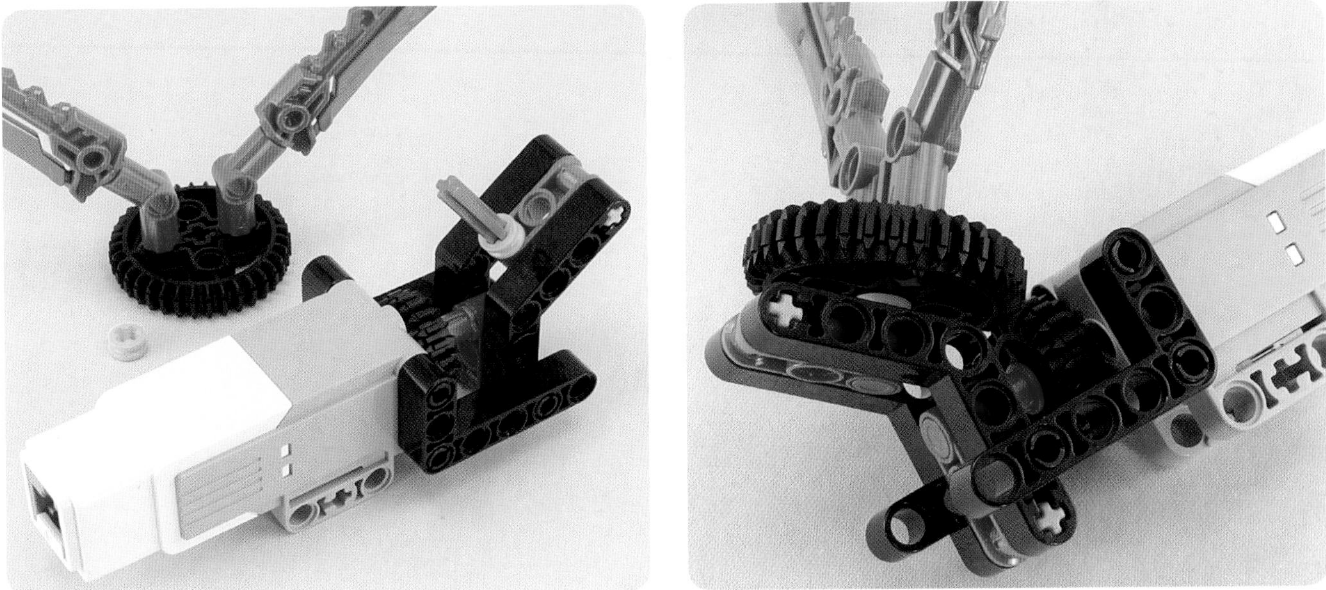

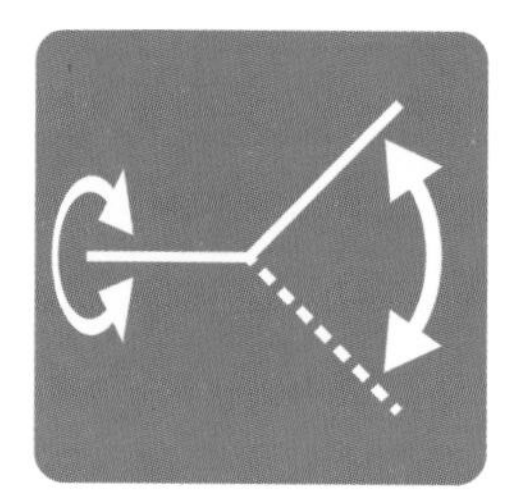

Changing the angle of rotation freely

#161

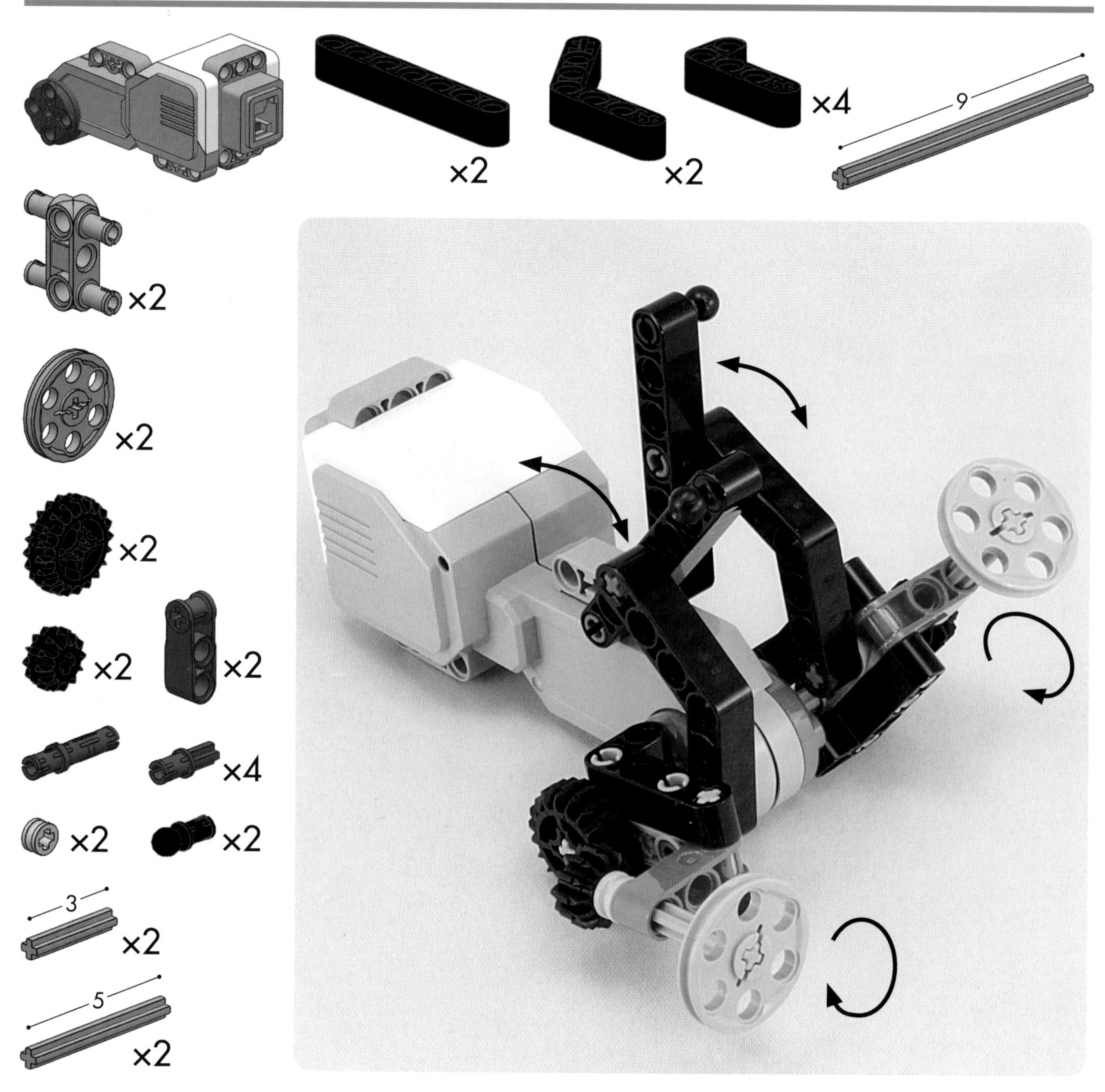
×2
×2
×4
9
×2
×2
×2
×2
×2
×4
×2
×2
3
×2
5
×2

#162

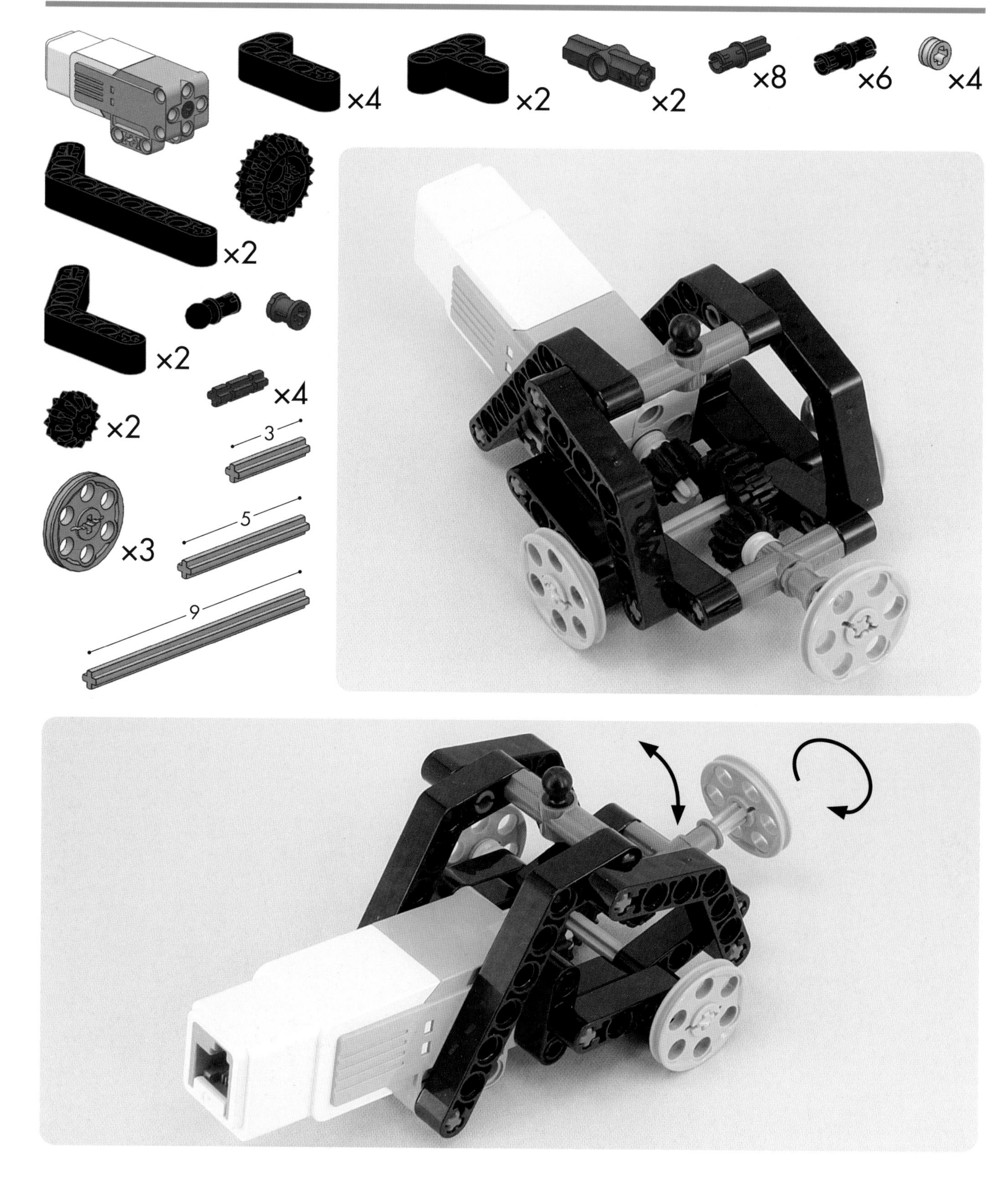
×4
×2
×2
×8
×6
×4
×2
×2
×4
×2
3
5
×3
9

#163

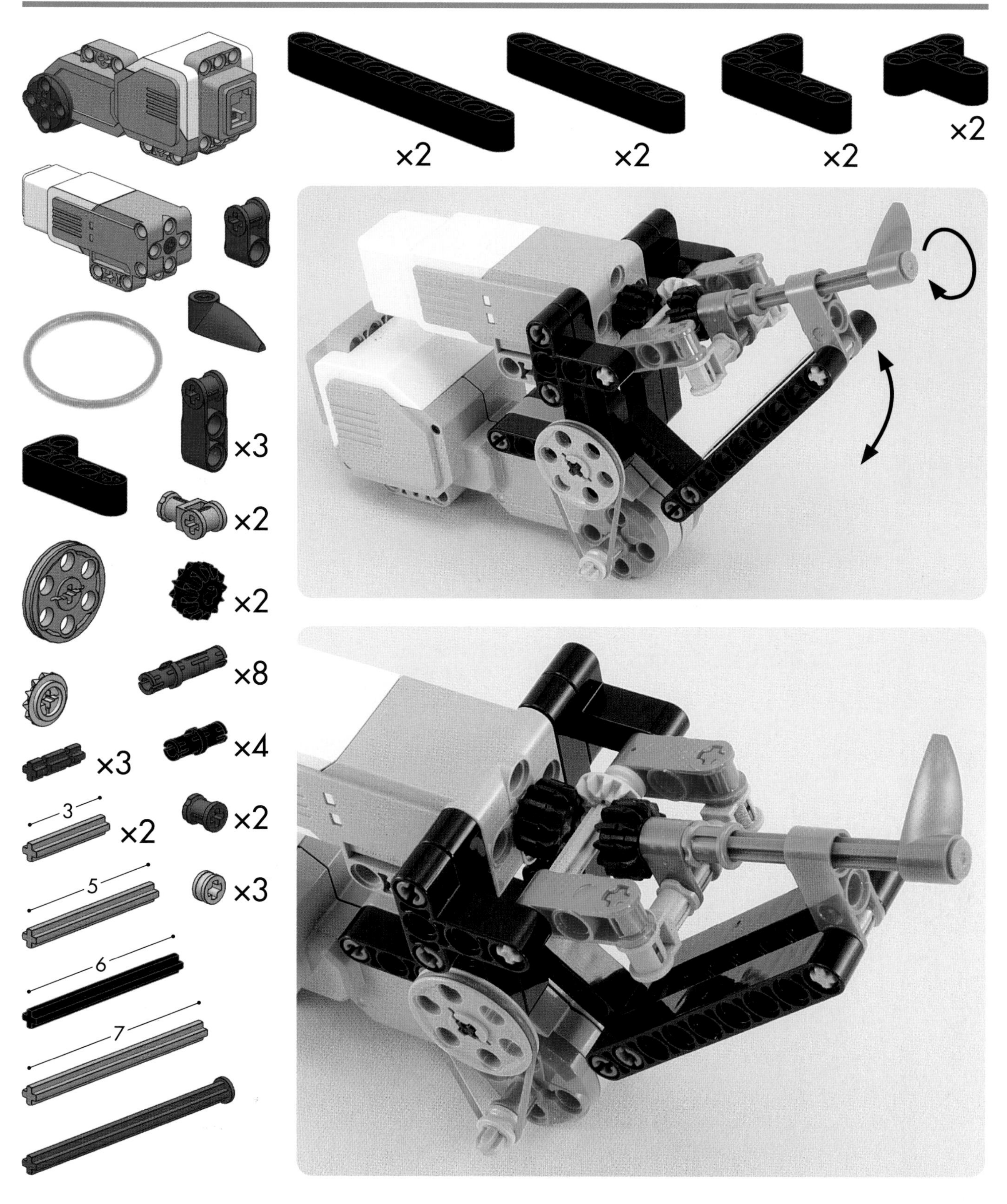
×2
×2
×2
×2
×3
×2
×2
×8
×3
×4
3
×2
×2
5
×3
6
7

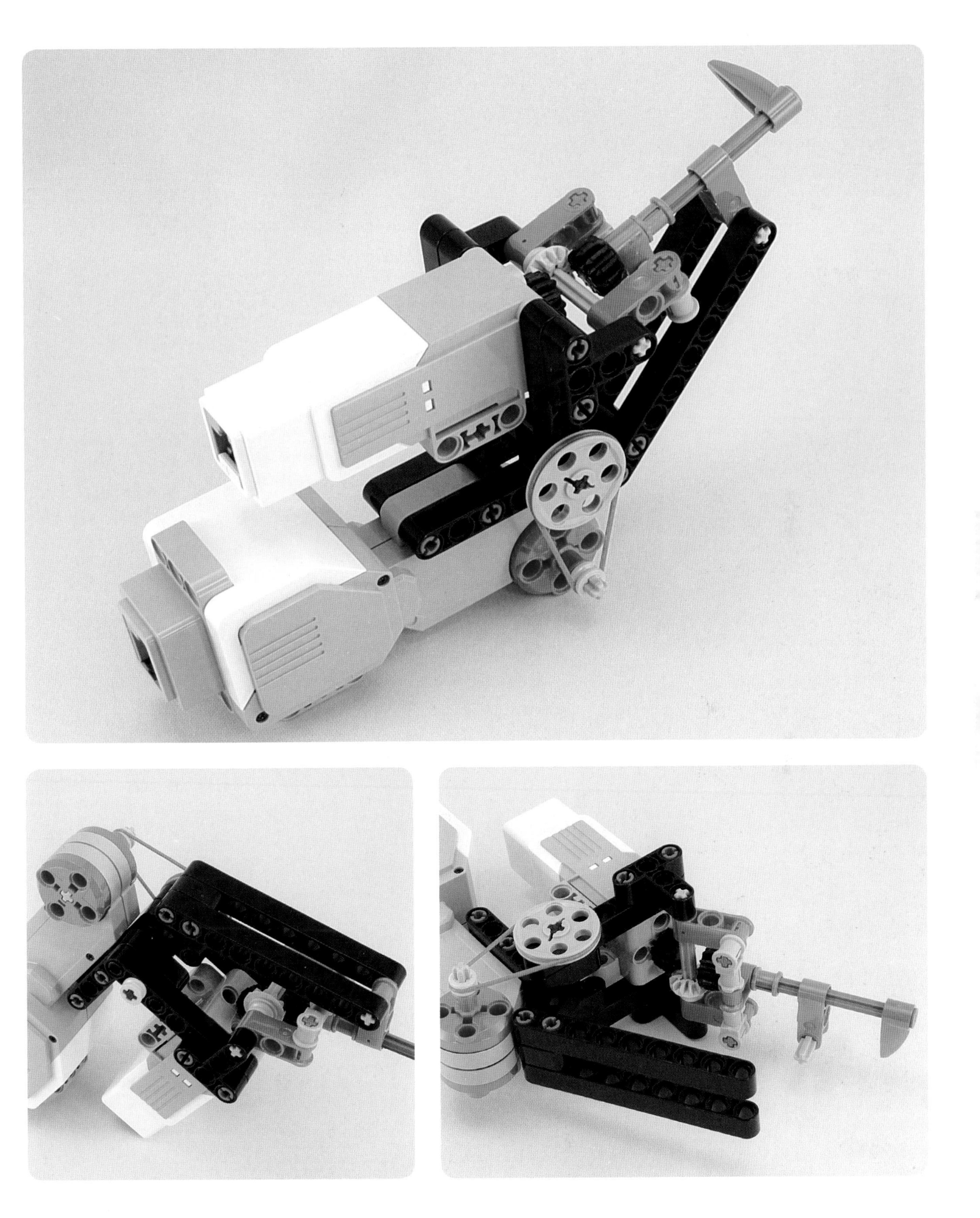

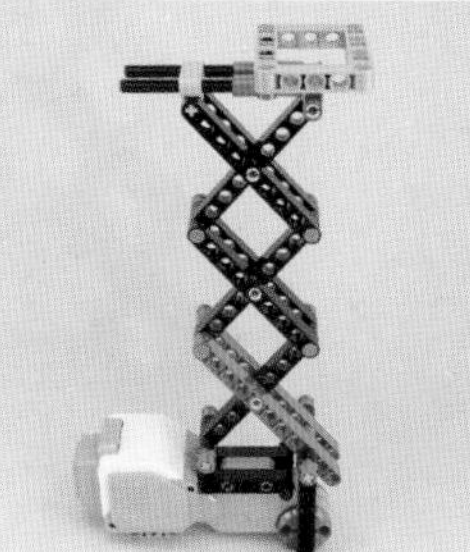

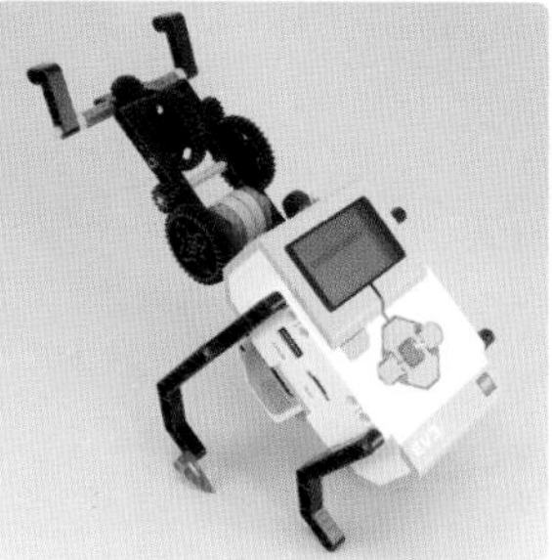

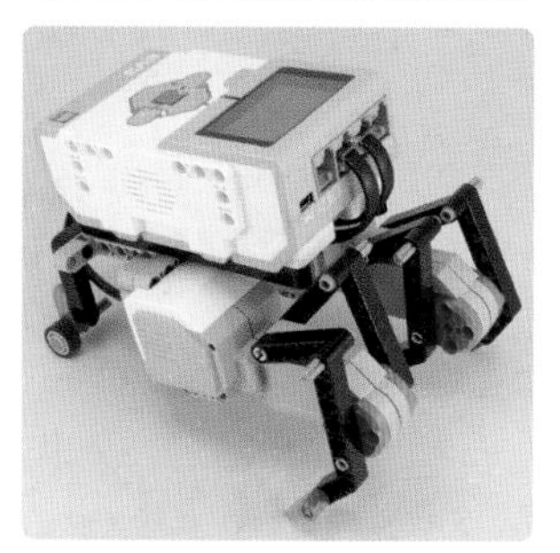

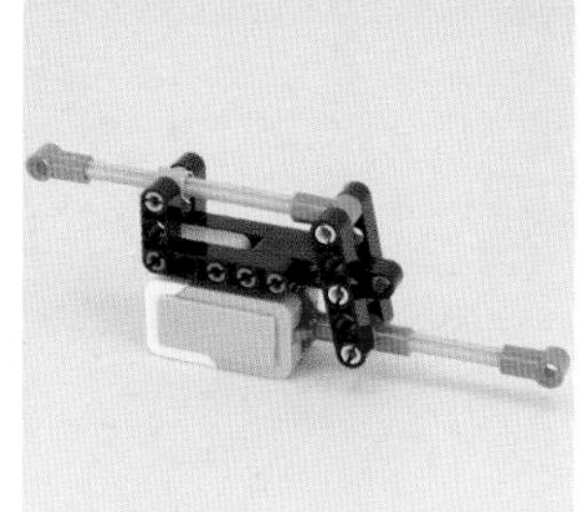

PART 5

Sensors

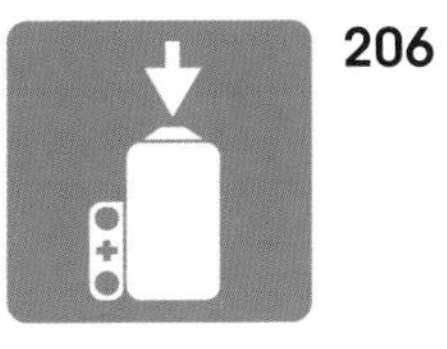

206

214

216

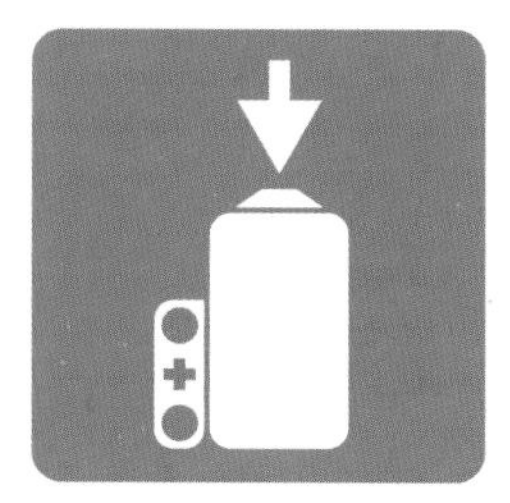

Ideas for using the touch sensor

#164

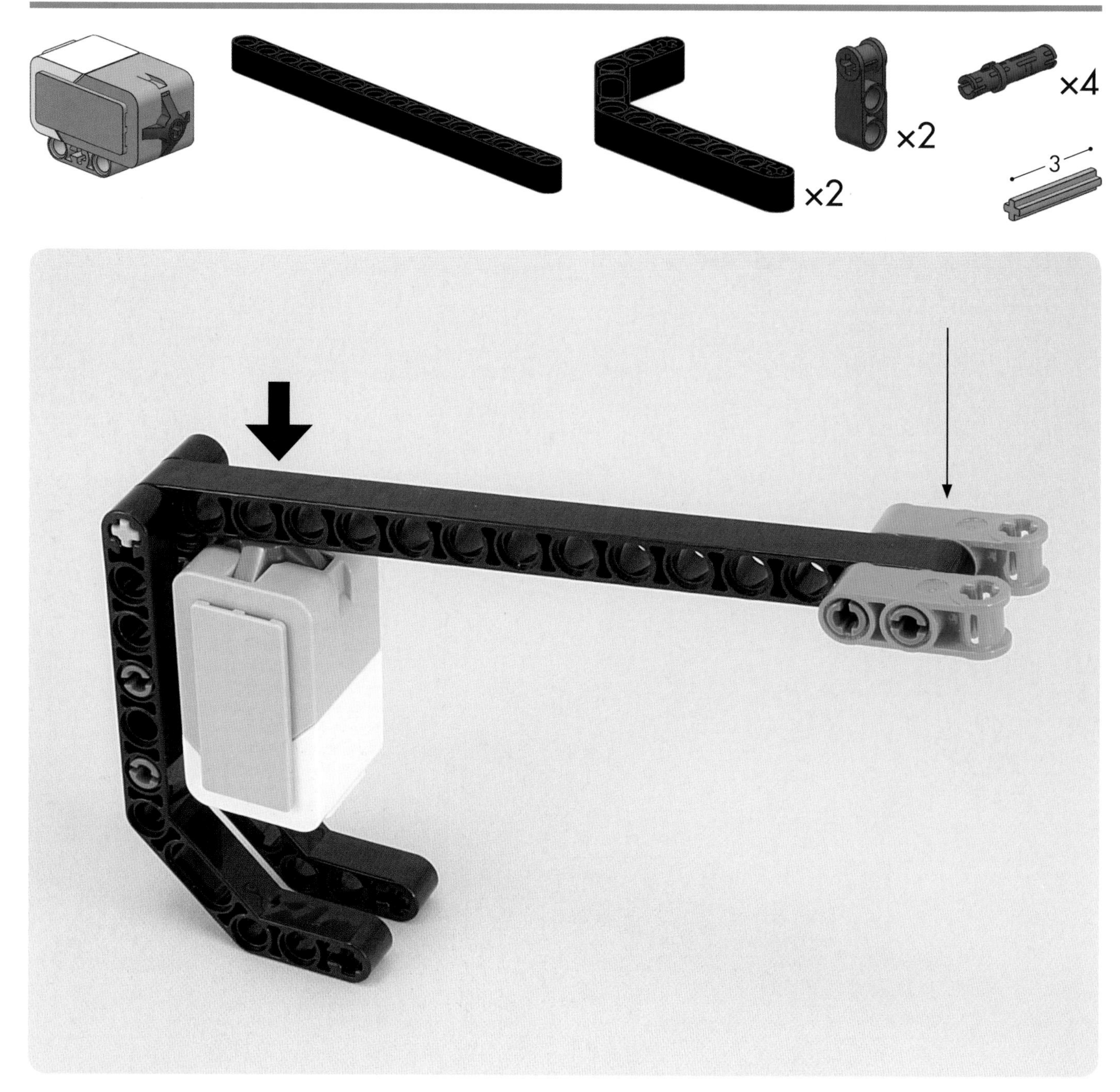
×2
×2
×4
3

#165

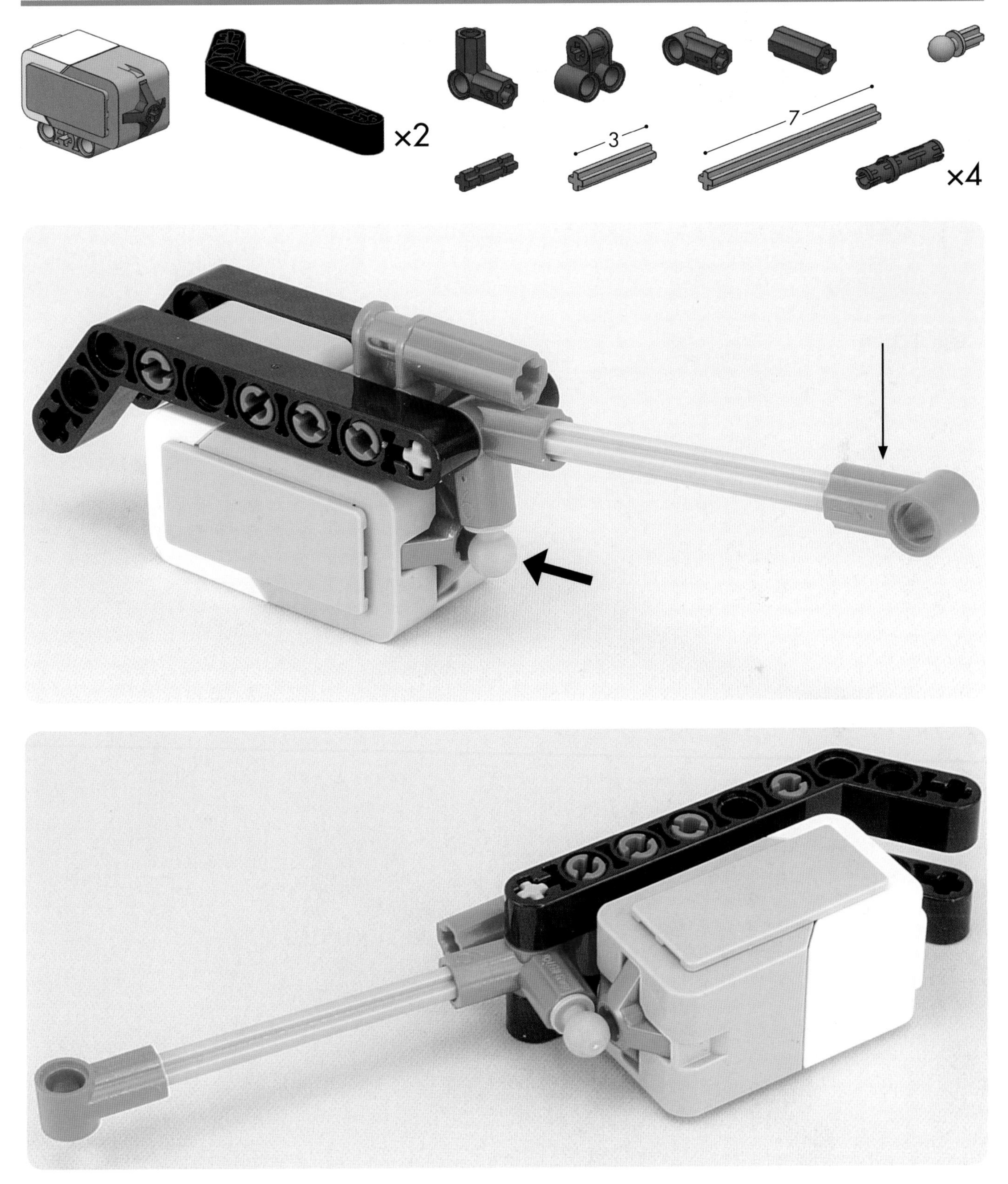
×2
3
7
×4

#166

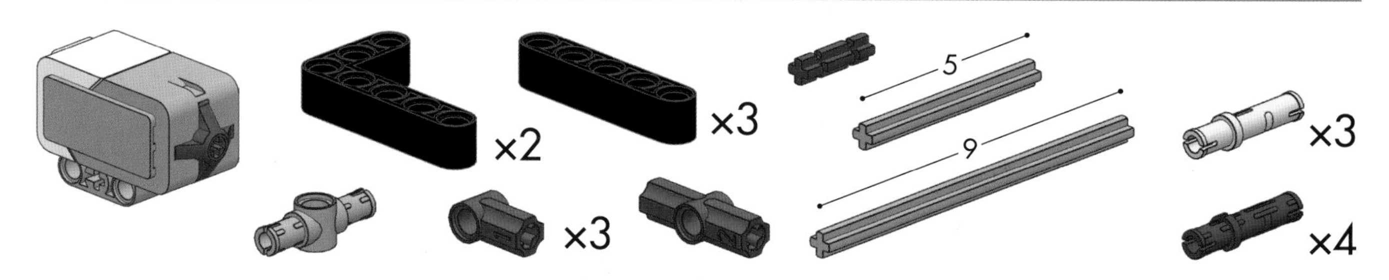
×2
×3
5
9
×3
×3
×4

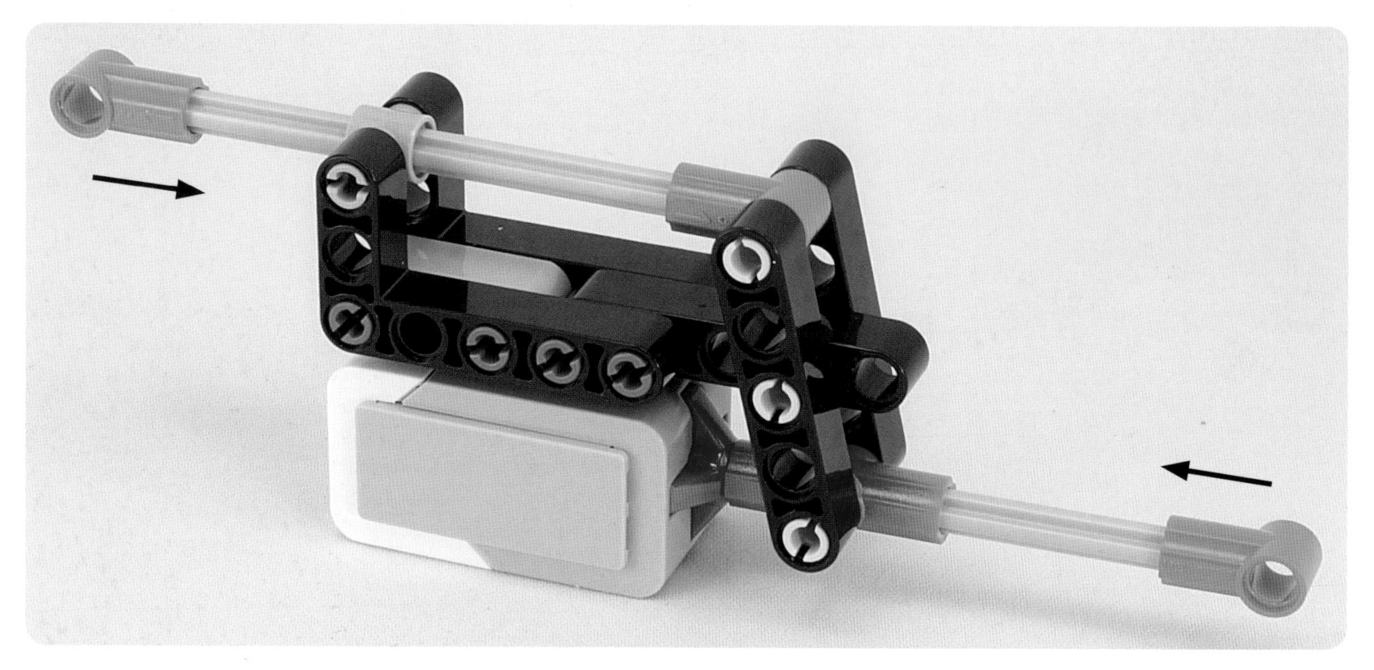

#167

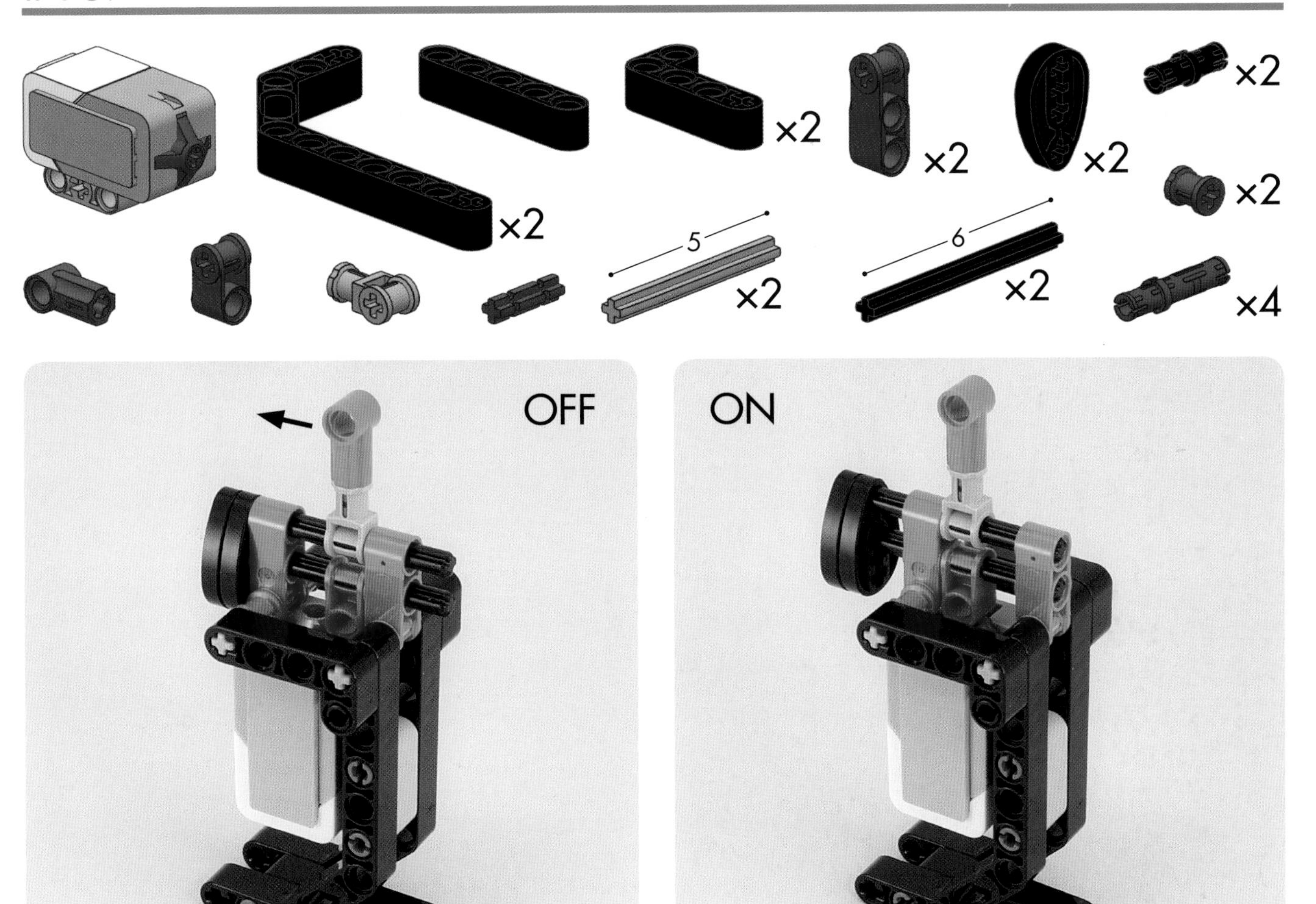
×2
×2
×2
×2
×2
×2
×2
5
×2
6
×2
×4
OFF
ON

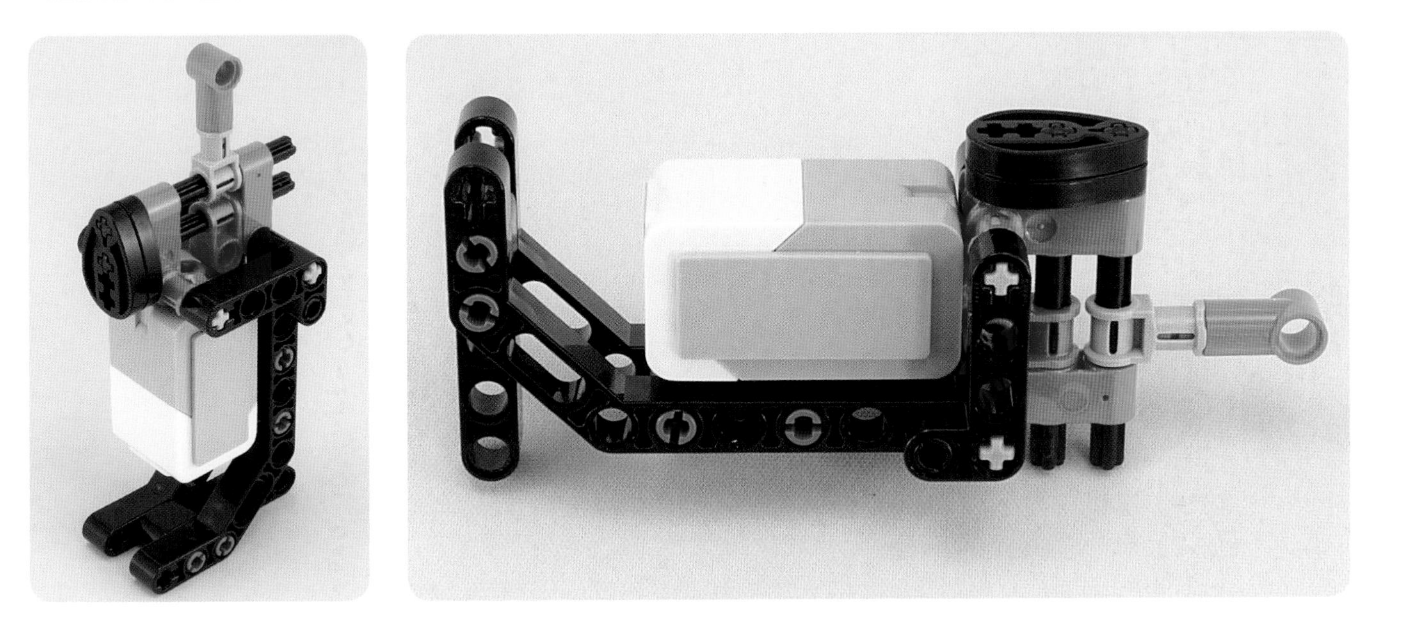

#168

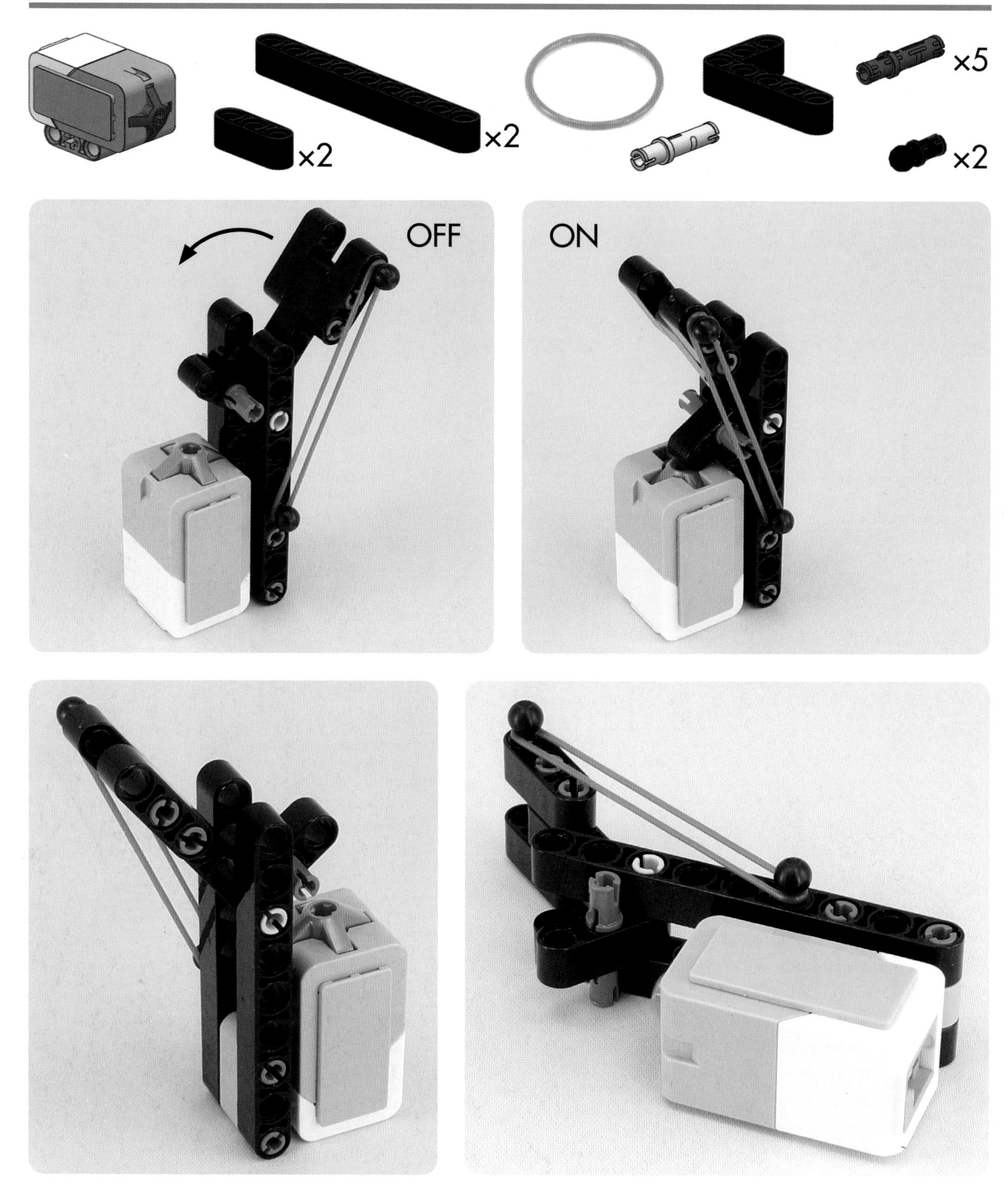

×2
×2
×5
×2
OFF
ON

#169

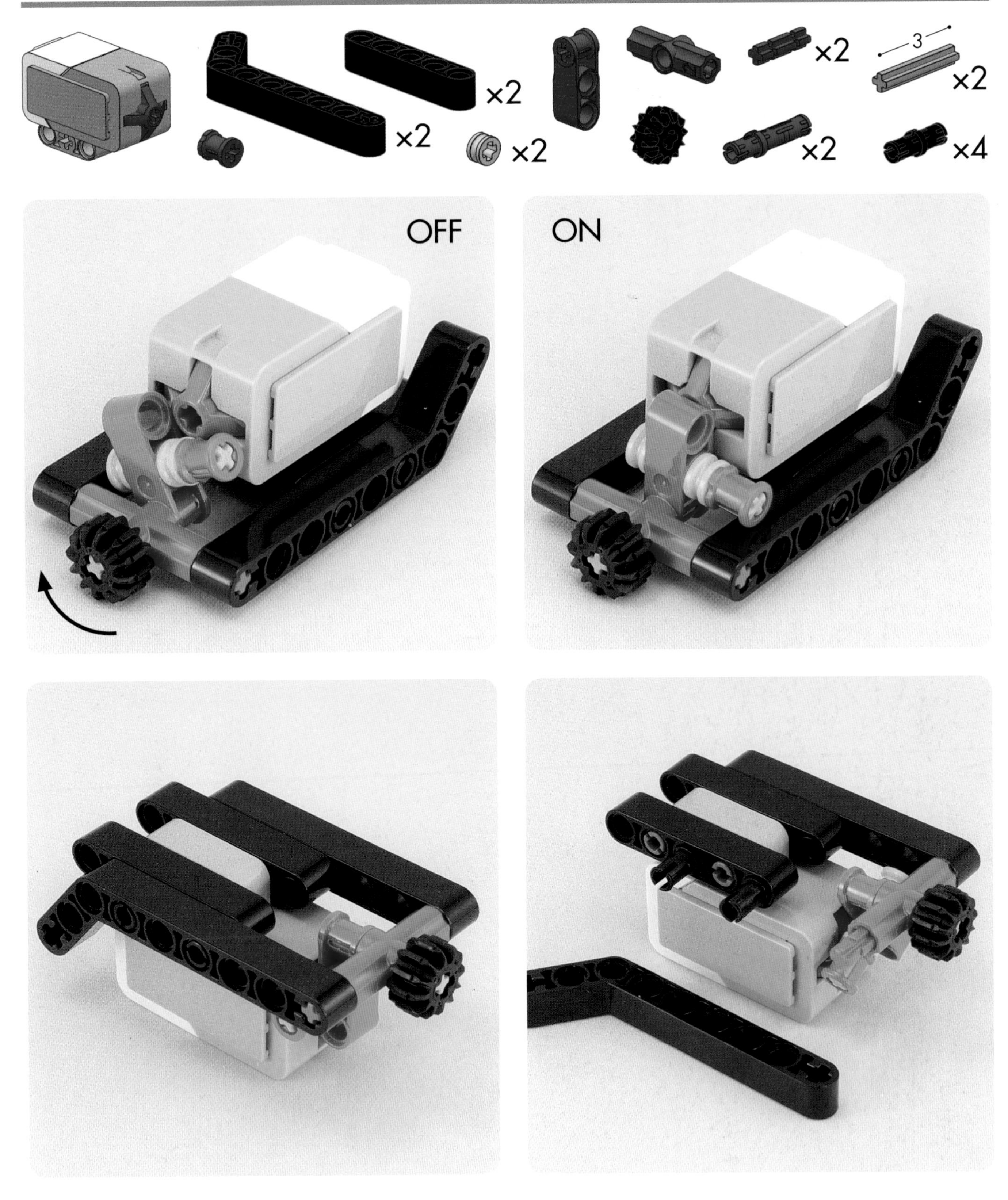
×2
×2
×2
×2
3
×2
×2
×2
×4
OFF
ON

#170

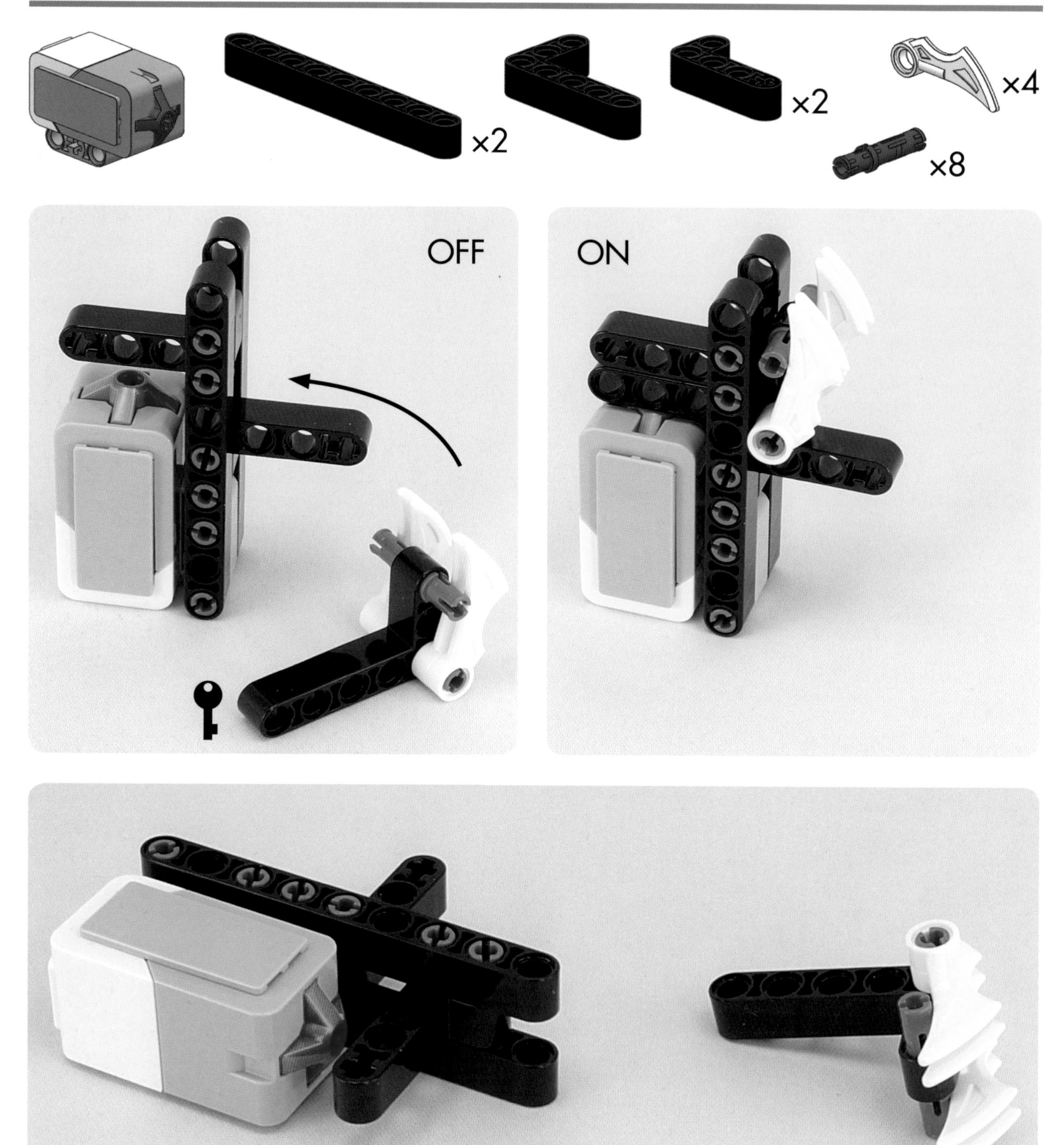

×2
×2
×4
×8
OFF
ON

#171

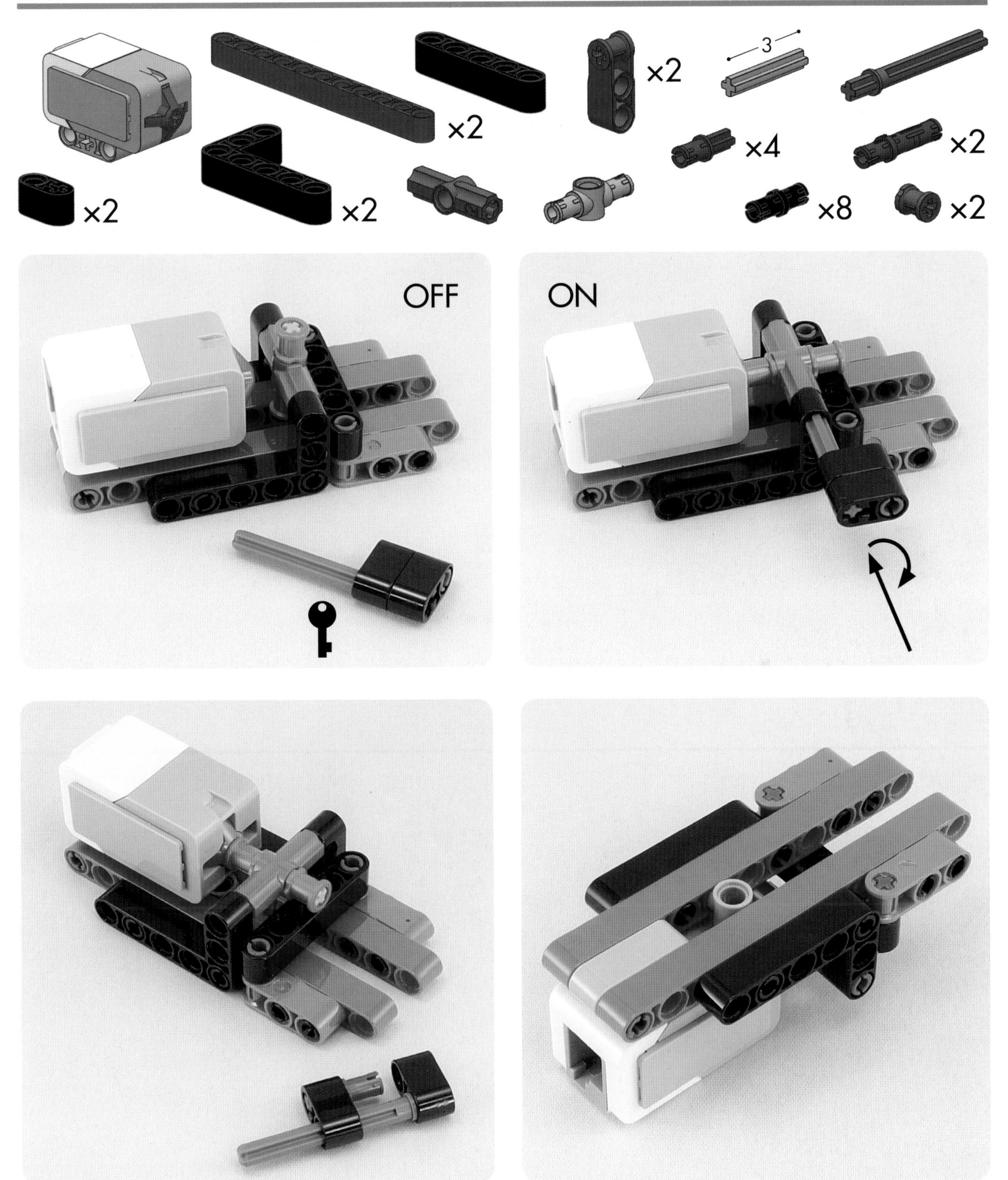

3
×2
×2
×2
×4
×2
×2
×2
×8
×2
OFF
ON

Ideas for using the buttons of the Intelligent EV3 Brick

#172

×2

×3

×2

×2

×8

×4

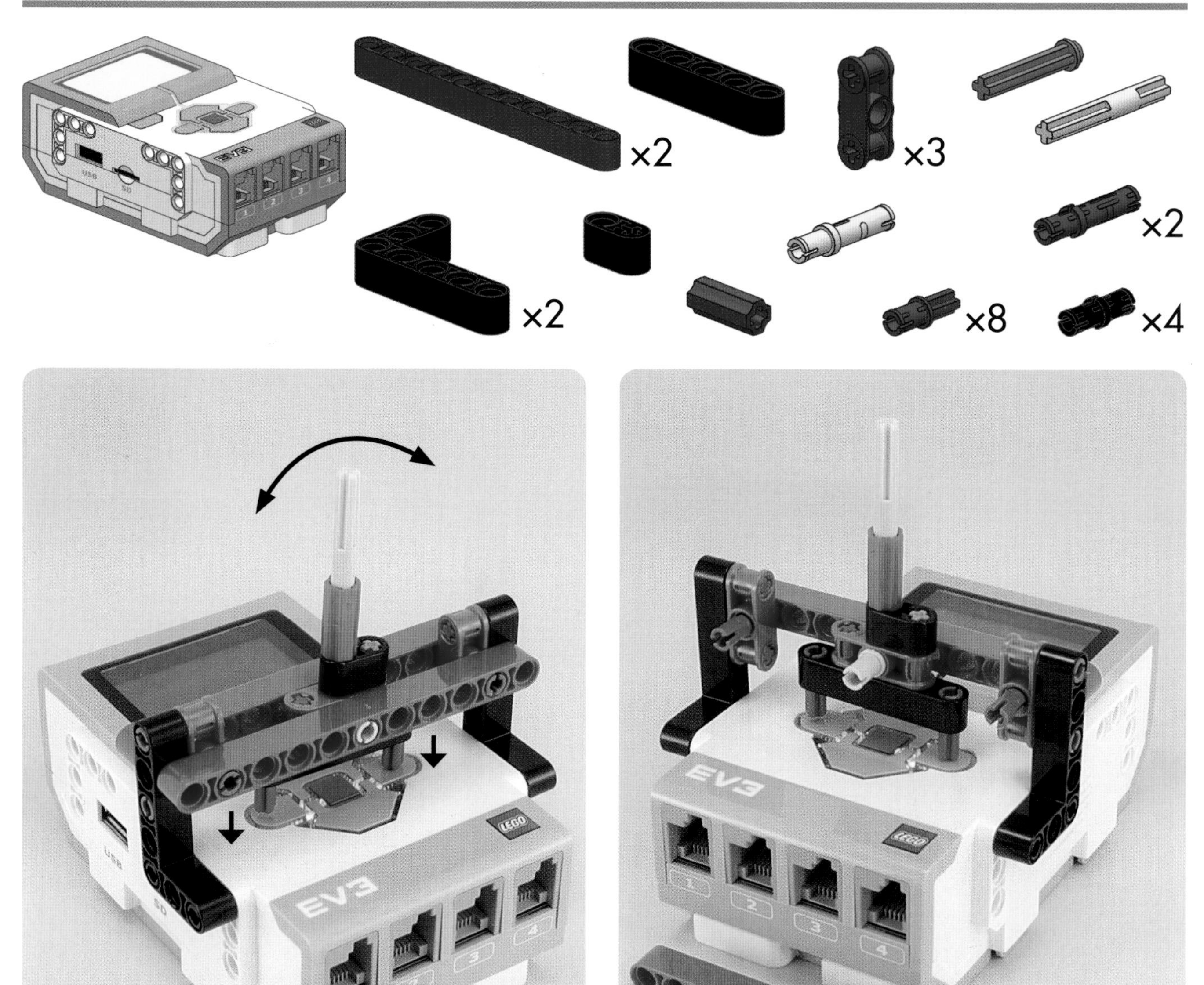

#173

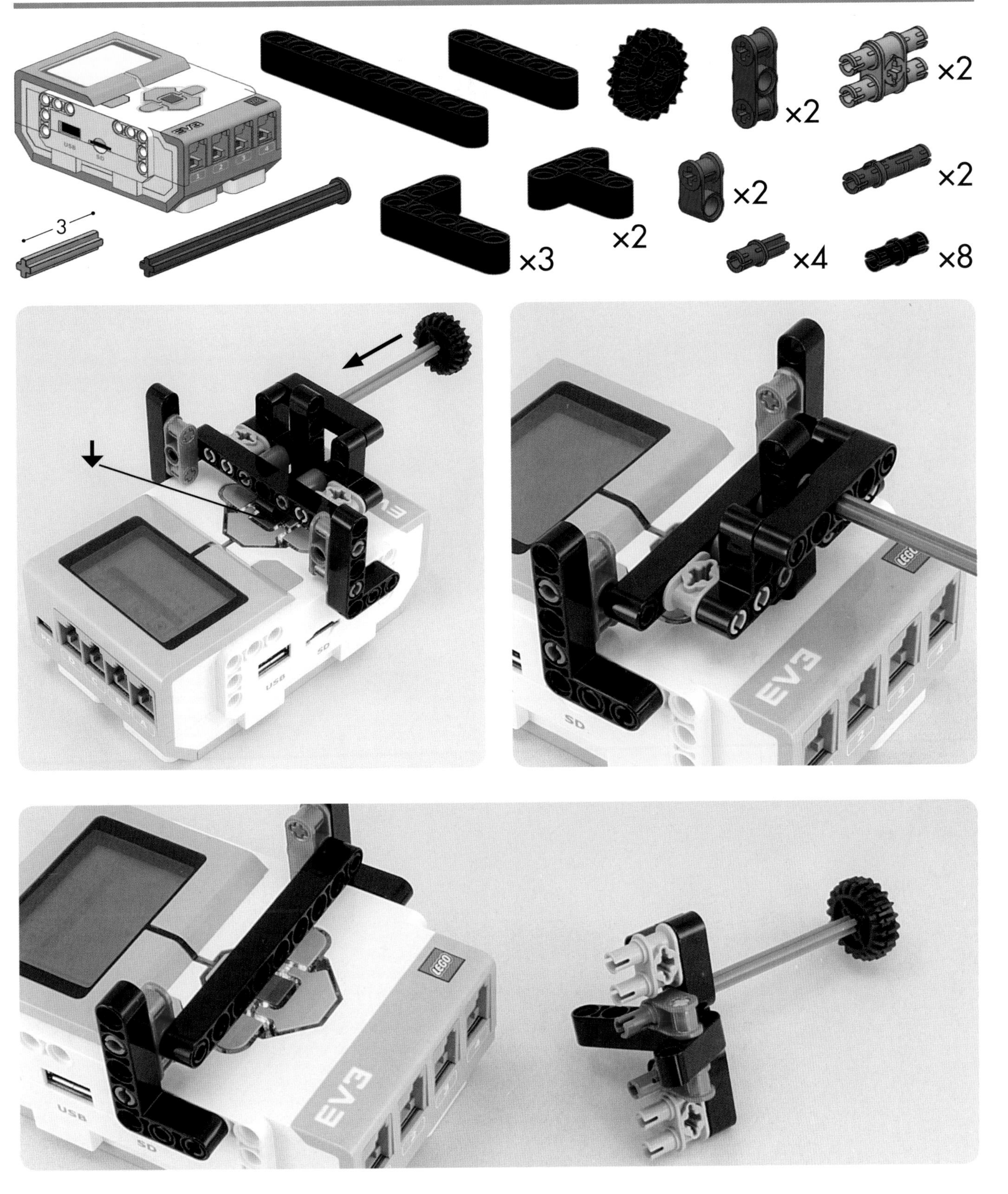
×2
×2
×2
×2
×3
×2
×4
×8
3
EV3
USB
SD

Ideas for using the color sensor

#174

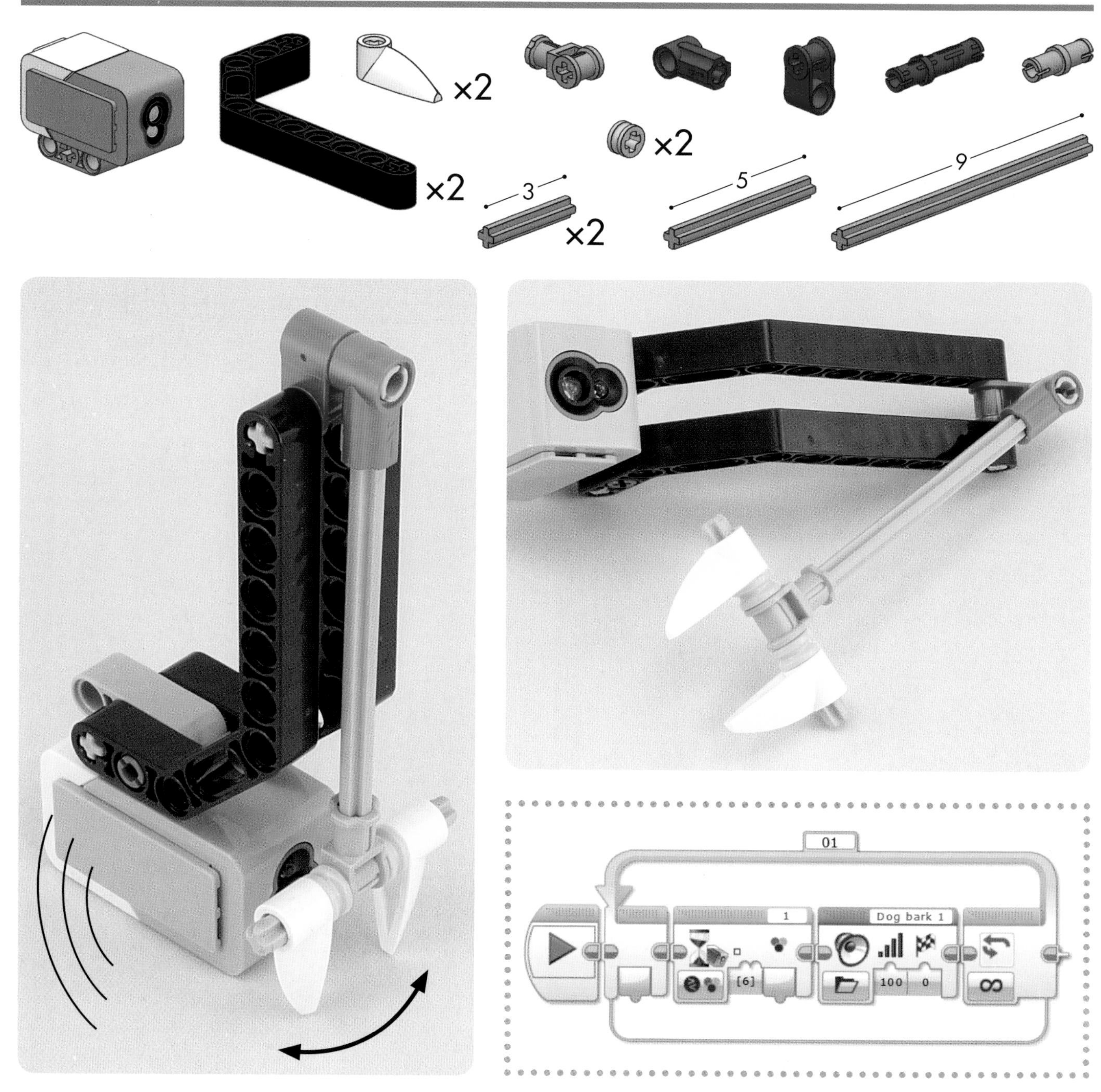

×2
×2
×2
3
×2
5
9
01
1
Dog bark 1
[6]
100
0

#175

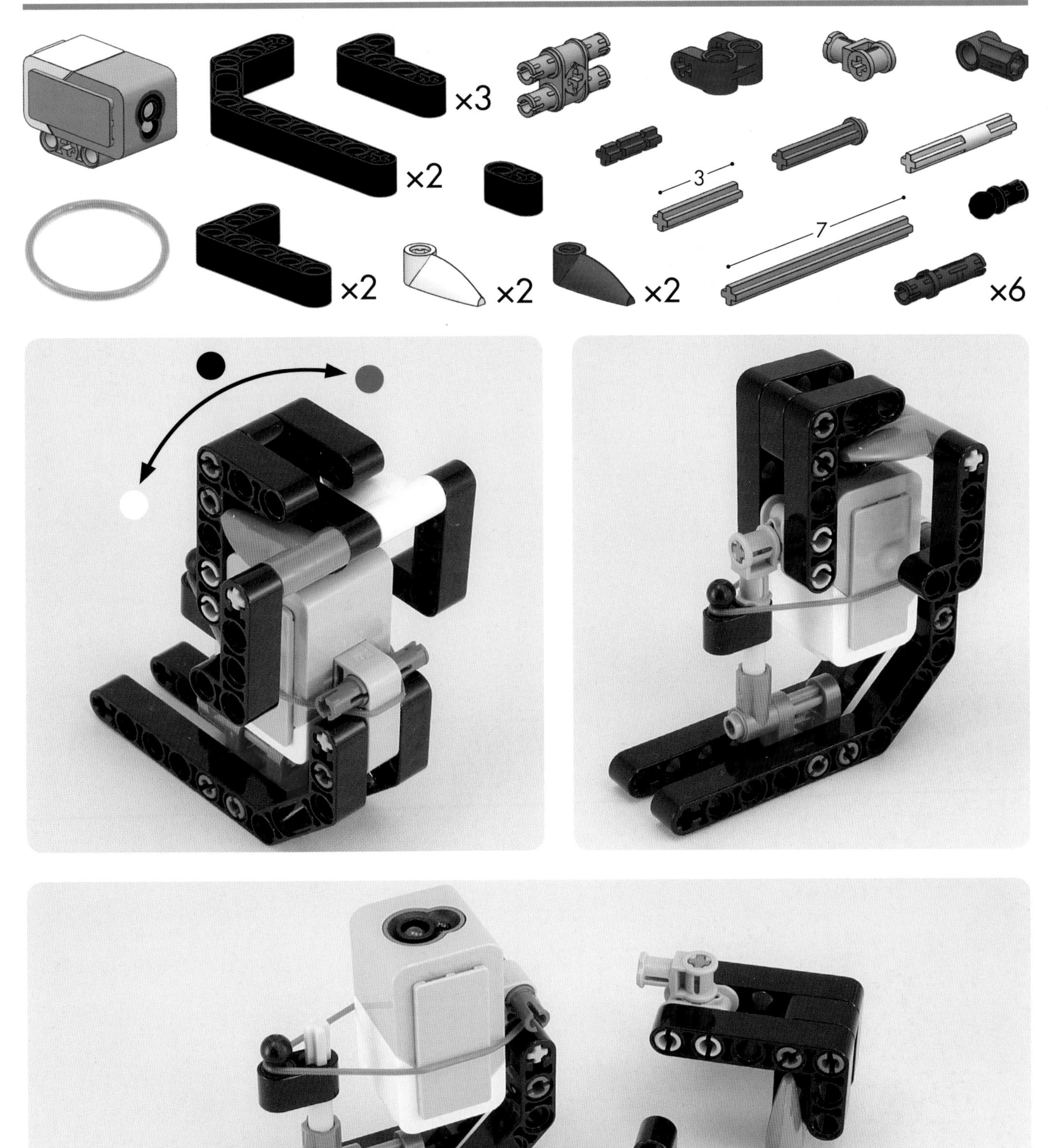

×3
×2
×2
×2
×2
3
7
×6

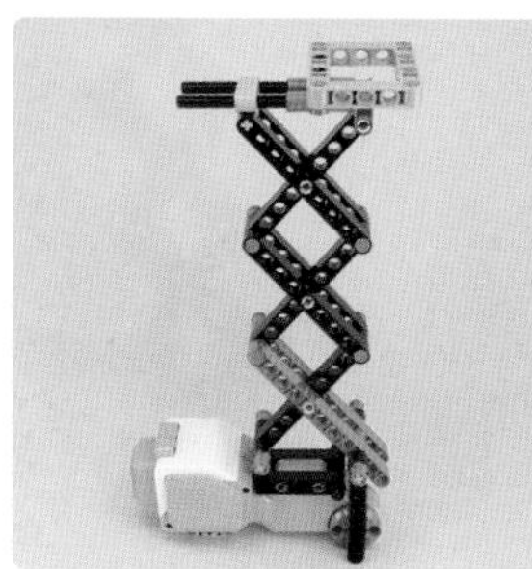
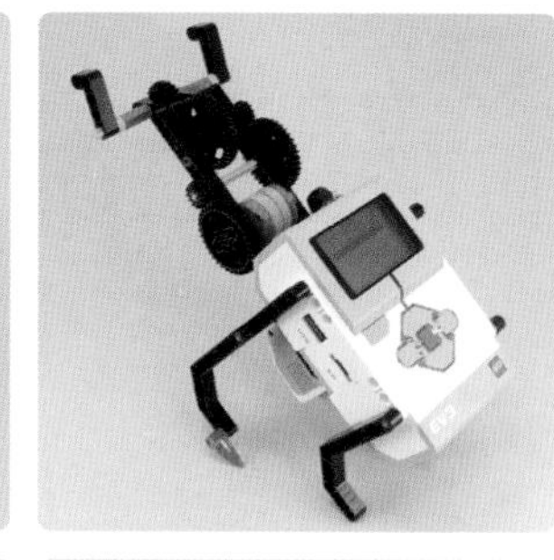
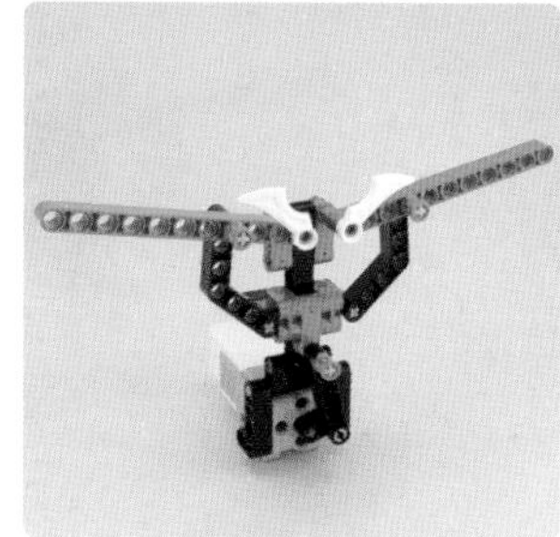
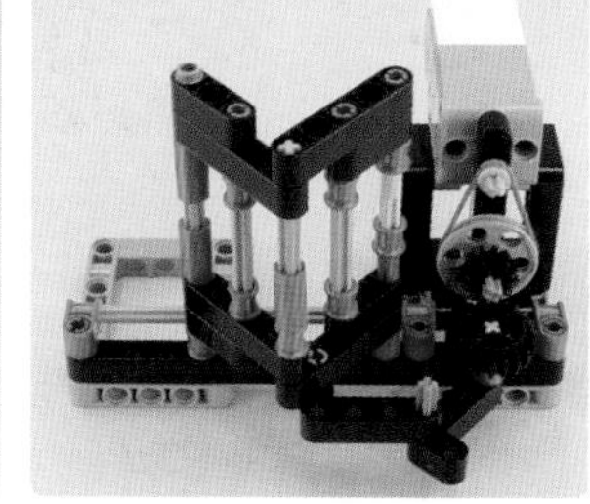

PART 6

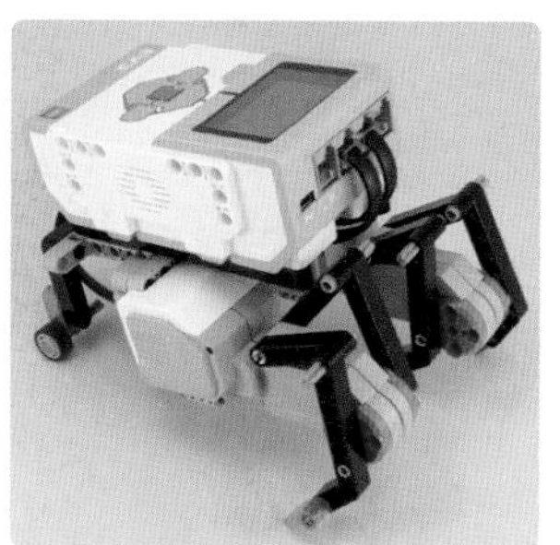

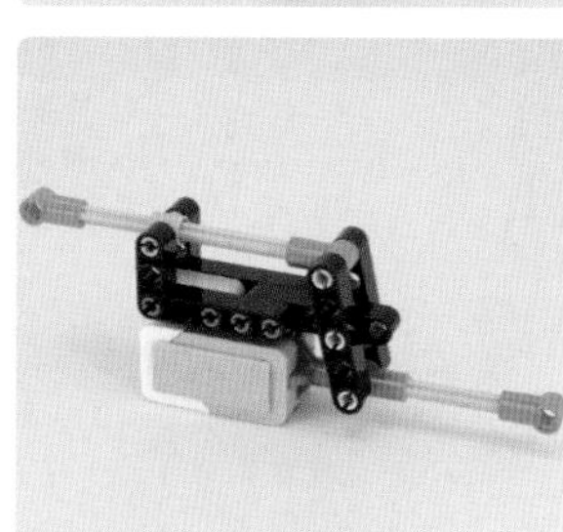

Something Extra

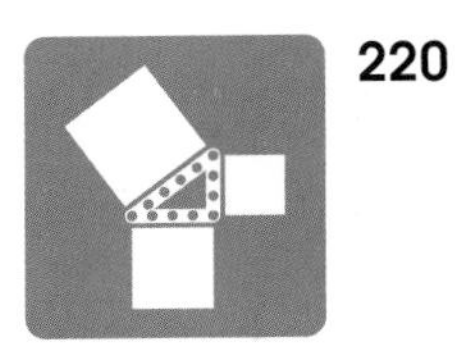

220

222

Using the Pythagorean theorem

#176

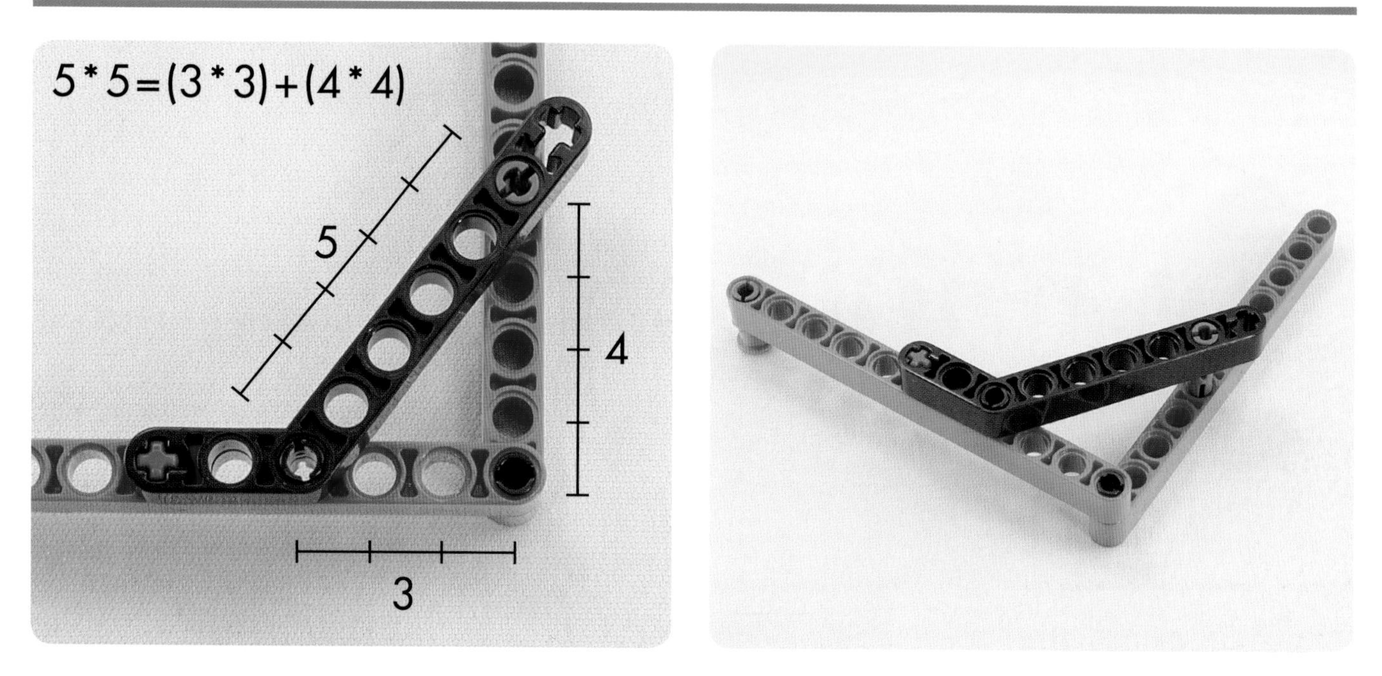
5*5=(3*3)+(4*4)
5
4
3

#177

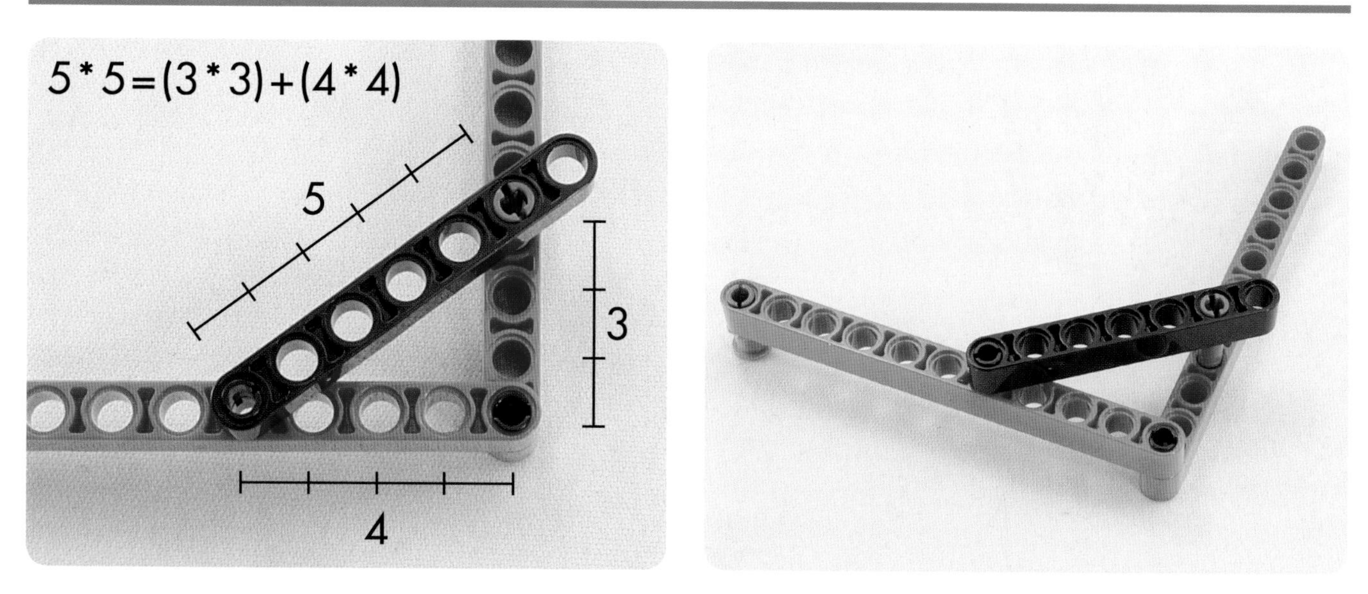
5*5=(3*3)+(4*4)
5
3
4

#178

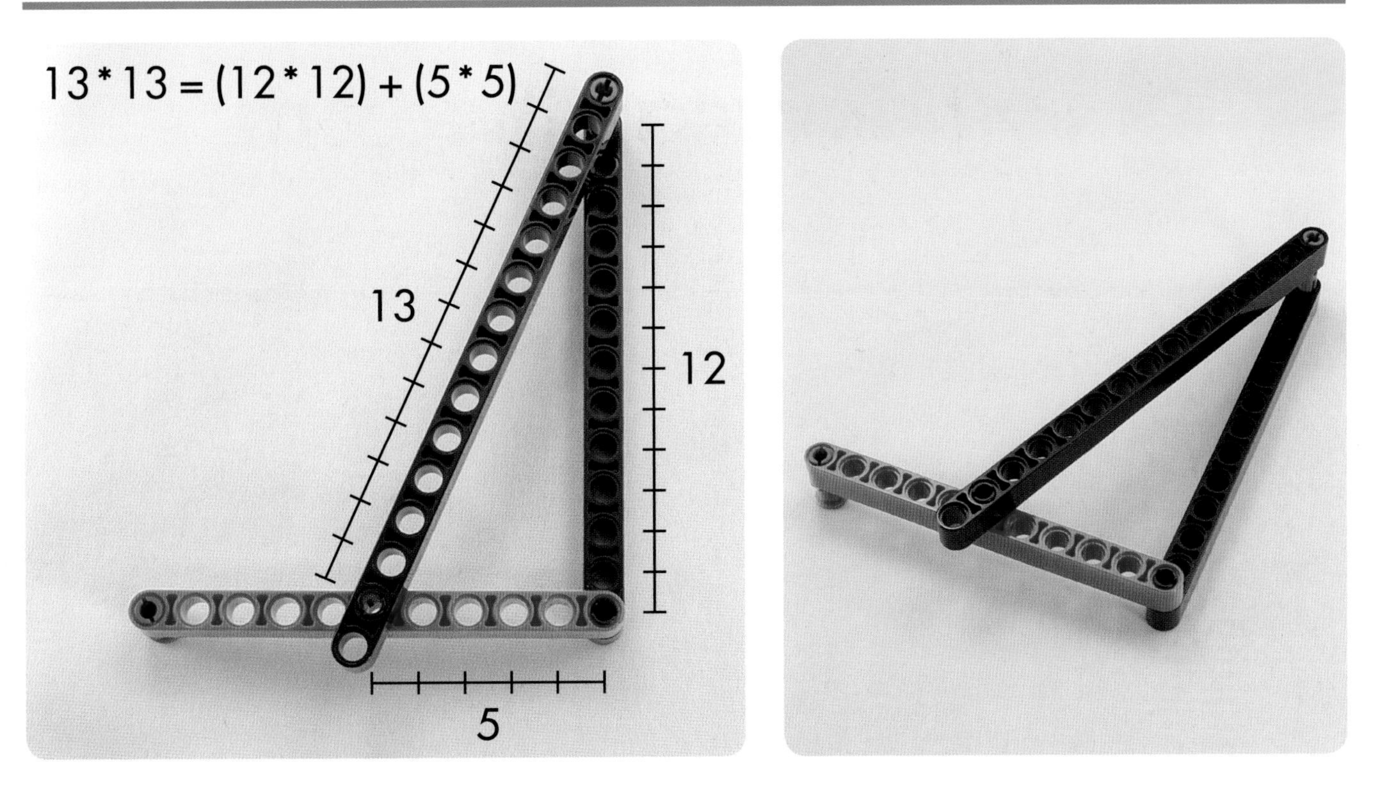
13*13=(12*12)+(5*5)
13
12
5

#179

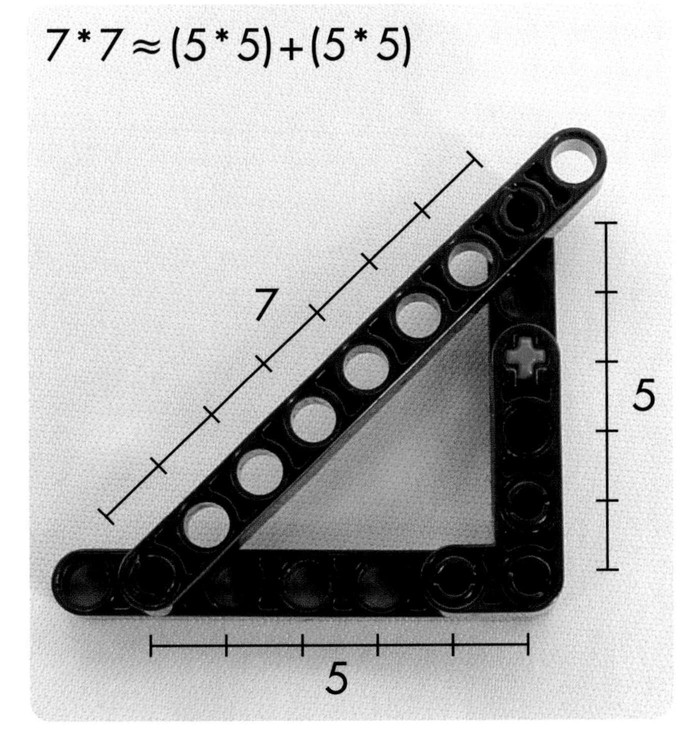
7*7≈(5*5)+(5*5)
7
5
5

#180

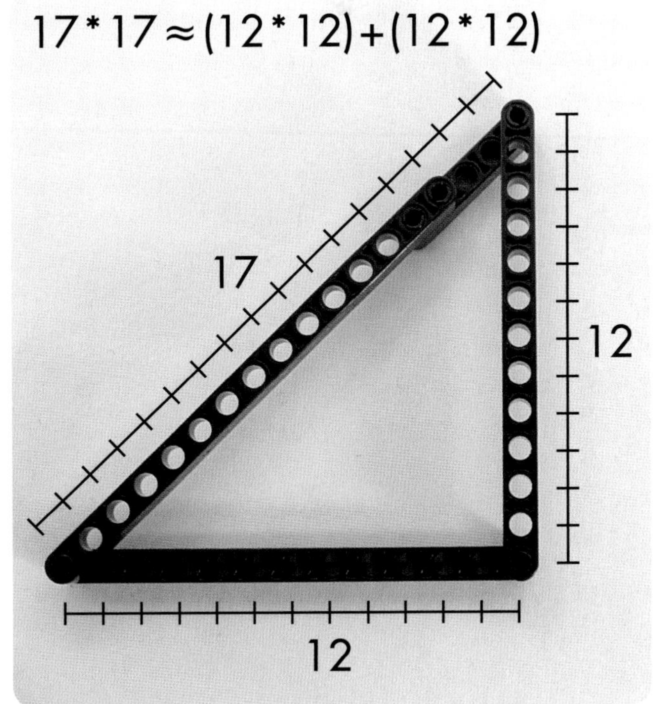
17*17≈(12*12)+(12*12)
17
12
12

Try building something handy!

#181

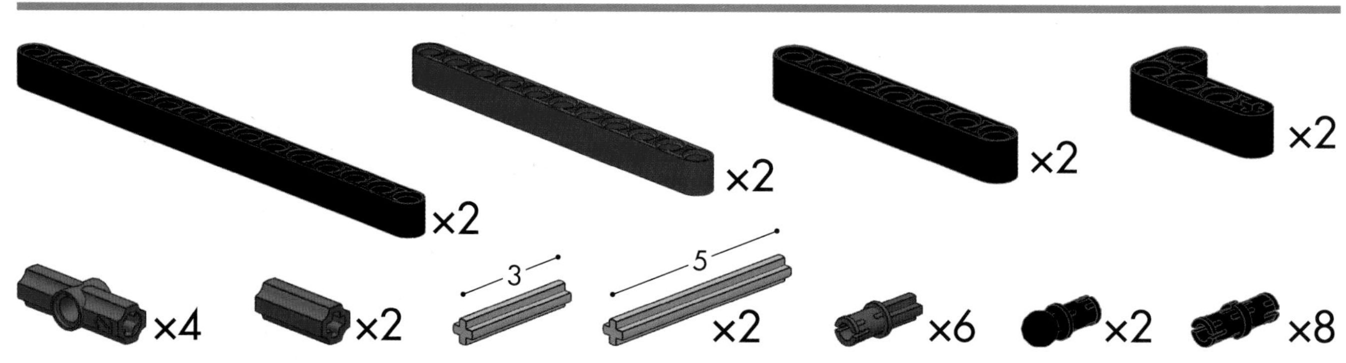
×2
×2
×2
×2
3
5
×4
×2
×2
×6
×2
×8

Updates

Visit *http://www.nostarch.com/ev3ideabook/* for updates, errata, and other information.

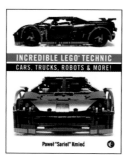